Physiological and Clinical Anatomy of the Domestic Mammals

Physiological and Clinical Anatomy of the Domestic Mammals

VOLUME 1

Central Nervous System

A. S. KING

Head of the Department of Veterinary Anatomy
The University of Liverpool

Oxford New York Tokyo
OXFORD UNIVERSITY PRESS

Oxford University Press, Walton Street, Oxford OX2 6DP
Oxford New York
Athens Auckland Bangkok Bombay
Calcutta Cape Town Dar es Salaam Delhi
Florence Hong Kong Istanbul Karachi
Kuala Lumpur Madras Madrid Melbourne
Mexico City Nairobi Paris Singapore
Taipei Tokyo Toronto
and associated companies in
Berlin Ibadan

Oxford is a trade mark of Oxford University Press

Published in the United States
by Oxford University Press Inc., New York

First published 1987
Reprinted 1993, 1994

British Library Cataloguing in Publication Data
King, A. S.
Physiological and clinical anatomy of the domestic mammals.
Vol. 1: Central nervous system
1. Mammals—Anatomy I. Title
599.04 QL739

Library of Congress Cataloging in Publication Data
King, Anthony Stuart.
Physiological and clinical anatomy of the domestic animals.
Bibliography: p. Includes index.
Contents: v. 1. Central nervous system.
1. Veterinary anatomy—Collected works. I. Title.
SF761.K564 1987 636.089'1 86–23693
ISBN 0 19 854187 2 (Pbk: v. 1)

Printed in Hong Kong

Preface

Nearly forty years ago, as a new Lecturer in Anatomy at the Royal Veterinary College, London, I had the good luck to hear E. C. Amoroso expounding to Honours B.Sc students the logic of somatotopic localization in the cuneate and gracile fascicles. At that time training veterinary students in neurology was considered unnecessary since euthanasia was still the standard treatment for neurological cases in veterinary practice. The publication of McGrath's *Neurologic examination of the dog* in 1956 initiated a series of books in English which clearly demonstrate the feasibility of analysing neurological conditions in domestic mammals despite the absence of active collaboration by the patient. When the opportunity came in 1962 to establish a totally new curriculum of veterinary anatomy at Liverpool, I included a course in neuroanatomy. The present book was founded then, in the form of stencilled summaries of individual lectures. By 1970 all the main components of the neuraxis had been covered, and this coincided with the appointment of A. M. Goldberg to the technical staff of my Department. His skill in reprographic technology enabled us to assemble these materials in 1971 into an in-house book. By cannibalizing several abandoned offset lithographic machines we were able to produce good quality print and coloured diagrams. The book then went through six editions (all of them identified by Penfield's motor homunculus on the cover), each one embodying many changes. The core material, which has always been distinguished by its own typeface, is the course taken by my own students (except for Chapters 18 and 21). The presence of other material should make the work useful to a wider range of readers. Certainly these books have been used by students in a number of veterinary schools in other universities during the course of the last fourteen years.

Not all veterinary educationalists would wish their students to receive such an extensive course of neuroanatomy as this, which does indeed require lectures, dissections, histology practical classes, discussions of video tapes, and examination of clinical cases by the students, totalling about 55 hours in all. It cannot be maintained, however, that neurological cases are encountered in veterinary practice so rarely that they can reasonably be ignored. Most animal owners expect at least to receive a diagnosis and more especially a prognosis, and treatment is by no means always out of the question. Intelligent owners of companion animals, particularly those with a biomedical background, may wish to discuss a case more fully. Therefore it seems reasonable to train the modern graduate to a level of basic competence in clinical neurology. If that be accepted, the process must start with neuroanatomy and neurophysiology, since it is only through a grasp of the basic principles of these subjects that the clinical signs of pathophysiology can be interpreted and the site of the disorder localized. It is here that the nub of the difficulty lies: the basic principles of *all* the main functional components of the nervous systems must be covered. If the course contents are limited to selected snapshots of neurology the clinician cannot be confident of interpreting clinical data logically. It follows that a foundation course of functional neuroanatomy has a critical mass below which one must not go. The core material in this book, which is printed in ordinary type (like this), strives both to attain and to remain within that critical mass. This enforces a great many severe anatomical simplifications which are essential to save the beginner time and labour, though without distorting the principles of function. However, in many places the reader is warned of the more substantial simplifications by passages in small type, which he can read or ignore according to the extent of his ambitions.

The text has been rewritten and reorganized under new headings, and nearly all of the diagrams have been redrawn. A determined attempt has been made to conform to the *Nomina Anatomica Veterinaria* (NAV) and *Nomina Histologica* (NH). The most substantial development is Chapter 21 by David Hogg on the topographical anatomy of the brain and spinal cord. Although presented in the same typeface as the core material elsewhere, this chapter is not intended to be core material. Its purpose is to provide a brief account of the location and connections of some of the multitude of smaller components which the more advanced reader is likely to encounter in the literature.

My thanks are due to many people: to G. C. Skerritt for commenting on passages in various chapters, as well as for contributing Chapter 20; to Dr David Bowsher for being an ever-available source of information; to Gary Martin for the artwork; to Mrs M. M. Thompson for the typing; and to Janet Cannell and especially to Merton Goldberg for the reprographic expertise on which the book was founded.

Liverpool A. S. K.
June 1986

Contents

1. Arterial supply to the central nervous system

Arterial supply to the brain

1.1 Basic pattern of the main arteries supplying the brain

Five pairs of arteries supply the brain (Fig. 1.1). The first four of these arise from the **cerebral arterial circle** on the ventral surface of the brain (the circle of Willis); the cerebral arterial circle roughly outlines the hypothalamus, with the stalk of the pituitary in its centre. The fifth arises from the basilar artery. The five pairs of arteries are: (i) the **rostral cerebral artery**; (ii) the **middle cerebral artery**, this being the largest cerebral artery in most mammals; (iii) the **caudal cerebral artery**; (iv) the **rostral cerebellar artery**; and (v) the **caudal cerebellar artery**.

There are also various smaller arteries which supply the medulla oblongata and pons.

Although there are minor species variations, these vessels occur in mammals generally. The cerebellar arteries are variable in number and origin even within the same species: for example in man and horse the rostral one may arise from the basilar artery. The three cerebral arteries are remarkably constant in amphibians and higher forms generally.

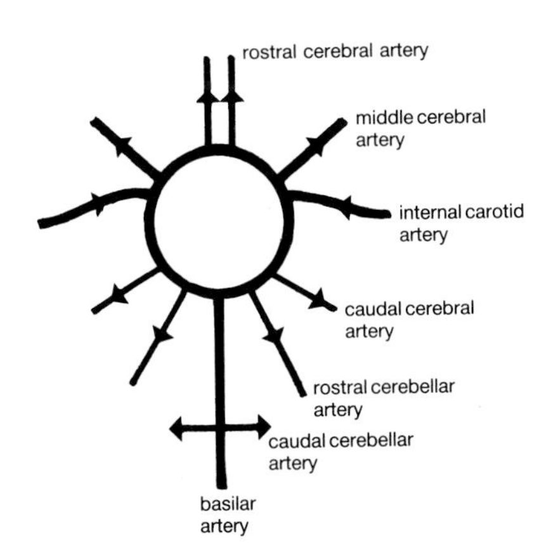

FIG. 1.1. **Diagram of the cerebral arterial circle and its outgoing branches.** The connection across the midline at the rostral end of the circle is inconstant in the dog and ruminants.

1.2 Basic pattern of incoming branches to the cerebral arterial circle

There are four **potential arterial channels** to the cerebral arterial circle in mammals generally (Fig. 1.2). **Channel 1**, the **internal carotid artery**. **Channel 2**, the **basilar artery**. This midline artery is a continuation rostrally of the ventral spinal artery. However, the blood that flows within the ventral spinal and basilar arteries has come from the vertebral artery via the segmental spinal arteries. **Channel 3**, the **maxillary artery**. The maxillary artery supplies the arterial circle by its so-called **anastomosing ramus**, which joins the maxillary artery to the internal carotid artery. **Channel 4**, the **vertebral artery**. The vertebral artery connects to the internal carotid artery, and in some species it supplies the arterial circle directly by this

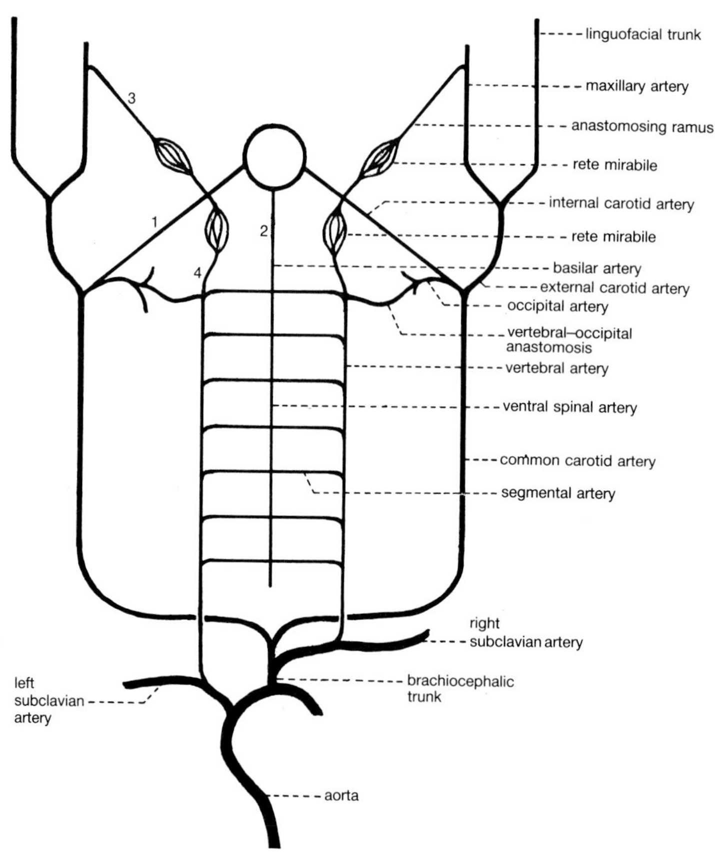

FIG. 1.2. Diagram showing the potential arterial channels to the cerebral arterial circle. There are four such channels, numbered 1 to 4 on the left: 1, the internal carotid artery; 2, the basilar artery; 3, the anastomosing ramus from the maxillary artery to the internal carotid artery; and 4, the connection of the vertebral artery to the internal carotid artery.

FIG. 1.3. Diagrams showing species variations in the sources of arterial blood to the brain. In each figure the upper diagram shows the distribution over the brain of internal carotid, vertebral, and maxillary blood in the intact live animal (see key); the lower diagram shows the anatomy which accounts for this distribution, based on the four potential arterial channels to the cerebral arterial circle. Arrows show the direction of flow in the basilar artery. The vertebral–occipital anastomosis (VO) can be disregarded in the intact animal. Channel 1, internal carotid artery; channel 2, basilar artery; channel 3, anastomosing ramus from maxillary artery to internal carotid artery; channel 4, connection of vertebral artery to internal carotid artery.

(a) Dog, man, and many other species. Channels 1 (internal carotid artery) and 2 (basilar artery) supply the arterial circle; the basilar artery carries blood **to** the arterial circle. Neither channel has a rete mirabile. Internal carotid blood reaches all of the cerebral hemisphere except its most caudal part. Vertebral blood supplies the remainder of the cerebral hemisphere, and all the rest of the brain.

(b) Sheep, cat. Only channel 3 (maxillary anastomosing ramus) supplies the arterial circle. It has a rete mirabile. Channel 2 (basilar artery) carries blood **away from** the arterial circle. Maxillary blood is distributed to all of the brain except the caudal part of the medulla oblongata, which is supplied by vertebral blood.

(c) Ox. Channels 3 (anastomosing ramus) and 4 (vertebral artery) both supply the arterial circle. Each has a rete mirabile. Channel 2 (basilar artery) carries blood **away from** the arterial circle. A mixture of maxillary and vertebral blood reaches all parts of the brain.

route. However, as stated under Channel 2, it may also supply the circle indirectly via the ventral spinal and hence the basilar artery.

Because of the anatomy of these four arterial channels, the blood which distributes itself over the brain may be **internal carotid blood**, **maxillary blood**, or **vertebral blood**, or a combination of these (Fig. 1.3).

1.3 Species variations

In no species of domestic mammal are all four of these potential arterial channels to the cerebral arterial circle fully developed. Some of the channels are reduced in calibre or even totally obliterated. The direction of flow in the remaining channels depends on the pressure gradients within the various vessels. The general relationships of these gradients have been worked out experimentally, thus establishing the direction of flow and the distribution of blood in each species. The following account applies these results to the intact live animal.

Dog, man, and mammals generally

These animals have what appears to be the most usual mammalian pattern of arterial supply to the brain (Fig. 1.3(a)). The blood reaching the rostral half of the brain is internal carotid blood, but the caudal half of the brain is supplied by vertebral blood. This is because the pressure gradients are such that the flow of blood in the basilar artery is **rostral**. Consequently, vertebral blood reaches not only the cerebellar arteries but also the caudal cerebral artery.

Channel 3 (the anastomosing ramus of the maxillary artery) is much reduced in these species. In the dog there is an anastomotic artery which connects the internal carotid to the external ophthalmic artery, the latter being a branch of the maxillary artery; this anastomosis could possibly serve as a channel 3. In man, the maxillary artery anastomoses with the internal carotid via the sphenopalatine artery and via the middle meningeal artery.

Sheep and cat

In these species the lumen of the proximal two-thirds of the internal carotid artery becomes obliterated in the weeks or months after birth. (At birth, however, the internal carotid is fully functional.) The **whole** of the adult brain is supplied by maxillary blood, via channel 3, the anastomosing ramus of the maxillary artery (Fig. 1.3(b)). (A **rete mirabile** develops on the anastomosing ramus in these species.) The direction of flow in the basilar artery is **caudal**. Consequently scarcely any blood from the vertebral artery reaches the brain.

The pressure gradients actually do allow a rostral flow in the **caudal** end of the basilar artery. As a result, vertebral blood does reach the **caudal half** of the medulla oblongata (Fig. 1.3(b)). It is certain, however, that **no** vertebral blood reaches the cerebral hemispheres.

Ox

In this species, the lumen of the **proximal** two-thirds of the internal carotid artery is obliterated by 18 months of age, but the **distal** third remains intact (Fig. 1.3(c)). The supply to the **whole** of the brain is by a **mixture** of maxillary and vertebral blood (Fig. 1.3(c)). This takes place by channels 3 (anastomosing ramus of the maxillary artery) and 4 (vertebral artery), both of which connect directly to the distal third of the internal carotid artery. (A **rete mirabile** develops on both of these channels.) The direction of flow in the basilar artery is again **caudal**, but as just stated vertebral blood already gains access to the brain through the distal remnant of the internal carotid artery.

Summary of species variations

An inverse relationship exists between the degree of development of the internal carotid and that of the anastomosing ramus of the maxillary artery: when one is large the other is small. In no species are both of these channels fully developed.

In all species a vertebral–occipital anastomosis (Fig. 1.2) is present and quite well developed. Despite the fairly substantial calibre of this pathway, it seems to have very little functional importance in the intact animal. It does become of interest, however, if the common carotid artery is cut.

1.4 Summary of the significance of the vertebral artery as a source of blood to the brain

In the normal sheep and cat little or no blood from the vertebral artery reaches the brain. In the normal ox, vertebral blood reaches **all** parts of the brain, and it is this point that becomes particularly important in the problem of humane slaughter. In other species vertebral blood reaches only the cerebellum and the caudal part of the cerebrum.

1.5 Humane slaughter

The essential characteristic of the Jewish and Mohammedan methods of slaughter is that they consist of a single rapid cut of the neck with an extremely sharp knife, severing among other things the common carotid arteries and the jugular veins on both sides of the neck, but not the vertebral arteries which are protected by the bony walls of the transverse foramina. Attempts have been made to apply anatomical and physiological knowledge to determine whether cattle and sheep slaughtered in this way do in fact lose consciousness with sufficient speed to be spared unnecessary suffering. The interpretation of the physiological evidence is controversial, but it appears to indicate that consciousness may be lost between 3 and 10 seconds after the neck is cut; the direct arterial pathway to the cerebral cortex via the vertebral artery in the ox but not in the sheep, gives a warning that the time taken in the ox may be relatively long (towards 10 seconds) though in the sheep it may be at the lower end of the range (towards 3 seconds). The captive bolt and electrical stunning used in other methods of slaughter are held to disrupt brain function completely within a fraction of one second, if properly used.

In Great Britain the Slaughter of Animals Act 1958, lays down three conditions for the slaughter of horses, cattle, sheep, pigs, and goats in abattoirs: all animals must be either (i) instantaneously slaughtered by a mechanically operated instrument, or (ii) stunned mechanically or electrically so as to be instantly insensible to pain until dead, or (iii) slaughtered by other means specified by the Minister, provided again that the animals are rendered insensible to pain until death supervenes. The Jewish and Mohammedan methods of slaughter are exempted from items (i) and (ii), but there is an overall proviso that they must not inflict unnecessary suffering.

In the preceding sections it has been shown how the brain is supplied with blood, and the events which follow the cutting of the neck can now be considered. Immediately after the common carotid arteries have been cut, blood will flow from both their cardiac and their cranial stumps. The blood which escapes from the cut cranial stumps will come from the vertebral arteries, via the occipito–vertebral anastomoses into the common carotid arteries. In the ox the vertebral arteries could continue to supply the brain via channel 4 (Fig. 1.3(c)). However, since the pressure should fall markedly at the cut cranial ends of the common carotid arteries, it is to be expected that most of the vertebral blood will flow down the steeper pressure gradient from the vertebral arteries to the open cranial stumps of the common carotid arteries, i.e. most of the blood from the vertebral arteries should escape from the cut cranial stumps of the common carotid arteries. In the sheep there is no channel 4 (Fig. 1.3(b)), so there is less opportunity for blood from the vertebral arteries to reach the brain. But even in the sheep, blood could reach the cerebral arterial circle via the occipito–vertebral anastomoses and channel 3, or by the reversal of flow in channel 2, though once again most of the flow should follow the steep pressure gradient to the cut cranial stumps of the common carotid arteries. One further factor is the possibility that the elastic fibres in the wall of the common carotid arteries may cause rapid retraction of the arterial wall with resultant sealing of the cut stumps. If this were to happen to the cranial stumps of the carotid arteries, the vertebral artery would then supply a closed system of arterial vessels including those to the brain.

Attempts have been made to measure the time taken for the animal to lose consciousness. It has been found that in standing sheep the electroencephalogram begins to change very quickly (about 2 seconds) after cutting the common carotid arteries. In sheep and goats the corneal reflex is lost on average in about 3 seconds. In cattle, the EEG becomes sleep-like in 3 to 5 seconds, and standing cattle fall within 8 to 10 seconds.

1.6 Rete mirabile

A maxillary rete mirabile (rete mirabile means 'marvellous network') occurs on the anastomosing ramus of the maxillary artery in all species in which channel 3 (from the maxillary artery) is well developed. Thus it occurs in the cat, and in the ox and sheep. In the ruminants (but not the cat) it is immersed in the **cavernous sinus** (see Section 3.7). The last part of the internal carotid artery passes through the cavernous sinus.

The function of the rete mirabile has long been debated. It was thought that it might eliminate pulsation before the blood reaches the brain itself. However, more recent observations indicate that the rete is involved in thermoregulation. The blood in the cavernous sinus comes partly from the nasal mucosa, where evaporative cooling occurs; consequently the blood in the cavernous sinus is cooler than the arterial blood coming direct from the hot core of the body. Presumably heat exchange occurs between the rete blood and cavernous blood, so that rete blood is somewhat cooled before it reaches the brain.

In the sheep differences of about 1 °C have been recorded between the body core and blood in the hypothalamus. It has been suggested that the object of this temperature difference is to protect the brain against rising temperatures. A notable example occurs in certain gazelles, which possess an intracranial rete mirabile. These animals can tolerate a body temperature of 46.5 °C for long periods; when the body temperature is driven up to 46.5 °C by exercise, the temperature of the brain still remains about 2.9 °C lower than the body core, this lower temperature probably being due to heat exchange between the rete blood and the cavernous blood.

The rete mirabile arises developmentally (Fig. 1.4(a), (b)) from many nutrient arterial twigs which form (i) from the terminal part of the internal carotid near the cerebral arterial circle and (ii) from the maxillary artery, and supply the root of the

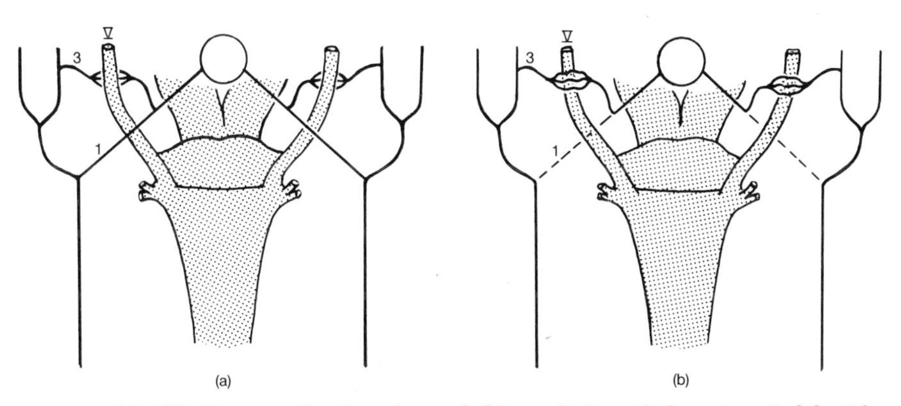

FIG. 1.4. (a), (b). Diagram showing the probable evolution of the rete mirabile. The trigeminal nerve (V) receives small nutrient branches from the internal carotid artery (1) and maxillary artery (3), in mammals generally (a). When the internal carotid is obliterated in ruminants and the cat these nutrient vessels anastomose across the trigeminal nerve and so form the rete mirabile (b).

trigeminal nerve. These nutrient twigs to the Vth nerve are present in mammals generally. In the few species in which the proximal part of the internal carotid becomes obliterated (sheep, ox, cat) these twigs proliferate, and those from the distal remnant of the internal carotid anastomose with those from the maxillary artery. These anastomoses then form a continuous arterial pathway from the maxillary artery to the cerebral arterial circle, complete with a basket-like plexus, the rete mirabile.

Superficial arteries of the spinal cord

1.7 Main trunks

Although species variations occur, the basic pattern in mammals is founded on three longitudinal trunks (Fig. 1.5):

Dorsolateral arteries

A pair of small dorsolateral arteries runs along the dorsolateral aspect of the spinal cord. These arteries are ill-defined and in some mammalian species, including the dog, cannot be identified as distinct vessels.

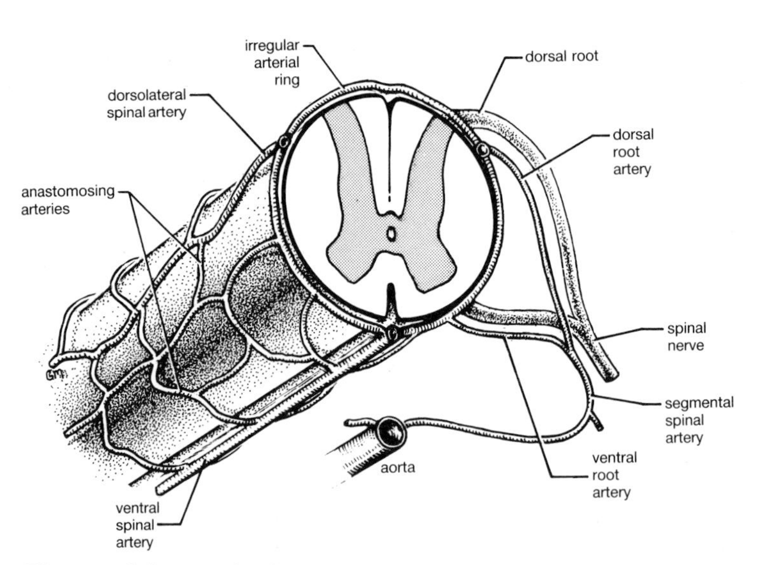

FIG. 1.5. Diagram of the superficial arteries of the spinal cord in a hypothetical mammal. Only the ventral spinal artery is constant and relatively large in mammals generally. The paired dorsolateral arteries, the anastomosing arterial network connecting these to the ventral spinal artery, and the arterial ring at the level of each intervertebral foramen are all inconstant and irregular in disposition depending on the species. These superficial arteries are supplied by paired segmental spinal arteries, which enter the vertebral canal as the dorsal root artery and ventral root artery on each side.

In man there tend to be two of these arteries on each side, one dorsal and one ventral to the dorsal roots of the spinal nerves.

Ventral spinal artery

A much larger and more important midline ventral spinal artery lies in the ventral fissure of the spinal cord.

1.8 Anastomosing arteries

A sparse and very irregular network of arteries connects the dorsolateral and ventral spinal arteries. However, these anastomoses are inadequate as an alternative pathway. If the ventral spinal artery is blocked the spinal cord will nearly always be damaged.

At the level of each intervertebral foramen, the anastomosing network tends to be reinforced by an irregular and somewhat incomplete **arterial ring** surrounding the cord (Figs 1.5, 1.6). This ring receives the incoming dorsal root and ventral root arteries.

1.9 Segmental arteries to the spinal cord

At each segment of the body two arteries on the left, and two on the right side, supply the spinal cord. These are the **dorsal root artery** and **ventral root artery** on each side (Fig. 1.5). These arteries arise from paired **spinal arteries** which in turn spring from the paired lumbar arteries of the aorta, from the intercostal arteries, and from the vertebral arteries.

In some species the dorsal and ventral root arteries are erratically absent from some or many segments of the body.

Besides interrupting the sensory and motor rootlets of that particular segment, injury to the roots of a spinal nerve can interfere with its arteries and thus cause degeneration within the spinal cord itself.

Deep arteries of the neuraxis

The general richness of vascularity of the neuraxis (brain and spinal cord) is a notable feature. Although the brain forms only about 2 per cent of the total body weight, it receives about 16 per cent of the total cardiac output, and accounts for about 20 per cent of the total O_2 consumption of the whole body.

1.10 General principles governing the distribution of arteries below the surface of the neuraxis

The requirements of axons, cell bodies, and synapses

These have increasing O_2 consumption in that order. Grey matter is therefore more vascular than white.

The function of the region

Motor areas, sensory areas, and association areas have increasing vascularity in that order. This reflects the functional importance of the association areas of the brain (see Section 17.1).

The phylogenetic age of the region

The ancient parts of the neuraxis typically are less vascular than the recent parts. Thus the spinal cord and the rhinencephalon (the latter being phylogenetically the oldest part of the brain, see Section 8.6) are less vascular than the cerebral cortex and cerebellar cortex. An exception is the hypothalamus which has the richest blood supply in the brain in spite of being phylogenetically ancient, but this is probably because it depends directly on a high vascularity to carry out some of its functions such as thermoregulation (see Section 16.3).

1.11 The deep arteries of the spinal cord

Two groups of arteries penetrate the surface of the spinal cord.

Vertical arteries

A series of vertical arteries arise from the ventral spinal artery like the teeth of a comb and pass, in the ventral fissure, towards the centre of the spinal cord (Fig. 1.6). They supply most of the grey matter, and reach peripherally into the white matter also.

In man, each vertical artery passes either to the right or to the left, supplying a territory on either side of the midline.

Radial arteries

These arise from all the other arteries on the surface of the cord. They supply the white matter and the outer regions of the grey matter (Fig. 1.6).

The spinal cord therefore has three vascular zones (Fig. 1.6). (i) The **inner vascular zone** is supplied solely by the vertical arteries of the ventral spinal artery. (ii) The **middle vascular zone** is supplied both by the vertical arteries of the ventral spinal artery, and by the radial arteries from all the other surface arteries. (iii) The **outer vascular zone** is supplied by the radial arteries alone.

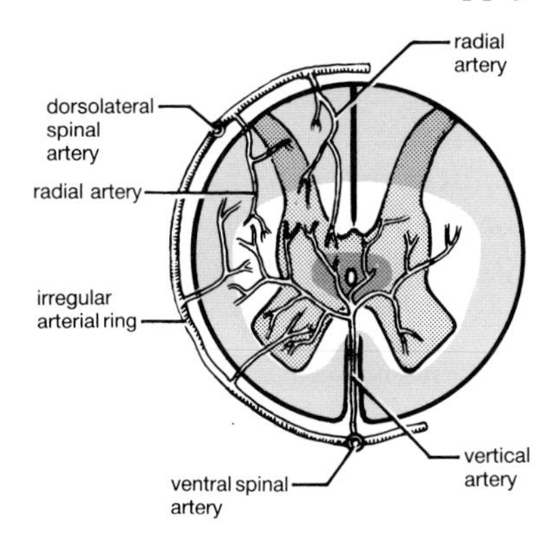

FIG. 1.6. Diagram showing the deep arteries of the spinal cord. The inner zone (dark green) is supplied by vertical arteries only. The middle zone (uncoloured) is supplied by both vertical and radial arteries. The outer zone (light green) is supplied by the radial arteries only.

1.12 The problem of pulsation

It appears that large pulsating arteries cannot be tolerated below the surface of the neuraxis. Since the fluid nature of brain tissue renders it incompressible, large pulsating arteries would generate pressure waves below the surface. This may be why no large arteries travel below the surface of the brain and spinal cord.

1.13 Arterial anastomoses of the neuraxis

On the surface of the neuraxis

Profuse anastomoses occur on the surface of the brain. These can sometimes provide effective alternative channels around an obstruction. The similar anastomoses on the surface of the spinal cord are finer, and generally inadequate as alternative pathways.

Below the surface of the neuraxis

The deep arteries rarely anastomose above capillary level. Some workers claim that there are no anastomoses at all above the capillary level; others believe in occasional anastomoses at arteriolar level. Functionally it matters little, since these arteries below the surface are all **functional end arteries**; thus total obstruction of any artery below the surface produces death of all neurons in its territory within about 8 minutes.

Failure of the blood supply to the neuraxis

The arteries of the brain and spinal cord may be blocked by small blood clots (thrombi) and emboli. Such obstruction results in irrevocable damage.

The arteries which penetrate and supply the depth of the neuraxis are relatively small and tend to arise at right angles from the large parent vessels on the surface. Normally the flow through these small penetrating vessels is already reduced to the minimal effective level; any further lowering of pressure through any local mechanical interference can reduce the flow below critical levels, causing anoxia and damage within the territory of such vessels.

In man the arteries which are most often involved in occlusion or haemorrhage ('**stroke**') are a group (the striate arteries) arising from the **middle cerebral artery** and supplying the basal nuclei (basal ganglia) and parts of the internal capsule. Haemorrhage into the brain interferes directly with the function of the territory supplied by the ruptured artery by causing anoxia of the tissues. It also interferes indirectly with the function of adjoining areas by means of pressure from the haemorrhage and oedema. The clinical neurological signs of stroke are surveyed in Section 11.6.

Cardiac arrest during anaesthesia may cause severe damage to the brain, even though the heart beat may be restored so that the subject survives the operation. There tends to be ischaemic necrosis of the cerebral cortex in the territory of the rostral and middle cerebral arteries. The dorsal regions of the cerebral cortex, including the visual area (Fig. 8.2(a)), are the most severely affected, and then the lateral regions including the primary motor area. Blindness and extensor rigidity may result.

2. Meninges and cerebrospinal fluid

Meninges

2.1 General anatomy of the cranial and spinal meninges

The brain and spinal cord are enclosed within three membranes, the dura mater, arachnoid, and pia mater (Figs. 2.1–2.3). Of these, the pia mater is the innermost layer and is in contact with the brain and spinal cord. The dura mater is the outermost layer. The arachnoid lies between the other two membranes.

Dura mater

In the cranial cavity the dura mater adheres to the periosteum lining the inner surface of the cranial wall. In the vertebral canal it is separated from the periosteum by the epidural space. The dura is the thickest of the

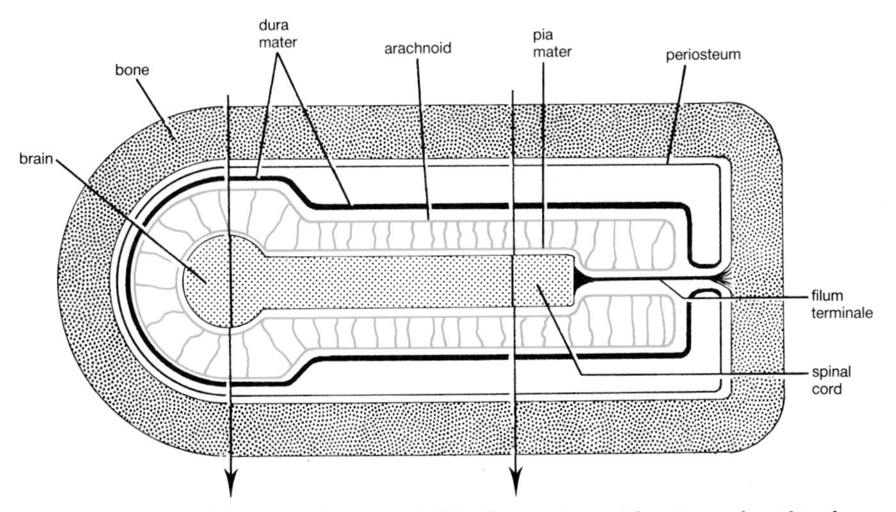

FIG. 2.1. Diagram of the neuraxis suspended in the meninges. The pia–arachnoid is shown in green. The left arrow indicates a transverse section through the brain and its meninges (Fig. 2.2). The right arrow indicates a transverse section through the spinal cord (Fig. 2-3). The filum terminale tethers the caudal end of the spinal cord to the coccygeal vertebrae.

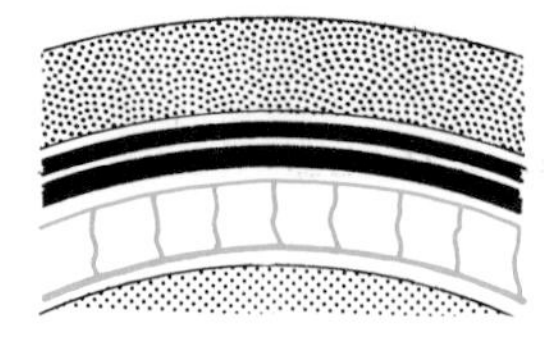

FIG. 2.2. Diagrammatic. transvere section through the brain and its meninges. The section is cut at the level of the left arrow in Fig. 2.1. The dura mater fuses with the periosteum of the cranial wall, thus eliminating the epidural space in the cranial cavity. The meninges are as labelled in Fig. 2.3, the pia–arachnoid being green.

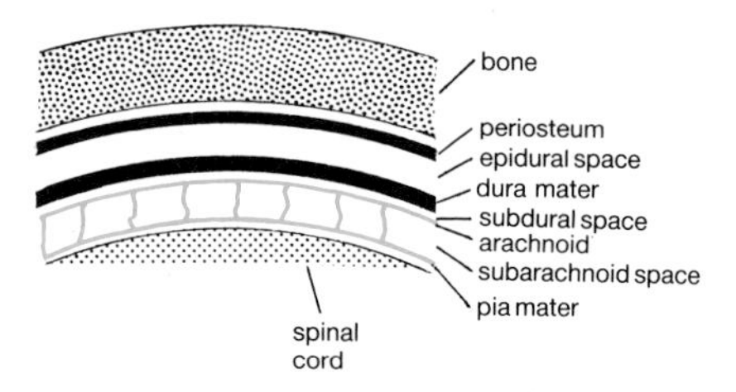

FIG. 2.3. Diagrammatic transverse section through the spinal cord and its meninges. The section is cut at the level of the right arrow in Fig. 2.1. The dura mater and periosteum are separated by the epidural space. The pia–arachnoid is shown in green.

meninges. Histologically, it consists mainly of dense connective tissue. The surface facing the arachnoid is covered by simple squamous epithelial cells.

Arachnoid

The arachnoid is a very thin membrane which is pressed against the dura mater by the pressure of the cerebrospinal fluid. Reflected from it are numerous fine filaments, which blend with the pia mater. These filaments resemble a spider's web, hence the term 'arachnoid'. Histologically, the arachnoid consists of a lamina of delicate collagenous connective tissue, coated on both sides by simple squamous epithelial cells.

Pia mater

The pia mater is thicker than the arachnoid but thinner than the dura mater. Its outer surface blends with the filaments of the arachnoid; its inner surface is fused to the brain and spinal cord. Histologically, it consists of connective tissue covered on its outer aspect by simple squamous epithelial cells. Midway between successive spinal roots the pia mater and arachnoid are firmly attached to the dura mater, forming the **denticulate ligament**; this ligament suspends the spinal cord within the dura mater.

Because they are continuous with each other by means of their fine filaments (and at the filum terminale), the pia mater and arachnoid are often collectively named the **pia–arachnoid** and regarded as a single entity.

2.2 Anatomy of the meninges at the roots of spinal and cranial nerves

The dura mater and arachnoid are extended around both the dorsal and the ventral roots of the spinal nerves in the form of tubular sleeves which prolong the subarachnoid space and therefore contain cerebrospinal fluid. In the dog, these tubular sleeves reach to the intervertebral foramina. The meninges also provide tubular sheaths for the cranial nerves, as they pass through their foramina.

Just outside the skull these sheaths generally fuse with the epineurium, but the sheaths of the olfactory and optic nerves extend to their peripheral ends. Thus the subarachnoid space connects with olfactory perineural spaces in the nasal cavity, creating a possible channel for the spread of nasal infections into the cranial cavity. The optic nerve carries the subarachnoid space to the sclera at the back of the eyeball, so that infections of the depth of the orbit can lead to meningitis.

2.3 The spaces around the meninges

Three spaces are associated with the meninges, namely the epidural space, the subdural space, and the subarachnoid space (Figs. 2.1–2.3).

Epidural space

This space is present in the vertebral canal but not in the cranial cavity. In the vertebral canal, it lies between the dura mater and the periosteum. It contains loose alveolar connective tissue, with fat which is semifluid at body temperature. The epidural space also contains the **longitudinal spinal venous sinuses**.

Subdural space

The subdural space lies between the dura mater and the arachnoid. In life this is really only a potential space containing no more than a thin film of clear yellow fluid.

Subarachnoid space

The subarachnoid space lies between the arachnoid and pia mater, and contains cerebrospinal fluid. This space is continued for a short distance around the larger blood vessels where they penetrate the surface of the brain and spinal cord (Fig. 2.5).

2.4 Collection of cerebrospinal fluid

The subarachnoid space is narrow, but there are two main sites where cerebrospinal fluid (CSF) can be withdrawn in the live animal.

The cerebellomedullary cistern (cisterna magna)

This is a large expansion of the subarachnoid space between the caudal aspect of the cerebellum and the dorsal surface of the medulla oblongata. This cistern is used in the dog (Fig. 2.4) for obtaining samples of cerebrospinal fluid with a needle and for myelography (see Section 20.8), and can also be used for sampling the CSF (usually under general anaesthesia) in the ox, sheep, horse, and cat.

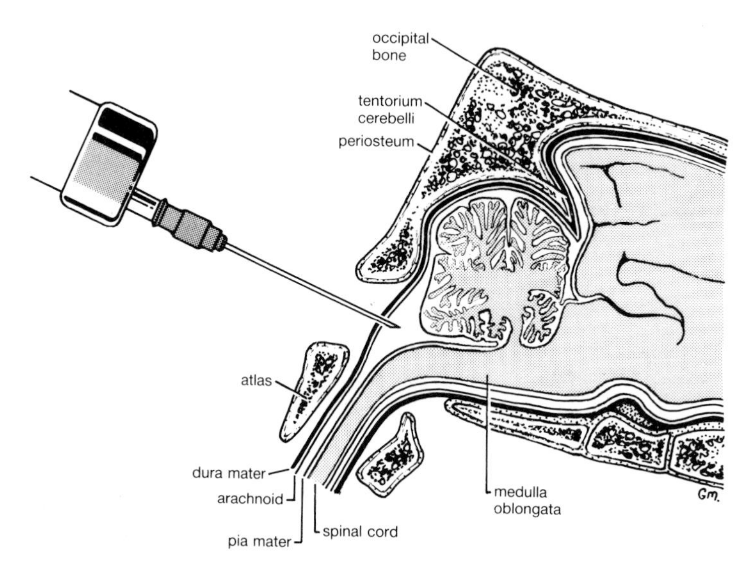

FIG. 2.4. Diagram showing the collection of cerebrospinal fluid. A needle is inserted in the cerebellomedullary cistern (cisterna magna) at the level of the foramen magnum of a dog. The head is flexed at about a right angle to the vertebral column.

The lumbosacral region

In the ox, the subarachnoid space can be entered at the lumbosacral junction (Fig. 4.1), where it is about 3 to 4 mm deep and provides an adequate volume of CSF for testing. **Lumbosacral puncture** in the dog yields only a few drops of CSF, and not enough for analysis. In man the subarachnoid space can be entered between the third and fourth or fourth and fifth lumbar vertebrae and provides sufficient volume for testing (see Section 4.4).

The chemical and cellular constituents of the cerebrospinal fluid, and also its pressure, may change in disease of the neuraxis and meninges. Analysis of these changes may be helpful in diagnosis.

2.5 Relationship of blood vessels to the meninges

Some of the smaller arteries and veins travelling on the surface of the brain and spinal cord are suspended in the filaments which pass between the pia mater and arachnoid (Fig. 2.5). Other larger ones are embedded in the pia mater on the surface of the neuraxis.

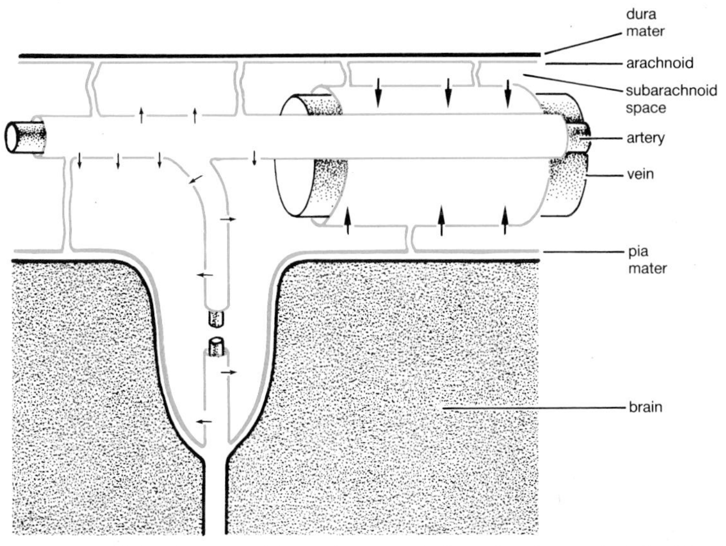

FIG. 2.5. Diagram of the intracranial meninges and their blood vessels. The blood vessels on the surface of the brain are suspended in the pia–arachnoid, which is shown in green. The formation of CSF from a small artery and an arteriole is indicated by the small arrows. Absorption of CSF into a small vein is indicated by the large arrows.

2.6 The filum terminale

The filum is a slender strand of tissue which tethers the end of the spinal cord to the coccygeal (caudal) vertebrae (Figs. 2.1, 4.1).

It has no practical importance, but is of minor theoretical interest because it causes the dura mater to be continuous with the periosteum, and the arachnoid to be continuous with the pia mater (Fig. 2.1).

2.7 The falx cerebri and membranous tentorium cerebelli

These are major folds of dura mater. The falx cerebri is a longitudinal sickle-shaped fold between the cerebral hemispheres (Fig. 2.8). The membranous tentorium cerebelli is a transverse fold separating the cerebral hemipsheres from the cerebellum (Fig. 2.4); it extends from the **bony tentorium cerebelli**, a leaf of bone which projects transversely from the parietal bone. The falx cerebri helps to protect the brain against tearing of nervous tissue during violent rotatory or sideways movements of the skull, as for example when the jaw is struck transversely by a heavy blow. The tentorium cerebelli has a similar function though in the rostrocaudal direction, as when the back of the head strikes the floor during falling over (see Section 2.13).

Cerebrospinal fluid

2.8 Formation of cerebrospinal fluid

Cerebrospinal fluid (CSF) is formed from two main sources: (i) from the small arteries, arterioles, and capillaries suspended in the filaments of the pia mater and arachnoid (Fig. 2.5); and (ii) from the choroid plexuses. Direct observations on the living tissues have shown that CSF accumulates in drops on an exposed choroid plexus.

2.9 The choroid plexuses

There are four choroid plexuses, located in the left and right lateral ventricles, the third ventricle, and the fourth ventricle (Fig. 21.14(b)). Of these, the plexus in the fourth ventricle adds a relatively large volume.

A choroid plexus (Fig. 2.6) comprises (i) a tuft of arterioles, (ii) a coat of pia–arachnoid suspending these vessels, and (iii) a lining of non-nervous epithelium consisting of cuboidal glandular cells, thrown into extensive folds to increase the surface area. The single layer of epithelial cells represents a persistence of the neural tube in its embryonic form.

2.10 Mechanism of formation of cerebrospinal fluid

CSF is probably produced mainly by dialysis, with some active secretion by the cuboidal glandular cells of the choroid plexus. The process of dialysis resembles the formation of tissue fluid, aqueous humour, synovial fluid, and glomerular filtrate. The dialysed fluid from a choroid plexus has to pass through the lining of cuboidal epithelium of the plexus (in order to enter a ventricle of the brain) and may then be modified by secretory processes of these glandular cells.

It is difficult to estimate the total volume accurately, but a figure of 200 ml has been given for the horse. In humans with a damaged cranium and an exposed subarachnoid space, volumes of up to 200 ml have been collected daily.

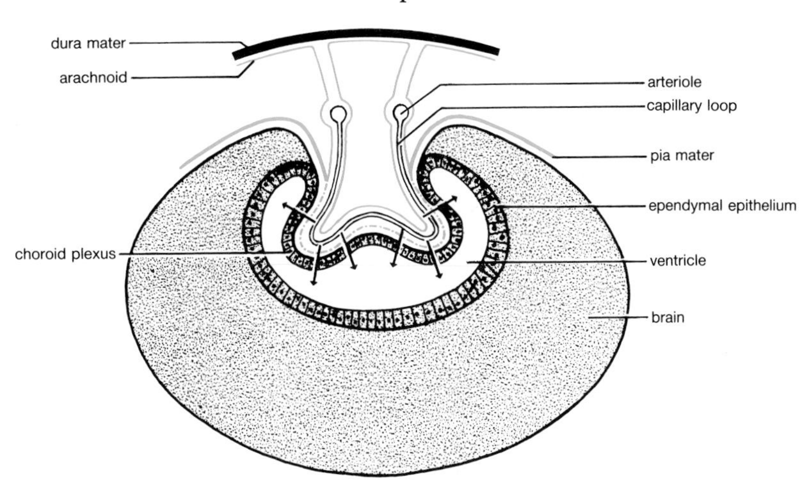

FIG. 2.6. Highly diagrammatic transverse section of a ventricle of the brain and its choroid plexus. It shows the choroid plexus invaginated into the lumen of the ventricle. A capillary loop is contained within the choroid plexus. The pia mater extends around the loop as a basal lamina (broken line). The ependymal epithelium is modified into cuboidal glandular cells which have their bases resting on the basal lamina. The formation of CSF is indicated by arrows. The pia mater is shown in green.

2.11 Circulation of cerebrospinal fluid

The cerebrospinal fluid formed by the choroid plexus of the **lateral ventricle** escapes via the **interventricular foramen** into the third ventricle (Figs. 2.7, 21.14(b)). Here the flow is augmented by the plexus of the **third ventricle**. After passing caudally through the **mesencephalic aqueduct** (cerebral aqueduct of Sylvius), the flow is again augmented, this time to a relatively greater extent, by the product of the plexus of the **fourth ventricle** (Fig. 21.14(b)).

The cerebrospinal fluid in the fourth ventricle may enter the central canal of the spinal cord. Most of it, however, escapes into the subarachnoid space through a pair of openings in the lateral walls of the fourth ventricle; these openings are called the left and right **lateral apertures** of the fourth ventricle (Figs. 2.7, 21.14(b)). Having thus entered the subarachnoid space around the medulla oblongata, the CSF circulates in all directions throughout the subarachnoid space of the cranial and spinal meninges.

In the higher primates, including man, there is a third opening in the midline; this is the **median aperture** of the fourth ventricle (Fig. 2.7).

Obstruction to the flow of CSF can occur by narrowing of the interventricular foramen, of the mesencephalic aqueduct, or of the lateral (and median, in man) apertures of the fourth ventricle. This causes **internal hydrocephalus**, which is distension of the ventricles rostral to the obstruction. Distension can gradually destroy the surrounding brain tissue by pressure from within (pressure atrophy).

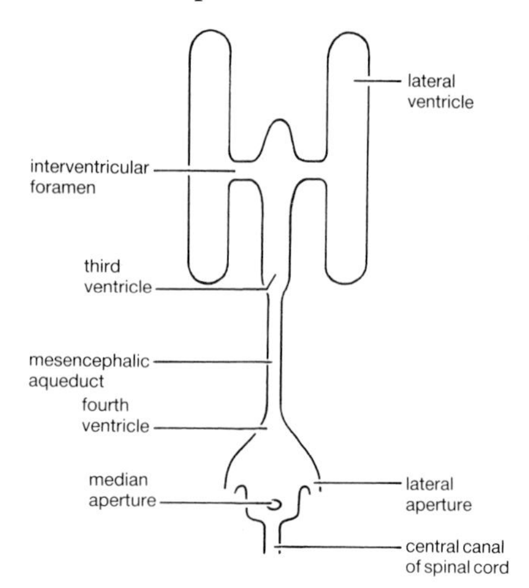

FIG. 2.7. Dorsal view diagram of the ventricles of the brain. The median aperture of the fourth ventricle occurs in man but not in the domestic mammals. The apertures of the fourth ventricle connect the ventricle to the subarachnoid space.

2.12 Drainage of cerebrospinal fluid

There are no lymphatics within the tissues of the brain and spinal cord. Cerebrospinal fluid is removed by three main routes, its speed of removal being rapid; a dye injected into the CSF appears in the venous blood within 30 seconds.

Venules of subarachnoid space

Some CSF is removed by absorption into the venules which are suspended in the subarachnoid space by the filaments of the pia mater and arachnoid (Fig. 2.5). This process is governed by essentially the same forces as those controlling the absorption of tissue fluid by the venous end of the capillary bed; it is therefore due primarily to the osmotic pressure of the venous blood.

Arachnoid villi

It is generally believed that the arachnoid villi take part in the absorption of CSF. Arachnoid villi are evaginations of the arachnoid into the venous sinuses of the dura mater (Fig. 2.8). In an arachnoid villus the barrier between CSF and blood is reduced to the arachnoid plus the endothelial lining of the sinus.

The importance of the arachnoid villi in removal of CSF is controversial. The presence of villi has been confirmed in man, sheep, and some other mammals, but it is uncertain whether villi are typical of mammals generally. It is commonly thought that they exist in the dog, but this has been disputed. In some species which definitely do have arachnoid villi, the electron microscope has revealed valve-like perforations in the wall of the villi, apparently bringing the CSF into direct continuity with venous blood.

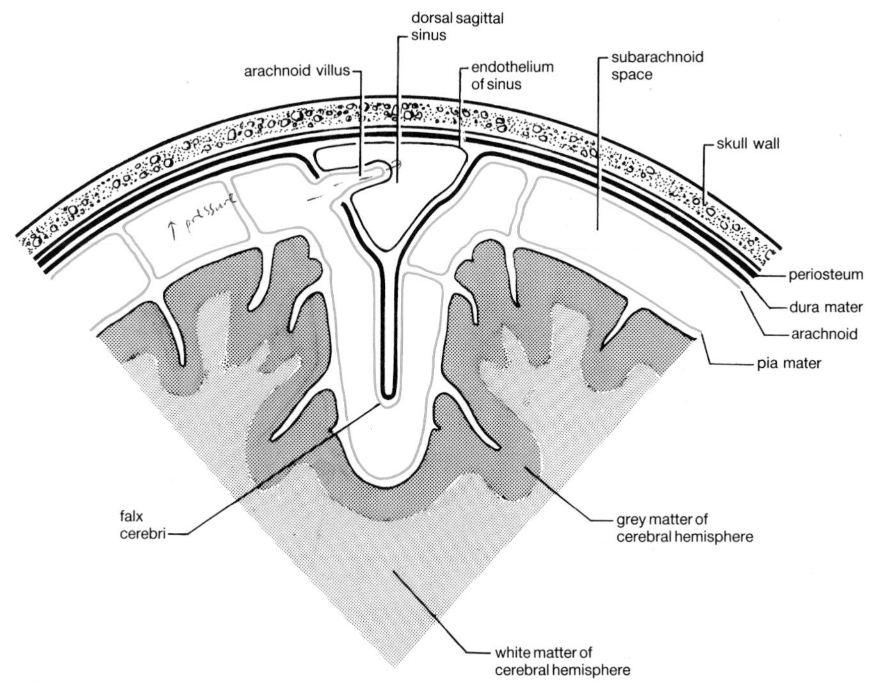

FIG. 2.8. Diagrammatic transverse section through the falx cerebri. The dorsal sagittal venous sinus is suspended in the falx, and an arachnoid villus projects into the dorsal sagittal sinus. In an arachnoid villus the CSF and blood are separated only by the arachnoid and the endothelial lining of the venous sinus. In some species, including man and horse, there is a ventral sagittal sinus in the ventral edge of the falx cerebri.

Lymphatics of cranial and spinal nerves

A less important route of drainage is into the lymphatic vessels on the roots of the cranial and spinal nerves.

Failure of the drainage mechanisms produces accumulation of cerebrospinal fluid in the subarachnoid space, leading to **external hydrocephalus**. This may be caused by damage to arachnoid villi by meningitis.

2.13 Functions of cerebrospinal fluid

Cerebrospinal fluid has three main functions: (i) **Protection** of the brain and spinal cord against impact upon their surrounding bony walls. (ii) **Nutrition** of the brain and cord. CSF may have significant metabolic functions. (iii) To permit **variations in the volume of blood** inside the cranial cavity. Since the cranial cavity has bony walls and is therefore not distensible, an increase in blood volume can only occur if some other content of the cranial cavity is proportionately reduced in volume; reduction of CSF permits a reciprocal increase of blood volume, and vice versa.

Protection of the neuraxis against impact

An unfixed brain, once removed from the skull, is as soft and vulnerable as a blancmange. When tipped out of its container, a very soft blancmange readily loses its shape. So also does the fresh brain; lay it on the table and it distorts, and may even rupture, as a result of the forces arising from its own weight. Without the support of the meninges and cerebrospinal fluid the brain within the skull could be damaged under the ordinary conditions of everyday life.

The protective function of the CSF depends mainly on buoyancy. A human brain and spinal cord weigh about 1500 g outside the body, but have a net weight *in situ* of only about 50 g. This great reduction in virtual weight renders effective the support given by the many delicate filaments of the pia–arachnoid, the denticulate ligaments, the falx cerebri, and tentorium cerebelli, and the nerves and blood vessels running to and from the neuraxis. The thin film of fluid in the subdural space allows the brain some freedom to slide within the cranial cavity, thereby imparting a certain degree of 'give' to the system, and yet surface tension prevents the dura mater and arachnoid from separating from each other. The importance of the cerebrospinal fluid in supporting the neuraxis by buoyancy has been shown by pneumoencephalography, or ventriculography using gas, procedures in which radiographic contrast is increased by substituting air for CSF (see Section 20.5). Without the support of the CSF the full weight of the neuraxis rests upon its delicate meningeal, nervous, and vascular supports. In man these techniques are painful, and the slightest jarring of the head greatly intensifies the pain.

Under severe impact these protective mechanisms can fail. Apart from direct injury to the brain by fractures of the skull, sudden acceleration and/or deceleration of the head (as caused by a blow to the side of the jaw, or falling over backwards and striking the back of the head) can cause shearing of nervous tissue, collision of the brain against the skull wall, or crushing by the ring of the tentorium cerebelli or foramen magnum. **Subarachnoid haemorrhage** (bleeding into the subarachnoid space) introduces large amounts of plasma proteins into the cerebrospinal fluid, which is normally nearly free of protein; this increases the osmotic pressure of the CSF and can lead to increased volume of CSF with a dangerous rise in intracranial pressure. Since the three meninges (including the subarachnoid space) extend along the optic nerve and attach to the sclera, a rise in intracranial pressure compresses the lymphatic channels of the optic nerve and the central vein travelling within the nerve; the resulting lymphatic engorgement causes **papilloedema** (swelling of the optic disc), and the venous obstruction dilates the retinal veins, changes which may be recognizable with the ophthalmoscope.

2.14 Blood–brain barrier

Some substances which normally pass freely from the blood through the tissue fluid in the body generally, are unable to penetrate the capillary walls in the brain. This fact has led to the hypothesis of the blood–brain barrier. The barrier is obviously not total since many substances like water, gases, electrolytes (Na^+, K^+, Cl^-), glucose, and amino acids do pass freely across the tissues of the brain.

Substances which are able to escape freely from the blood and into the tissue fluid in the body but cannot pass through the capillary walls in the brain include dyes such as trypan blue and large molecules such as ferritin.

The mechanism of the barrier is not fully understood. The cellular components of the central nervous system are so tightly packed together that the extracellular space is thought to be unusually small but there is quite enough room for substances in solution to travel freely. The main elements of the barrier seem to lie in the capillaries of the brain. In contrast to capillaries elsewhere in the body, their endothelial cells have very few micropinocytotic vesicles and their overlapping edges are sealed by continuous tight junctions (true zonulae occludentes) which leave no gaps anywhere. In addition the outer surface of the basal lamina of nearly all the capillaries of the brain is covered by the perivascular feet of the astrocytes (about 85 per cent of the total capillary surface is believed to be covered in this way), and this may contribute to the barrier (see Section 5.20). The function of the barrier may be to protect the brain from invasion by toxic compounds.

3. Venous drainage of the spinal cord and brain

The venous sinuses of the neuraxis have two components, namely a spinal system and a cranial system. The spinal system drains the spinal cord and the vertebrae. The cranial system drains the brain and certain extracranial regions. These two components joint to form one continuous system of venous sinuses, in which valves are poorly developed and scarce, or even absent altogether. Transverse connections from the left to the right side of this system of sinuses are common. They include (i) the midline cavernous sinus, (ii) the transverse sinus, and (iii) the anastomoses of the spinal system across the midline.

All of the venous sinuses are outside the dura mater. In the head they are sandwiched between the dura mater and the periosteum. In the vertebral canal they are in the epidural space.

The spinal system of venous sinuses

3.1 General plan

The main component is a pair of **longitudinal spinal sinuses**, running parallel with each other along the whole length of the vertebral canal (Fig. 3.1). They lie on the ventral floor of the vertebral canal. Valves are absent or poorly developed. Transverse connections from one side to the other occur erratically at the junctions between the vertebrae.

The left and right longitudinal spinal sinuses are sometimes called the left and right **vertebral sinuses**. In the Nomina Anatomica Veterinaria they are named the **internal vertebral plexus**.

3.2 Connections to the cranial system of sinuses

At the foramen magnum there is direct continuity, end to end, of the longitudinal spinal sinuses with the connecting system of cranial sinuses (Fig. 3.1).

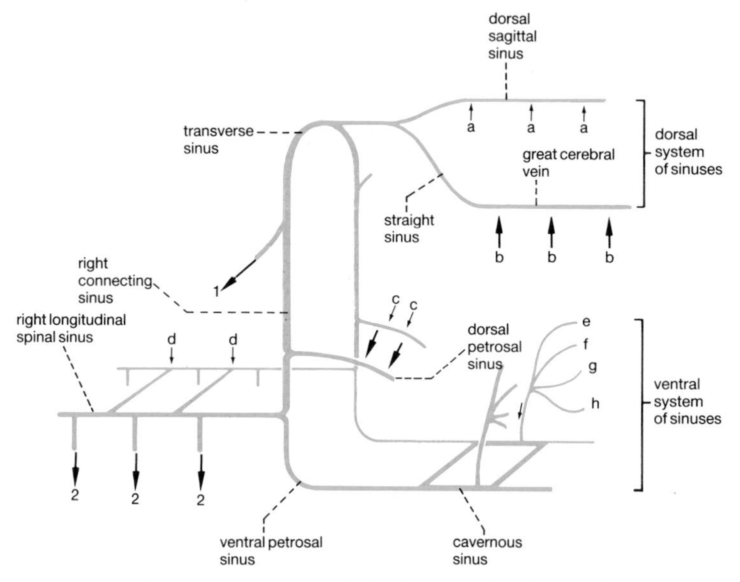

FIG. 3.1. **Diagram of the venous sinuses of the brain and spinal cord.** The structures are viewed from the right side. Arrows indicate direction of blood flow. 1, drainage **from** the cranial sinuses into the maxillary vein; 2, drainage **from** the spinal sinuses into the intervertebral veins. a to h, drainage **into** the sinuses; a, from the dorsal surface of the forebrain; b, from the dorsal deep parts of the forebrain; c, from the ventral surface of the forebrain; d, from the spinal cord; e, from the face; f, from the orbit; g, from the nasal cavity; h, from the upper teeth.

3.3 Territory drained by the spinal system of sinuses

The spinal sinuses drain the whole length of the spinal cord, and the cauda equina.

The veins of the spinal cord empty into the **ventral spinal vein** (or veins), which accompanies the ventral spinal artery along the ventral fissure of the spinal cord. Other veins drain from the spinal cord into the network of veins lying on the surface of the cord. Veins accompanying the roots of the spinal nerves drain the ventral vein and the venous network into the longitudinal spinal sinuses. The spinal sinuses also drain the vertebral bodies including their haemopoietic tissues.

3.4 Drainage of the spinal sinuses into the systemic circulation

The longitudinal spinal sinuses drain by **intervertebral veins** (Fig. 3.1); these are segmental and escape from the vertebral canal by passing through the intervertebral foramina alongside the spinal nerves. In the neck they empty into the vertebral veins. In the thorax they drain into the azygos vein. In the abdomen they discharge mainly into the caudal vena cava and also into the azygos vein. Valves direct the flow away from the spinal cord, but distension of these intervertebral veins can apparently render their valves incompetent (see Section 3.10 for the pathological significance of this).

The cranial system of venous sinuses

3.5 General plan

The cranial system of venous sinuses can be subdivided into three systems of sinuses:

Dorsal system of sinuses

This group of sinuses drains the dorsal areas of the forebrain, by the **dorsal cerebral veins**; these areas include most of the cortex of the cerebral hemispheres. The dorsal system of sinuses also drains deep dorsal parts of the forebrain by the **great cerebral vein** (Fig. 3.2).

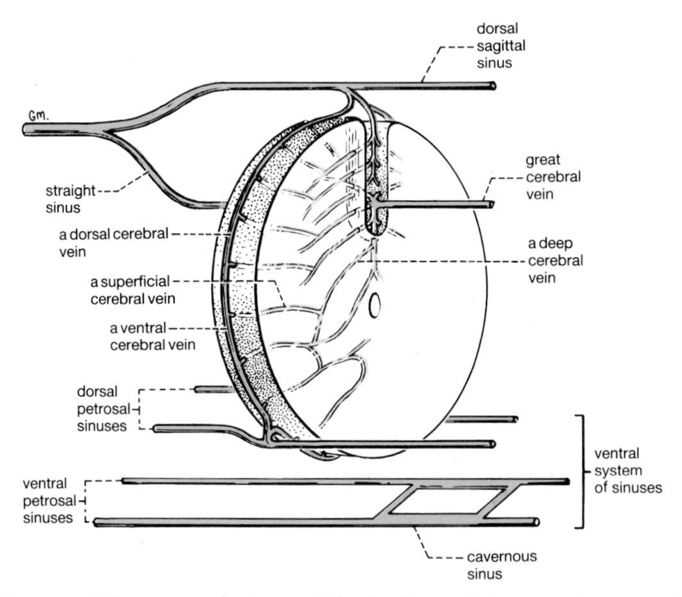

FIG. 3.2. **Diagram of the venous drainage of the forebrain.** It shows a diagrammatic slice cut transversely through the cerebral hemispheres, and illustrates the principles of the venous drainage of the forebrain into the dorsal and ventral systems of the cranial venous sinuses, viewed from the right side.

Ventral system of sinuses

This system drains (i) the ventral areas of the forebrain, by the **ventral cerebral veins** (Fig. 3.2), and (ii) the face, nasal cavity, orbit, and upper teeth (Fig. 3.1). The orbital drainage includes the **angularis oculi vein** which drains the skin in the region of the upper eyelid and is used for cavernous sinus venography (see Section 20.5).

This vein, which is valveless, empties at one end into the facial vein and at the other end into the dorsal external ophthalmic vein and thence into the cavernous sinus. The nasal drainage is important in protecting the brain against a rise in body temperature (see Section 1.6).

Connecting system of sinuses

The connecting system consists of a (paired) **connecting sinus** (known as the **sigmoid sinus** in the *Nomina Anatomica Veterinaria*). The connecting sinus is the focal point of the cranial venous sinuses. It receives the dorsal system via the transverse sinus, and the ventral system via the petrosal sinuses, and connects directly with the spinal system of sinus (Fig. 3.1). It also provides the principal drainage of the cranial venous sinuses by emptying into the maxillary vein (1, in Fig. 3.1).

3.6 The components of the dorsal system of sinuses

The dorsal system comprises three principal sinuses (Figs. 3.1, 3.2):

Dorsal sagittal sinus

This sinus lies in the falx cerebri (Fig. 2.8) and is the principal site of arachnoid villi. It empties at the junction in the midline of the left and right transverse sinuses.

Straight sinus

The straight sinus drains the great cerebral vein into the dorsal sagittal sinus (Figs. 3.1, 3.2). It lies in the caudal part of the ventral edge of the falx cerebri.

Transverse sinus

The transverse sinus is considered to be paired. The left and right sinuses are joined in the dorsal midline, where they receive the dorsal sagittal sinus (Fig. 3.1). They continue laterally into the connecting sinuses.

3.7 The components of the ventral system of sinuses

The ventral system contains three main sinuses:

Cavernous sinus

This is strictly a pair of sinuses, but the pair are cross-connected and can therefore be regarded as a median sinus (Figs. 3.1, 3.2).

Dorsal petrosal and ventral petrosal sinuses

The dorsal petrosal sinus drains the ventral regions of the brain (Figs. 3.1, 3.2). The ventral petrosal sinus is buried in bone, and simply joins the cavernous sinus to the connecting sinus (Fig. 3.1).

3.8 Drainage of the cranial sinuses into the systemic circulation

The cranial system of venous sinuses drains mainly via the connecting sinus into the **maxillary vein** (Fig. 3.1).

Strictly it is the transverse sinus that drains into the maxillary vein, via the **temporal sinus**. Other **emissary veins** drain the cranial system into the occipital vein and internal jugular vein.

The veins of the brain

The veins of the brain are unusual in having no valves. Also they have no smooth muscle in their walls and are therefore very thin-walled. Anastomoses between the veins of the brain occur relatively commonly (as compared with the arteries), particularly on the surface of the brain. Unlike the arteries there are also some anastomoses within the substance of the brain. Since the veins have no valves, there is the possibility that alternative channels may develop if a vein is obstructed.

The veins of the more caudal part of the brainstem correspond essentially to the arteries. The **cerebral veins**, however, are not satellites of the arteries. These veins are arranged in two groups, superficial and deep. The **superficial cerebral veins** drain by **dorsal cerebral veins** (four or five large ones) into the **dorsal** system of cranial venous sinuses, and by **ventral cerebral veins** into the **ventral** system via the dorsal petrosal sinus (Fig. 3.2). The **deep cerebral veins** drain the deep dorsal parts of the hemispheres; these deep veins empty into the **great cerebral vein**, and thence into the **straight sinus** (Fig. 3.2).

Clinical significance of the venous drainage of the neuraxis

3.9 Spread of infection in the head

Infective material which enters the veins of the face, nasal cavity, orbit, and upper teeth, can drain directly into the ventral system of cranial sinuses. Bacterial attack on the endothelium of the sinuses may cause a thrombus (clot), and this can lead to a sudden obstruction of venous drainage in the sinus system with serious if not fatal results.

3.10 Paradoxical embolism

The continuity of the valveless (or effectively valveless) sinus system throughout the whole length of the neuraxis enables bacteria or tumour cells to travel from the thorax or abdomen to the head, or from the head to the thorax or abdomen. For instance, abdominal straining can produce a pressure gradient within the sinus system, capable of driving blood from an abdominal organ such as the kidney into the spinal sinuses. Such a flow would initially have to take place **against** the valves which guard the intervertebral veins connecting the venous sinuses with the caudal vena cava; however, during 'straining' the pressure rises in these veins and distends them, and this is evidently sufficient to render their valves incompetent. Abdominal blood can then enter the longitudinal spinal sinuses and travel to the head, or drain into the vertebral vein in the neck, or into the azygos vein in the thorax. In this way microorganisms from an infected kidney can reach the head, neck, or thorax. These routes produce unexpected movements of infection or of tumour cells, a phenomenon known as paradoxical embolism.

3.11 Venous obstruction

Probably because of the frequent occurrence of venous anastomoses, even below the surface of the brain, lesions resulting from venous occlusion appear to be much less important clinically than those arising from arterial obstruction. Nevertheless, from general principles it can be predicted that a sudden blockage of a substantial vein may cause capillary haemorrhage in the area which it drains; extensive occlusion (by thrombosis) of a considerable length of such a vein will certainly cause severe damage to the brain.

The actual recognition of lesions resulting from intracranial venous obstruction has been slow, in both man and lower mammals. Few observations have been made, but it is known that in man surgical interference with the great cerebral vein can lead to coma and muscular rigidity. It has been suggested that lesions of the veins, if they were but looked for, might turn out to be the basis of certain cerebral vascular accidents in man. There is some evidence that some of the slight but distinct neurological defects which are found in young children are due to hypoxic lesions, mainly in the cerebral cortex, which have arisen basically from thrombosis of the dural sinuses and cerebral veins; such thromboses appear to arise from prenatal circulatory failure, which is commonly due to placental defects, especially premature detachment. There is also evidence that tearing of the venous sinuses at the junction of the falx cerebri and membranous tentorium cerebelli, due to excessive distortion of the head during delivery, is a common injury in infants which can sometimes lead to fatal subdural haemorrhage into the posterior fossa.

3.12 Angiography for diagnosis

The pattern of the venous sinuses and veins inside the cranial cavity is clinically important, since cerebral venography is used to diagnose disorders of the CNS. Most of the intracranial sinuses and veins of the brain which have been mentioned above are visible in venograms (see Section 20.5).

4. Applied anatomy of the vertebral canal

The anatomy of epidural anaesthesia and lumbar puncture

4.1 The vertebrae

Numbers of lumbar and sacral vertebrae

In the ox and horse there are six lumbar and five sacral vertebrae. The sheep and pig have six or seven lumbar and four sacral vertebrae. In the dog and cat there are seven lumbar and three sacral vertebrae. Man has five lumbar and five sacral vertebrae. Thus apart from the ox and horse which have 11, there are 10 lumbar and sacral vertebrae altogether.

Coccygeal vertebrae (caudal vertebrae)

In the ox the first coccygeal vertebra is firmly attached to the sacrum, forming a relatively firm and immobile joint. The joint between the first and second coccygeal vertebrae is much more flexible and mobile, forming the first easily movable joint of the tail. The dorsal aspect of this articulation between Co1 and Co2 is a useful site for epidural injection in the ox (Fig. 4.1), a common and important technique in this species.

Ligaments

The spinous processes of the vertebrae are joined together by two ligaments, (i) the **supraspinous ligament**, covering the dorsal summits of the spines, and (ii) the **interspinous ligament**, passing between the spines (Fig. 4.1). The **interarcuate ligament** (ligamentum flavum) is a thin elastic sheet joining the arches of adjacent vertebrae (Fig. 4.1).

4.2 Spinal cord

Filum terminale and cauda equina

The filum terminale is supported by a leash of sacral and coccygeal spinal nerve roots called the cauda equina (horse's tail). The filum and cauda equina escape from the subarachnoid space and dural tube at about the

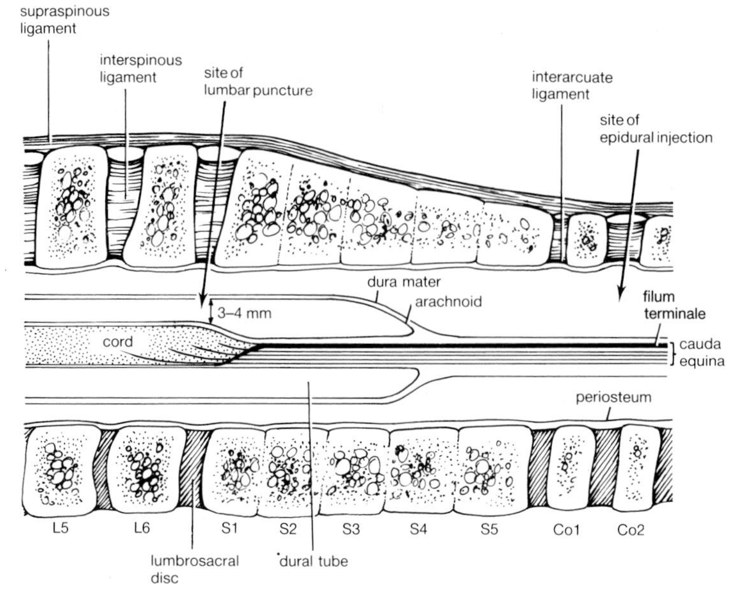

FIG. 4.1. Diagrammatic longitudinal section of the lumbosacral vertebral canal in the ox. The diagram shows sites of lumbar puncture and epidural anaesthesia. The spinal cord ends at about S1. The dural tube extends to about S3.

fourth sacral segment (S4) in the ox (Fig. 4.1). Caudal to this point they are covered only by dura mater, which in fact now becomes the epineurium.

Strictly, the filum terminale combines with the dura mater to form the **filum of the dura mater**.

The end of the spinal cord

In the domestic mammals the spinal cord ends at about the lumbosacral junction.

In the adult ox it ends at vertebra S1; in the calf it reaches S3. It ends at S1 or S2 in the horse, L7 in the dog (Fig. 20.6), L7 in the cat and pig, and S1 in the sheep. In man the spinal cord is relatively much shorter, ending at about L1.

4.3 Meninges

The **dural tube** (and within it the subarachnoid space) continues caudally beyond the end of the spinal cord, ending in the sacrum (Fig. 4.1). It contains the beginning of the cauda equina and filum terminale.

The dural tube reaches as far as vertebra S4 in the ox and horse, and S1 or S2 in the dog and cat.

4.4 Lumbar puncture

In the ox the lumbar subarachnoid space between L6 and S4 is relatively voluminous, enabling CSF to be withdrawn by lumbar puncture between L6 and S1 (Fig. 4.1).

In the average adult ox the subarachnoid space measures about 3 to 4 mm dorsoventrally at this point, the distance from the skin to the spinal cord being about 7 cm. Lumbosacral puncture is also feasible in the sheep, pig, and horse. See also Section 2.4 for discussion of the cerebellomedullary cistern. In man CSF can be obtained relatively easily by lumbar puncture (between L3–4 or L4–5), since the spinal cord ends cranial to these levels.

4.5 Epidural anaesthesia in the ox

Site

Epidural anaesthesia in the ox is usually carried out by injecting between Co1 and Co2 (Fig. 4.1), but can be done at L6–S1. The principle is to inject local anaesthetic into the epidural space. The anaesthetic spreads gradually through the semifluid fat, blocking the roots of the spinal nerves as it goes. Observations on the ox, dog, and man suggest that the dura is slightly permeable to local anaesthetics and that diffusion into the CSF causes some if not all of the effects of epidural injection.

Objectives

A common obstetrical reason for use of epidural anaesthesia in the ox is to prevent 'straining' during parturition. Epidural anaesthesia eliminates these expulsive efforts by interrupting the **sensory** part of the reflex arc which arises from the stimulus to the vaginal mucosa when the fetal membranes or fetus enter the vagina. The mother reflexly responds to these objects as though they were a foreign body, and tries to force them out. The sensory pathway from the vagina is the **pudendal nerve**, which projects into the CNS via sacral spinal nerves, S2 to S4 in the ox. In the ox sufficient anaesthetic can be injected epidurally to block the dorsal and ventral roots as far cranially as S2 only. Provided that the nerves at S1 and L6 are not blocked as well, the sciatic nerve is not paralysed and the animal will remain standing.

Anatomical hazards

There are anatomical hazards in epidural anaesthesia. (i) It is desirable not to pierce the cauda equina in domestic animals, as this may impair the nerve supply to the tail. (ii) Further overshooting may insert the needle into the longitudinal spinal sinus; local anaesthetic should not be injected into a vein. (iii) If the site of injection is Co1–2 it is impossible to inject into the CSF, but if the injection is at the lumbosacral junction or cranial to this level, this danger exists; since local anaesthetics mix freely with CSF the anaesthetic spreads much further cranially and gives over-extensive anaesthesia.

Epidural anaesthesia in other species

Epidural anaesthesia is also used in the horse for obstetrical purposes and general surgery in the perineal region. In the sheep the anatomy of epidural anaesthesia and lumbar puncture is essentially the same as in the ox, except that there are only four sacral vertebrae and the dural tube ends in segment S3. In man epidural anaesthesia is obtained by injecting between vertebrae L3–4, or L4–5.

4.6 Injuries to the root of the tail

In the pig, the tail may be bitten off and this can introduce infection into the vertebral canal and longitudinal venous sinuses. In the cat a bite at the root of the tail can introduce infection and lead to abscess formation in the epidural space.

The anatomy of the intervertebral disc

4.7 The components of the disc

The main components of the intervertebral disc are the anulus fibrosus and the nucleus pulposus. The disc is also supported by the dorsal and ventral longitudinal ligaments.

Anulus fibrosus

The anulus fibrosus consists of a series of about 25 to 35 concentric laminae, one inside the other like the layers of an onion (Fig. 4.2). Each

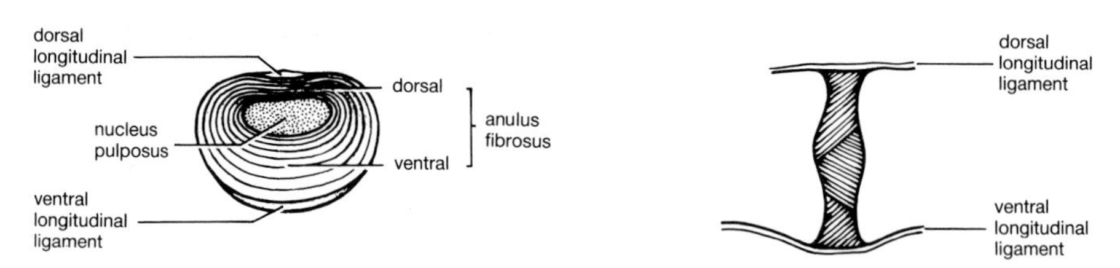

FIG. 4.2. (left) Normal intervertebral disc in transverse section of a dog. The rings represent the 25 to 30 laminae of the anulus fibrosus. The laminae are much thinner dorsally than ventrally. The nucleus pulposus is dorsally eccentric in position.

FIG. 4.3. (right) Diagrammatic lateral view of an intervertebral disc. The outermost lamina has been partly removed, exposing two successively deeper laminae with their collagen fibres passing obliquely to each other, in alternating directions. The collagen fibres are embedded in the bony epiphysis at each end.

lamina consists of collagen fibres. The direction of the fibres alternates in successive laminae (Fig. 4.3). The collagen fibres of each lamina pass diagonally from the bony epiphysis of one vertebra to the bony epiphysis of the succeeding vertebra (Fig. 4.3), being deeply and firmly embedded in bone at each end.

Nucleus pulposus

The nucleus pulposus consists of an embryonic jelly-like tissue with few cells. Its role is to distribute forces evenly throughout the disc (Figs. 4.4(a), (b)) in the same manner as the fluid-filled bag in the centre of a golf ball. It lies somewhat dorsally in the disc, and this makes the dorsal part of the anulus fibrosus much thinner and weaker than the ventral part (Fig. 4.2). The nucleus pulposus is a remnant of the notochord.

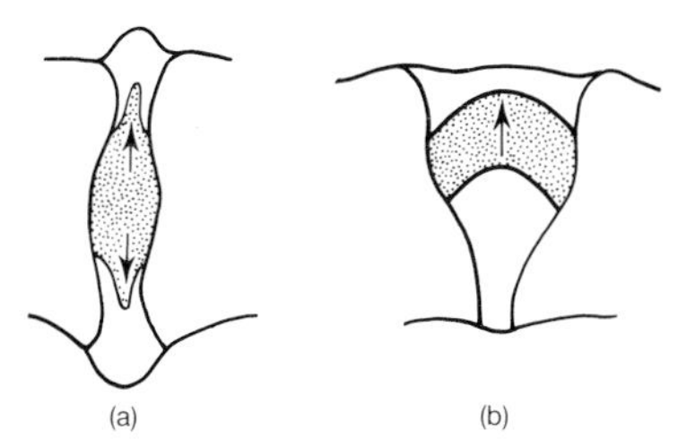

FIG. 4.4. **(a) Diagram showing the effect of craniocaudal compression of an intervertebral disc.** The arrows indicate forces in the nucleus pulposus. The anulus fibrosus is distended both dorsally and ventrally, but the thinness of the anulus dorsally favours dorsal protrusion of the nucleus pulposus.

(b) Diagram showing the effect of flexion of an intervertebral joint. The arrow indicates forces in the nucleus pulposus. The anulus fibrosus is stretched dorsally and the nucleus pulposus is being forced dorsally, thus predisposing to dorsal disc protrusion.

The longitudinal ligaments

The dorsal and ventral longitudinal ligaments blend with the outermost laminae of the intervertebral discs, and thus reinforce the discs dorsally and ventrally (Figs. 4.2, 4.3).

The ventral ligament is much reduced or absent between the midthoracic region and head in the dog, but reaches the head in man; the dorsal ligament extends throughout the whole length of the vertebral column in both man and dog.

Blood and nerve supply

The whole disc is avascular, the lumbosacral disc being the largest avascular structure in the body. The laminae of the outer half (and possibly the deeper laminae also) of the anulus fibrosus are innervated, and this may be enough to make small ruptures of the disc painful (although it is not known for certain that this is so).

4.8 Senile changes

Senile changes begin to overtake the disc relatively early in life, when the rest of the body is otherwise in its prime (by about 20 years in man, or 7 years in the dog). These age changes consist essentially of a loosening and disruption of the collagen fibres in the laminae, and conversion of the jelly-like nucleus pulposus into fibrocartilage. Once the nucleus pulposus has undergone this change it can no longer distribute forces evenly throughout the disc. Great forces can now develop at particular points in the anulus fibrosus. The dorsal part of the anulus fibrosus is especially vulnerable, notably when the vertebral column is flexed ventrally (Fig. 4.4(b)). These forces in the dorsal part of the anulus become maximal in man when lifting a heavy weight with the back flexed.

In some breeds of dog, notably Dachshunds, the senile changes are advanced by the time the animal is only 2 years of age, but this is a pathological state within these breeds (*chondrodystrophia*).

4.9 Disc protrusion

General principles

Once the normal senile changes have begun, there is a tendency for the collagen fibres of the **dorsal** part of the anulus fibrosus to begin to rupture. This may happen gradually, or many fibres may rupture abruptly. The tissue of the senile fibrocartilaginous nucleus pulposus then escapes through the ruptured fibres in the dorsal part of the anulus. The escape may be gradual, the outermost lamellae remaining intact but bulging outwards (a **type 2 protrusion**); alternatively the anulus fibrosus may be completely perforated (a **type 1 protrusion**). Type 2 protrusions can become very large and may then compress the spinal cord from the ventral direction. Type 1 protrusions tend to occur explosively and quite often strike the ventral aspect of the cord a violent blow, capable of causing both severe mechanical damage directly to the spinal cord, and also major injury to the blood vessels of the cord thus leading to extensive ischaemic necrosis (death of neurons through reduced blood flow). Even when the volume of protruded tissue is small, or even quite minute, a type 1 protrusion can be extremely destructive since it may be ejected against the spinal cord with projectile-like force. Escape of tissue of the nucleus pulposus is known to the layman as a 'slipped disc', but obviously it is impossible for the entire disc to slip between its two vertebrae. If a protrusion escapes dorsolaterally, it may compress the spinal nerve in the intervertebral foramen and cause acute pain.

Regional distribution in domestic animals

The commonest site of disc protrusion in dogs is at the thoracolumbar junction. Here it can cause severe damage to the spinal cord, resulting in paralysis caudal to the injury. This is not uncommon in Dachshunds by 3 years of age. The nine discs between the first and tenth thoracic vertebrae are protected dorsally by the strong **intercapital** (conjugal) **ligament**, which joins the head of each pair of ribs from the second to the tenth rib inclusive (Fig. 4.5). Protrusions scarcely ever occur from these nine discs in dogs. In cats cervical and mid-lumbar protrusions are common, but they seem to cause little trouble clinically. In bulls protrusions have been recorded particularly at the lumbosacral junction.

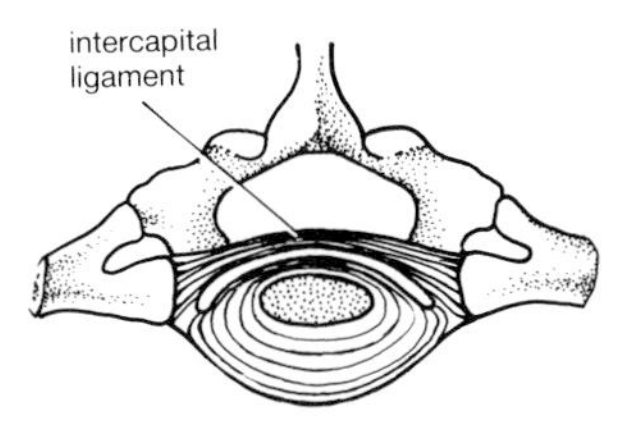

FIG. 4.5. Semidiagrammatic transverse section through a typical intervertebral disc between vertebrae T1 and T10 in the dog. The intercapital ligament joins the two ribs. This ligament almost completely prevents dorsal protrusions of the nine intervertebral discs between vertebrae T1 and T10 in the dog.

Regional distribution in man

The commonest site of disc protrusion in man is at the lumbosacral junction. Since the spinal cord ends at L1 in man, a protrusion at the lumbosacral junction strikes the cauda equina and not the spinal cord. Consequently disc protrusions rarely cause paralysis in man. Sometimes, however, they cause incapacitating pain through compression of the cauda equina or of spinal nerves in the intervertebral foramina.

4.10 Fibrocartilaginous embolism

Fragments of fibrocartilage have been found within the lumen of the arteries of the spinal cord. The resulting lesions in the spinal cord may produce neurological signs typical of an upper motor neuron or lower motor neuron disorder (see Sections 10.12 and 13.6). The cause appears to be emboli from an intervertebral disc, but the mode of entry of such an embolus into the arterial circulation is difficult to explain. This condition has been reported in the dog, cat, horse, and pig.

Malformation or malarticulation of vertebrae

4.11 The wobbler syndrome in the dog

This condition is well recognized in several breeds (e.g. Great Dane, Doberman Pinscher, Irish Wolfhound, Rottweiler, Saint Bernard). All these are large breeds with a massive head, and tend to have a high arching neck.

One of the more caudal of the cervical vertebrae (C5, C6, or C7, but typically C7) may become malformed with subluxation, so that the cranial end of the vertebra becomes displaced dorsally into the vertebral canal; a similar abnormality can occur in the Bassett Hound but then the abnormal vertebra is usually C3. In a lateral radiograph the affected vertebra appears to be out of alignment with the one in front, though this may only be obvious when the neck is flexed. Alternatively there may be a narrowing (stenosis) of the cranial opening of the vertebral canal of the affected vertebra. In many cases both of these abnormalities are present. In Great Danes such abnormalities seem often to be correlated with overnutrition, giving an excessive growth rate. Genetic factors may also be involved. The lesions in the spinal cord, and the resulting clinical signs, are discussed in Section 7.4.

4.12 The wobbler syndrome in the horse

Vertebral abnormalities similar to those found in dogs also occur in horses, except that the affected vertebrae tend to be more cranial in position (C3 to C6) (see Section 7.4).

4.13 Atlanto-axial subluxation in dogs

This condition involves dorsal luxation of the axis into the vertebral canal. The lesions include congenital malformation of the dens, tearing of the transverse ligament of the atlas, and factures of the atlas or axis (see Section 20.7).

4.14 Anomalous atlanto-occipital region in Arab horses

The atlas is typically fused to the occipital bone. The canal of the atlas is very narrow, leading to compression of the spinal cord with a resulting upper motor neuron disorder. The neck is extended and appears to be unusually immobile. This condition has been found only in Arabian horses and is believed to be genetic in origin.

4.15 Other vertebral abnormalities in dogs

Spina bifida, spondylosis, and discospondylitis, all of which are capable of causing neurological signs, are briefly discussed in Section 20.7.

5. The neuron

The anatomy of the neuron

5.1 General structure

A typical neuron has dendrites, a cell body (perikaryon), an axon, collateral branches, axon terminals, and synaptic bulbs.

Relationships of dendrites and axon to cell body

The most common arrangement of neurons lying within the neuraxis is for the dendrites to lead directly into the cell body, and the whole of the axon to follow the cell body (Fig. 5.1(a)). The position of the cell body can vary, however. A **primary afferent neuron** (the first neuron in a sensory pathway from the periphery) has its cell body in a **dorsal root ganglion** of a spinal nerve. In this type of neuron, the cell body has 'slid' along the axon and, in functional terms, is now near the end of the axon instead of its beginning (Fig. 5.1(b)). The peripheral end of a primary afferent neuron typically consists of fine branching **receptor terminals**, which correspond to the dendrites of an ordinary neuron; the extreme tip of a receptor terminal is

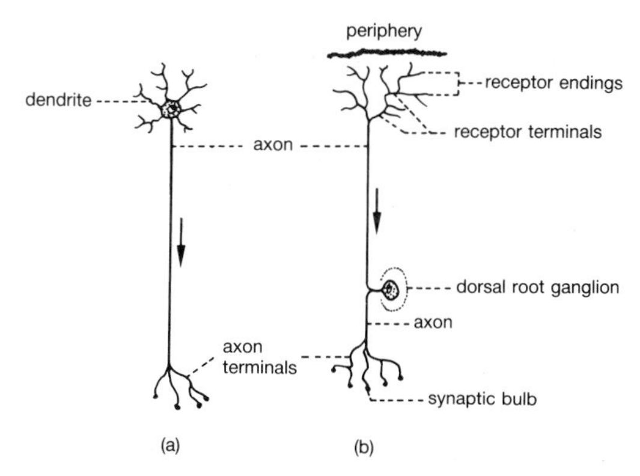

FIG. 5.1. Two types of neuron with the cell body in different positions on the axon. (a) Typical neuron within the neuraxis. The numerous dendrites lead directly into the cell body, and the whole of the axon follows the cell body. (b) Primary afferent pseudounipolar neuron with its cell body in a dorsal root ganglion. The cell body is now near the end, rather than the beginning of the axon. The cell body has only a single cell process, which immediately divides into two.

specialized, structurally and electrophysiologically, to form a **receptor ending**. The resiting of the neuronal cell body of a primary afferent neuron reflects the evolutionary advantage of removing the cell body as far as possible from the surface of the animal's body, where it would be too easily damaged. No matter what the position of the cell body, the nerve impulse normally travels from either the dendritic or the receptor endings towards the **axon terminals**, and ends at the synaptic bulbs of the axon terminals.

Branches of a neuron

A neuron typically has many branches. There are usually numerous **dendrites**, and each dendrite may itself have many fine branches. The zone of dendrites around one neuron is called its **dendritic field**. The axon ends in several branching **axon terminals**. The axon may also possess one or several **collateral branches**, each arising from a node of Ranvier (Fig. 5.2). Every collateral branch also has its own axon terminals which may be numerous. Each axon terminal, or branch of an axon terminal, ends in a **synaptic bulb** (Fig. 5.4). The term **neurite** covers the axon, collaterals, and dendrites.

Synaptic bulbs

The synaptic bulb (synaptic knob, terminal button) is an enlargement at the effector end of an axon terminal (Fig. 5.4). The surface of a typical neuronal cell body within the neuraxis is richly encrusted with synaptic bulbs; bulbs in this site are called **axosomatic bulbs** (Fig. 5.2). The dendrites are

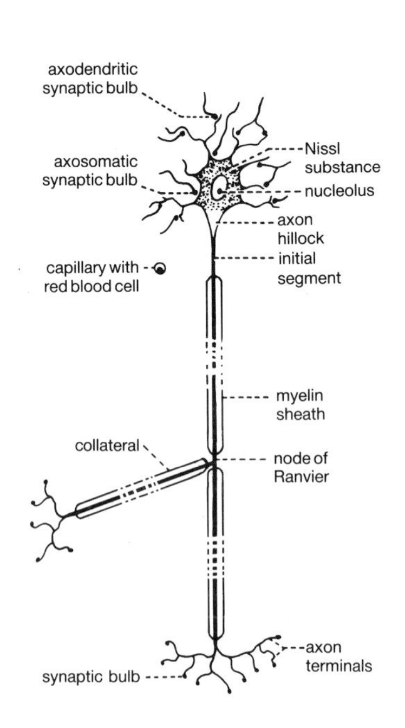

FIG. 5.2. The components of a neuron. The capillary and red blood cell indicate the very large size of the neuronal cell body, nucleus, and nucleolus. Otherwise the proportions are highly schematic. An axon cylinder 20 μm thick would have an internodal distance of up to 1500 μm, i.e. 1.5 mm. Broken lines indicate shortening.

similarly covered with bulbs, but the number of bulbs decreases as a dendrite divides into its fine terminal branches; the bulbs on dendrites are known as **axodendritic bulbs** (Fig. 5.2).

Some bulbs may end on the axon terminals of a neuron, rather than on the cell body or dendrites. Bulbs in this position have been named **axo-axonal bulbs**, but **termino-terminal bulbs** would be more descriptive of their position. These bulbs have been regarded as the anatomical basis of presynaptic inhibition (see Section 5.13).

The total number of synaptic bulbs received by a single neuron may be several thousand. Some individual neurons in the neuraxis are believed to receive up to 10 000 synaptic bulbs, and a large proportion of these may have come from separate neurons. Since the total number of neurons in the human neuraxis may be about 15 000 000 000 (considerably greater than the human population of the world), the total number of possible neuronal connections is truly prodigious.

Size and shape of cell body

The cell body of a typical neuron is very large indeed, the nucleus alone being bigger than the whole of many other individual cells and the nucleolus being only slightly smaller than a red blood cell (Fig. 5.2). If the axon is also taken into account the cell may be immensely long; for example a mechanoreceptor ending in the hindfoot of a thoroughbred horse leads into an axon which ascends the leg to a dorsal root ganglion and then continues along the spinal cord to end in the medulla oblongata, giving a total length of about 4 metres or more.

Most neurons are **multipolar**, with the dendrites and axon pointing in various directions (Fig. 5.2). **Pseudounipolar neurons** have only one cell process which immediately divides into two (Fig. 5.1(b)); these are primary afferent neurons with their cell body in a dorsal root ganglion. They arise in the embryo from a bipolar cell with two processes which fuse together. True **bipolar neurons** do occur in the adult animal, being spindle-shaped with a dendrite at one end and the axon at the other end; neurons of this type occur in the retina and the vestibulocochlear pathway.

Cell structure

A typical neuron contains a single voluminous nucleus with a very large nucleolus. The nucleus is markedly euchromatic and therefore pale-staining. Conspicuous clumps of basophilic material, known as **Nissl substance** (substantia chromatophilica), are dispersed throughout the cytoplasm including that of the beginning of the dendrites but not the axon hillock and axon (Fig. 5.2). The Nissl substance consists of aggregations of rough endoplasmic reticulum, with attached and free clusters of poly-ribosomes.

The membranes destined to form neurosecretory granules, or synaptic vesicles, or the cell membrane itself, are synthesized in the endoplasmic reticulum and processed in the well-developed Golgi apparatus. A major proportion of the cell's metabolism is devoted to the elaboration of membranes, since essentially all the membranes contained within the dendrites, axon, and axonal terminals have originated in the cell body, either fully assembled or in precursor form.

Neurotubules, **microfilaments**, and **neurofilaments** run through the cytoplasm into the dendrites and axon. There are usually numerous mitochondria. **Lysosomes** are common, and nearly all neurons contain membrane-bound granules of a brown 'wear-and-tear' pigment known as **lipofuscin** which are probably secondary lysosomes. The **cell membrane** (cytolemma) has the usual trilaminar structure, but certain of its integral protein molecules can reorientate themselves to create ion channels (see Section 5.10).

Storage diseases arise from deficiencies of specific lysosomal enzymes, and therefore result in the accummulation within lysosomes of the products of lipid, protein, or carbohydrate metabolism.

Protein synthesis

The euchromatic nucleus, large nucleolus, and abundant rough endoplasmic reticulum and polyribosomes, are all signs of active protein synthesis within the cell body, for which the many mitochondria provide a source of energy. The maintenance and repair of cytoplasmic proteins contained in the relatively enormous volume of cytoplasm and membrane proteins of a neuron demand a high rate of protein synthesis. The elaboration of transmitter substances to be released at the axon terminals also account for a substantial proportion of this protein synthesis.

5.2 The axon

The axon is also known as the **axis cylinder**. The term '**nerve fibre**' is sometimes used as a synonym for axon, but really applies to the axon together with its sheath, i.e. the Schwann cell sheath in the peripheral nervous system and the oligodendroglial sheath in the central nervous system. This definition of 'nerve fibre' holds good in the peripheral nervous system where all axons, whether myelinated or unmyelinated, have a Schwann cell sheath (except, sometimes, for the axon terminals); however, in the central nervous system unmyelinated fibres are totally bare and have no sheath at all, but not unreasonably are often referred to as nerve fibres, nevertheless. The term '**nerve**' should not be used to signify a single myelinated or unmyelinated fibre, but applies only to a bundle of such fibres.

The axis cylinder

The cytoplasm of an axon is known as **axoplasm**. About 80 per cent of the dry weight of axoplasm consists of proteins, and the rest is lipids and sugars. There are thousands of different proteins, some in solution, some embedded in membranes, and others assembled into elongated threads. Among those in solution are enzymes specific to the synthesis of neurotransmitter substances. The thread-like elements are **neurotubules**, **microfilaments**, and **neurofilaments** (Fig. 5.3); these constitute a cytoskeleton, running through the cell body and penetrating the whole length of the dendrites and axon without branching. There are also membrane-bound organelles including (i) vesicles containing neurotransmitters (or neurosecretory substances), (ii) occasional small cisternae of smooth endoplasmic reticulum for storing and transporting substances such as lipids and membrane glycoproteins required in the axon terminals, (iii) lysosomes, and (iv) mitochondria (Fig. 5.3). Ribosomes are absent, a fact which shows that essentially no synthesis of protein from amino acids can occur in an axon. The cell membrane of an axon is called the **axolemma** (Fig. 5.3). It resembles the cell membrane of the nerve cell body in possessing ion channels in its integral proteins.

Like all other varieties of microtubule, the neurotubules are formed by polymerization of subunits of the protein tubulin. Microfilaments consist chiefly of a neural

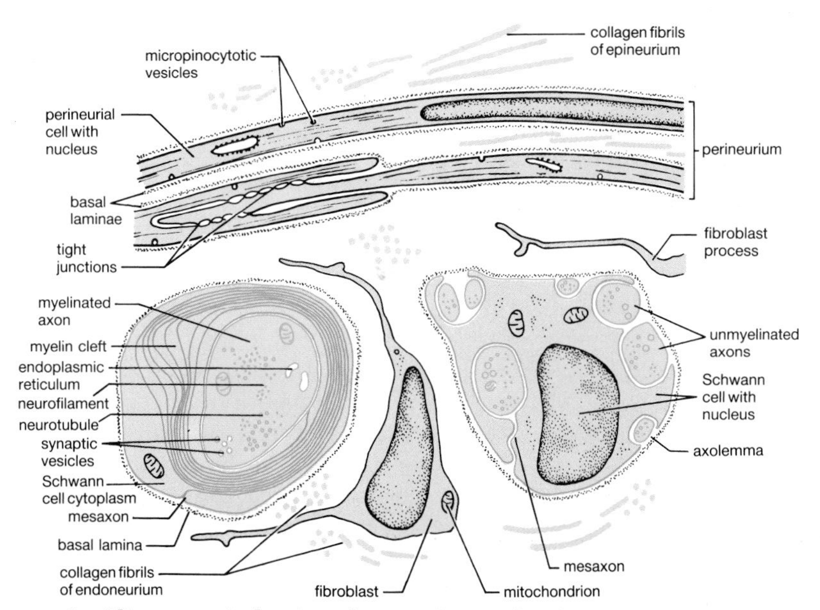

FIG. 5.3. Semidiagrammatic drawing of a part of a peripheral nerve. The illustration shows the structure of the epineurium, perineurium, and endoneurium, and of a myelinated and several unmyelinated nerve fibres. The axons and the cell membranes of the Schwann cells are green.

variant of the protein actin. Both microtubules and microfilaments are also present in most other cell types. Neurofilaments are characteristic of nerve cells and neuroglial cells, and are formed from controversial protein constituents. They are usually more numerous than the microtubules and microfilaments. Neurofilaments (about 10 nm) are about half the diameter of neurotubules, and neurotubules (about 20 nm) are about half the diameter of synaptic vesicles (about 40 nm) (Fig. 5.3).

Axonal transport

Various experiments including ligation of axons, labelling of proteins by incorporating radioactive amino acids, and direct observations in tissue culture, have shown that axoplasm is continuously on the move. Not only is it moving, but it goes in both directions at the same time. Thus in a ligated axon, vesicles and other axonal constituents, such as proteins, catecholamines, and phospholipids, accumulate on **both** sides of the ligature. The net transport, however, is from the cell body to the axon terminals, and this process is known as anterograde transport.

Anterograde transport occurs as two types of movement, one slow (about 1 mm per day) and the other fast (about 100 to 400 mm per day). **Slow transport** is the bulk flow of the whole semisolid axonal column, which is being produced continuously in the cell body and undergoing internal dissolution at the terminals. It seems that the entire axonal cytoskeleton (i.e. neurotubules, microfilaments, and neurofilaments) moves as an extended scaffolding rather than as subunits. Carried along with the scaffolding are numerous and diverse proteins, consisting mainly of many types of soluble enzymes. The function of slow transport is to renew the axoplasm continuously, or to supply new axoplasm in a regenerating axon. In terms of weight, most of the protein in axons moves by this method, which resembles extrusion from a tube of toothpaste. **Fast transport**, on the other hand, works by the movement of materials within the scaffolding of the axonal column. It carries new membrane material and membrane-bound organelles. Foremost among these are vesicles, or the precursors of vesicles, the former containing neurotransmitter substances. However, it is not necessary for all synaptic vesicles to be exported intact from the cell body. In most neurons, synaptic vesicles are formed largely by local recycling of membrane in the axon terminal. Thus in the majority of neurons the axon terminal contains enzymes for the local synthesis of neurotransmitters; nevertheless, there is still a continued need for fresh supplies of synthesized membrane and enzymes from the cell body. Much of the new membrane material exported by the cell body fails to reach the axon terminals, but is deposited instead along the axon to maintain the axolemma.

Transport from the axon terminals to the cell body is known as **retrograde transport**. This goes at about half the rate of fast anterograde transport. The organelles which travel by retrograde transport are again membrane-bound and consist mainly of lysosomes. Some of the contents of these lysosomes are worn out membrane materials from the axon terminals which are being returned to the cell body for either degradation or restoration. Other substances within lysosomes are acquired from the extracellular fluid, either in small quantities by endocytosis of solutes, or in much larger quantities by endocytosis of substances which bind to specific receptors on the axolemma and are then taken up, incorporated into lysosomes, and shipped rapidly to the cell body.

The latter category includes some dangerous toxins and particles, for example tetanus toxin (which is taken up and transported by virtually all kinds of neurons), and neurotropic viruses, notably those of herpes and rabies. The latter particles are able to multiply in non-neuronal cells, but at physiological concentrations of the virus they are effectively excluded through lack of appropriate surface receptors: on the other hand, neurons offer a sequence of mechanisms which promotes the rapid arrival of the virion at the cell body, where it proliferates.

The **mechanisms** of axonal transport are not understood, but the participation of contractile proteins and microtubules has been considered. The contractile proteins actin and myosin, the prime biological movers, certainly are present in neurons and in theory could be the source of the force needed for **fast transport**, the microtubules merely providing a passive skeletal framework. However, the evidence for this is not substantial. Alternatively the microtubules could provide a slow moving track on which organelles could undergo fast transport in an active stepwise manner. Movement would be provided by local, sequential, energy-dependent conformational changes, resulting from interaction of the organelle with a subunit of the track in a ratchet mechanism or as a sliding-filament mechanism. Tubulin (and actin and myosin) are polar molecules. Polar molecules could operate in either the anterograde or the retrograde direction, depending on their orientation. Therefore transport would be possible in both directions if some of the molecular tracks had one polarity and others had the opposite polarity, thus explaining the simultaneous occurrence of anterograde and retrograde transport within one axon.

Rapid axonal transport in both directions over distances of a metre or more through such a narrow tube as an axon is an amazing engineering achievement, but similar phenomena of intracellular motility occur in many other cells, e.g. the movement of the spindle during mitosis and amoeboid movement. Therefore axonal transport is really only an example, albeit highly specialized, of standard cellular activities, anterograde transport being a step in the biogenesis of membranes and retrograde transport being a step in the cycle of membrane turnover.

The myelin sheath

The myelin sheath of a myelinated fibre in the peripheral nervous system is formed from the rolled-up cell membrane of the **Schwann cell** (neurolemmocytus) (Fig. 5.3). In the central nervous system the oligodendroglial cells form myelin by throwing out cell processes (Fig. 5.14) which wrap round an axon (see Section 5.21). In the peripheral nervous system, the unmyelinated axons are suspended, several together, within one Schwann cell (Fig. 5.3). In the central nervous system, the unmyelinated fibres are bare and totally devoid of any supporting cells. Even a myelinated fibre has regions where the axon is bare of myelin. Such regions include the **initial segment** and the conical **axon hillock** (colliculus axonis) (Fig. 5.2), which together form the beginning of the axon in motoneurons and interneurons. The axon hillock and initial segment have great functional importance, since they constitute the **trigger region** of the neuron from which the volleys of action potentials originate. The **node of Ranvier** (nodus neurofibrae) is also bare of myelin; each node represents the junction of two adjacent Schwann cells (Fig. 5.2). The nodes of Ranvier are also of great functional significance, because they form the basis of **saltatory conduction** of the nerve impulse, which is characteristic of myelinated axons. The high resistance of myelin effectively seals the surface of the axon. Only at the nodes of Ranvier can the depolarizing ionic flux of the nerve impulse take place, and therefore the depolarizing process has to jump from node to node; hence the term saltatory conduction, which means 'leaping' conduction (L. *saltare*, to leap). The functional value of the leaps is shown by the fact that conduction in a myelinated fibre is about three times faster than in a non-myelinated fibre of the same external diameter.

Fibre diameter and conduction velocity

In vertebrates in general the diameter of the axis cylinder (axon) and the thickness of its myelin sheath tend to increase by similar proportions, as the thickness of the whole fibre increases.

In the higher mammals the external diameter of the largest myelinated fibres is about 20 μm and the smallest about 1.0 μm, including the myelin sheath in both instances. The largest unmyelinated fibres have a total external diameter of about 2.0 μm, and the smallest about 0.1 μm. Since the limit of resolution of the light microscope is only about 0.25 μm the smallest unmyelinated fibres cannot be studied with the light microscope. Moreover, the distance between even the larger unmyelinated fibres is often less than 0.25 μm and consequently these fibres can only be distinguished individually with the electron microscope.

The larger the diameter of the fibre the faster it conducts a nerve impulse, whether it be myelinated or unmyelinated. As a rough guide, the conduction velocity (in metres per second) of myelinated fibres can be estimated by multiplying the diameter (in μm) by a factor of 6. Thus the largest myelinated fibres of 20 μm diameter conduct at about 120 metres per second or 270 mph, and the smallest at about 6 metres per second.

As already stated, myelinated fibres conduct about three times faster than unmyelinated fibres of similar external diameter. The conduction (in m/s) of the larger unmyelinated fibres can be estimated by multiplying the diameter (in μm) by 2. Thus an unmyelinated fibre of 1.0 μm diameter conducts at about 2 m/s or 4.5 mph.

If myelinated fibres were not available to mammals would it be possible to utilize unmyelinated fibres which would be capable of conducting at adequate velocities? Theoretically the answer is 'yes', but these would have to be unmyelinated fibres of colossal diameter. A typical nerve innervating a mammalian muscle is about 1 mm in diameter and contains about 1000 myelinated fibres from 10 to 20 μm in diameter; if such a nerve were to consist of the same number of unmyelinated fibres, but with the same conduction velocities as the myelinated fibres, the fibres would have to be of such massive dimensions that the nerve would resemble a thick rope about 4 cm in diameter. If the human spinal cord contained only unmyelinated axons, it would need to be several metres in diameter in order to maintain its conduction velocities.

5.3 Epineurium, perineurium, and endoneurium

The epineurium and endoneurium are typical connective tissue elements. The **epineurium** provides an irregular collagenous coat around the outside of a peripheral nerve, enclosing all the nerve bundles and penetrating between them (Fig. 5.3). The **endoneurium** forms a delicate layer of collagen fibrils and fibrocytes around the individual nerve fibres within a nerve bundle (Fig. 5.3). It also forms thin septa which blend with the perineurium and enclose groups of nerve fibres. Included within the endoneurium are capillaries (and other small blood vessels) which are

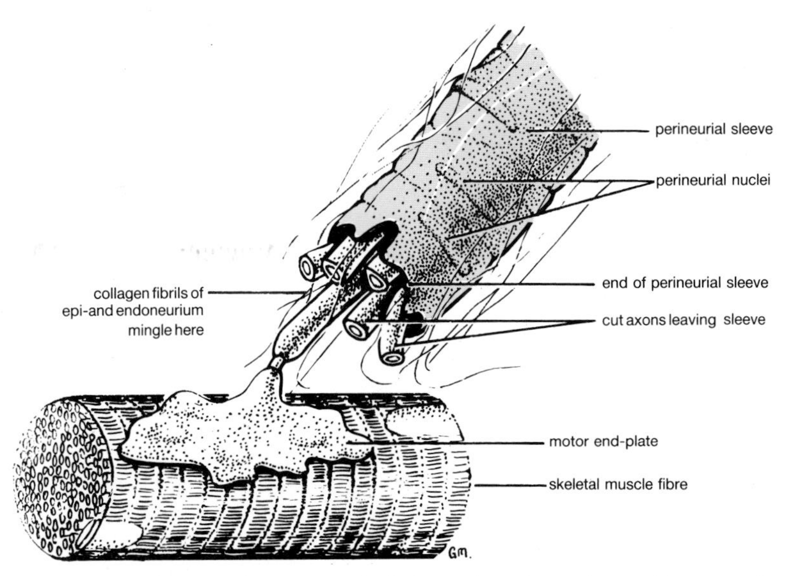

FIG. 5.4. Diagram showing the open end of the perineurium of a small nerve. The perineurium (green) is like the sleeve of a jacket. The nerve fibres within it reach their target tissues by emerging through the open end of the sleeve.

useful for estimating the diameter of nerve fibres in histological transverse sections of a nerve.

The **perineurium** is a much more specialized connective tissue structure, forming a thin but very well-defined sleeve around each bundle of nerve fibres in a nerve. It consists of several concentric layers of thin (epitheloid) cells and collagen fibrils. The cells are attached to each other by tight junctions (Fig. 5.3), which form an effective diffusion barrier to most macromolecules thus isolating nerve fibres from noxious materials in the surrounding tissue fluid. The cells also have many micropinocytotic vesicles. The perineurium is open at its end, like the sleeve of a jacket, thus allowing the nerve fibres access to their target tissues (Fig. 5.4). The opening also allows toxins and pathogens to enter the sleeve, and this has been suspected as another route by which the rabies virus reaches nerve cell bodies.

5.4 The synapse

Synapses have essentially the same components within all parts of the nervous system. The junctions between efferent neurons and their effector organs (e.g. between the axon terminal of a motoneuron and a skeletal muscle cell), and between receptor cells (such as those in the organ of Corti in the inner ear) and the axonal receptor endings which are applied to these cells, are also essentially similar.

A synapse within the neuraxis forms a good model, and has the following characteristics:

General structure of the synaptic bulb

A synaptic bulb is the enlarged end of an axon terminal (Fig. 5.5). It usually contains some mitochondria, and numerous small vesicles which are believed to enclose transmitter substance. **Cholinergic vesicles** are spherical and about 30–60 nm in diameter, and in electron micrographs

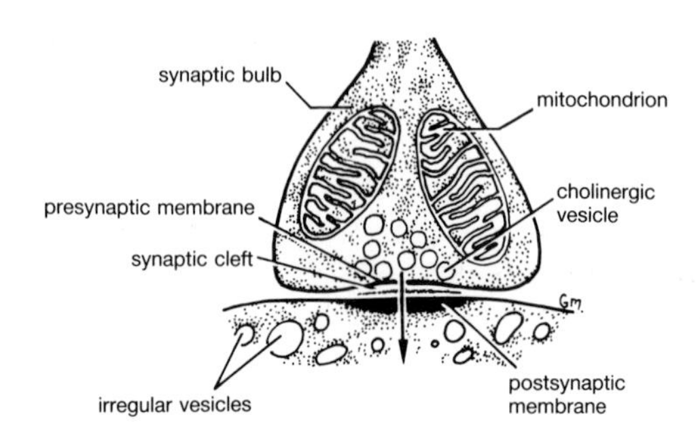

FIG. 5.5. A synaptic bulb, showing the components of a synapse. The direction of transmission (arrow) is indicated by the accumulation of synaptic vesicles on the presynaptic membrane, and by the relatively greater thickening of the postsynaptic membrane.

appear to be empty. **Noradrenergic** and **adrenergic vesicles** are also 30–60 nm in diameter, but contain a dense core and are therefore called small granular vesicles or small dense-cored vesicles.

The presynaptic membrane

The axolemma of a synaptic bulb, where it is applied to the cell membrane of the adjacent neuron, is flattened, and is called the presynaptic membrane (Fig. 5.5). In electron micrographs it usually appears thickened and electron-dense, though somewhat less so than the postsynaptic membrane. Transmitter vesicles tend to be massed along the presynaptic membrane. During synaptic transmission, vesicles fuse with the presynaptic membrane and open into the synaptic cleft, this being the same process of **exocytosis** as occurs during the release of secretory granules by the cells of exocrine and endocrine glands. The term presynaptic membrane is also applied to the cell membrane of a receptor cell (e.g. a carotid body type I cell), where it forms a synaptic contact with its sensory axonal (receptor) ending.

The synaptic cleft

The space, usually slightly widened, between the presynaptic and postsynaptic membranes is known as the synaptic cleft (Fig. 5.5).

As a rule it contains a narrow zone of dark material, which typically lies nearer to the postsynaptic membrane.

The postsynaptic membrane

This is the cell membrane (Fig. 5.5) of the adjacent neuron. In electron micrographs it usually appears thicker and denser than the presynaptic membrane.

The asymmetry of the pre- and postsynaptic membranes seems to be characteristic of excitatory synapses, but in inhibitory synapses the postsynaptic membrane is apparently no thicker than the presynaptic membrane.

Transmitter substances

In synapses between neurons there are essentially two varieties of transmitter substance, excitatory and inhibitory, which may be called **E-substance** and **I-substance** respectively. After release, transmitter substances are rapidly removed from the synaptic cleft. Acetylcholine and noradrenaline are common transmitter substances throughout the nervous system. The acetylcholine that is released at a synapse from cholinergic vesicles is rapidly inactivated, mainly by acetylcholinesterase which splits it into choline and acetic acid; the choline is reabsorbed and recycled within the synaptic bulb. The noradrenaline that is delivered into the synaptic cleft from noradrenergic vesicles is largely reabsorbed and recycled within the synaptic bulb. The release of transmitter substances, and the onset of their effects on the postsynaptic membrane, both take time. This pause in the transmission of nervous activity is known as **synaptic delay**.

At each synapse, synaptic delay is about 0.5 to 1.0 millisecond. With the advance of knowledge, a wide variety of possible **neurotransmitters** (transmitter substances) is beginning to emerge. **Dopamine** appears to be an important catecholamine within the neuraxis, and seems to be mainly inhibitory in action. The amine **serotonin** (5-hydroxytryptamine) is evidently an important synaptic transmitter in some parts of the brainstem and spinal cord. **Purines**, or their derivatives such as ATP, may form another group of neurotransmitters. Several amino acids are possible transmitter substances in the mammalian neuraxis. One of these is **GABA** (gamma aminobutyric acid), and this appears to be inhibitory; others are believed to be excitatory. There is some evidence that in the mammalian neuraxis the amino acids could be the major transmitters, while the better known transmitters such as acetylcholine, noradrenaline, dopamine, and serotonin, may only account for transmission at a small percentage of central synapses.

Finally, a group of **neuropeptides** has recently attracted much attention as possible transmitter substances. These are synthesized in the nerve cell body, transported to the axon terminals, stored there in dense-cored vesicles about 80–120 nm in diameter, and released on depolarization of the terminals. Notable among them are the opioid peptides, known as **enkephalins** and **endorphins**. These are naturally occurring (endogenous) substances, but their actions are similar to exogenous opiates like morphine; they appear to be involved in neuronal pathways regulating the conscious perception of pain (see Section 9.9, part (ii)). All opiate neuronal receptor sites are blocked by naloxone.

It has long been believed that each neuron is equipped with only one neurotransmitter, this concept being known as **Dale's hypothesis**. It is now clear that this concept is not valid. For example, many pre- and postganglionic parasympathetic neurons contain neuropeptides as well as acetylcholine, and so also do many preganglionic sympathetic neurons; likewise many postganglionic sympathetic neurons contain neuropeptides in addition to noradrenaline and adrenaline. When more than one of these substances is released from an axon terminal, it is postulated that they modulate each other's neurotransmitter activities.

The release of transmitter substance is affected by the concentration of **calcium** and **magnesium** ions in the extracellular fluid. In every known secretory process, depolarization allows Ca^{++} ions from the extracellular environment to enter a cell, via channels formed by protein molecules, causing an increase in intracellular Ca^{++}. The increase in Ca^{++} ions can also be achieved by the release of stored intracellular Ca^{++}. It seems likely that, in the synaptic bulb of a neuron, the Ca^{++} ions promote the fusion of synaptic vesicles with the presynaptic membrane, and hence induce the release of transmitter substance. Magnesium ions have the opposite effect, blocking the release of transmitter substance by exocytoxis. A disturbance in the balance of magnesium or calcium ions can produce pronounced neuromuscular disorders such as grass staggers (hypomagnesaemia) or milk fever (hypocalcaemia).

Curare and synthetic curare-like compounds (e.g. flaxedil) compete with acetylcholine for the receptor sites at the postsynaptic membrane of the motor end-plate; they produce a block without causing depolarization of the postsynaptic membrane.

Other drugs (e.g. succinylcholine) paralyse muscle by producing persistent depolarization of the postsynaptic membrane at the motor end-plate. **Botulinum toxin** paralyses by preventing the release of acetylcholine at the motor end-plate.

5.5 Phylogenetically primitive and advanced neurons

Phylogenetically primitive neuron

A 'primitive' neuron (Fig. 5.6(b)) has a long axon with many collaterals and many axon terminals. The terminals diverge to make synaptic connections with many other, widely separated, neurons thereby illustrating **divergence**. The dendrites of a primitive neuron, which usually have only one generation of branches, are straight rather than curved and therefore receive many synapses from relatively distant regions of the neuraxis. Because of their widespread and numerous connections, primitive neurons form extensive networks of interconnected neurons.

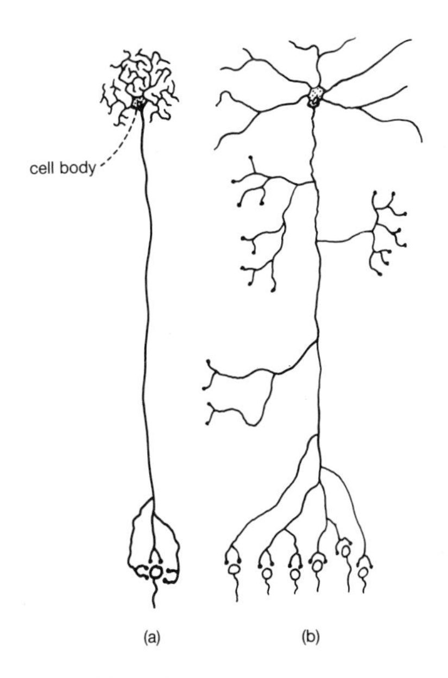

FIG. 5.6. **(a) An 'advanced' neuron.** This type of neuron is characterized by a long axon, few collaterals, and terminal filaments converging on only one or two other neurons. Its dendrites are curved and branching, and remain near the cell body.
(b) A 'primitive' neuron. This variety of neuron is typified by a long axon, many collaterals, and terminal filaments diverging to many neurons. The dendrites are straight, have only one generation of branches, and reach far from the cell body.

Phylogenetically advanced neuron

In contrast, a 'phylogenetically advanced' neuron (Fig. 5.6(a)) has a long axon with only a few collaterals and axon terminals. Moreover its axon terminals show **convergence**, i.e. they tend to converge on only one or two other neurons. The dendrites of this type of neuron branch repeatedly and are relatively long, but they are also curved rather than straight and therefore remain fairly close to their cell body; consequently they tend to receive synapses from regions nearby rather than far away. Neurons of this variety are characteristic of those which form so-called 'specific pathways'.

Specific and non-specific pathways

Neuronal pathways which conduct only a single and particular kind of information are known as **specific pathways**. An example is the visual system. **Non-specific pathways**, on the other hand, share a wide variety of information; an example is the ascending reticular formation, in which the neuronal pathways consist of diffuse networks of primitive neurons, shared by impulses arising from receptors of touch, pressure, pain, temperature, etc. In its most refined form, a specific pathway would consist of a one-to-one chain of neurons, thus enabling a single stimulus to be conveyed with maximum precision to the cerebral cortex where it could be accurately identified. One-to-one pathways, however, require a great wealth of neurons, and are a luxury which even the human neuraxis can seldom if ever afford.

For instance even in the eye, where specific pathways would seem to be especially advantageous, there are six to seven million cones but only one million axons in the human optic nerve; evidently several receptor cells converge on one bipolar cell, and several bipolar cells converge on one ganglion cell. However, it is apparently not uncommon in the neuraxis of the higher mammals for a specific neuron to make synapses with only two or three other neurons. This happens, for example, in the medial lemniscal sensory system of touch, pressure, and joint movement (kinaesthesia).

5.6 Axonal degeneration and regeneration in peripheral nerves

The responses of a peripheral nerve fibre to transection are shown in Fig. 5.7. The proximal stump (joined to the cell body) undergoes **ascending degeneration**, in which a short length of the axon degenerates and its myelin disintegrates. The cell body swells and the nucleus moves to an eccentric position away from the axon hillock. The Nissl substance undergoes dissolution with a resulting decrease in basophilia of the cytoplasm (**chromatolysis**). As soon as the debris in the proximal stump has been removed by macrophages, the axon sprouts a leash of fine filaments. Schwann cells begin to proliferate and move towards the distal stump, and

---

FIG. 5.7. **Diagrams of the degeneration and regeneration of a single myelinated peripheral nerve fibre.** The fibre was transected at the arrow.

(a) Normal motor nerve fibre innervating a skeletal muscle fibre via a motor end-plate. Basal lamina, nerve cell body, and axon, are green.

(b) Changes during first week after transection. In the proximal stump (towards the cell body) the axon and myelin sheath die back a short distance (ascending degeneration), but the Schwann cell just above the cut survives and begins to proliferate (2). The neuronal cell body swells, the nucleus becomes eccentric away from the axon hillock, and the Nissl substance disintegrates (chromatolysis). The axon sprouts a leash (1) of fine filaments. In the distal stump the axon (3) and its myelin sheath (6) degenerate (descending degeneration), the debris

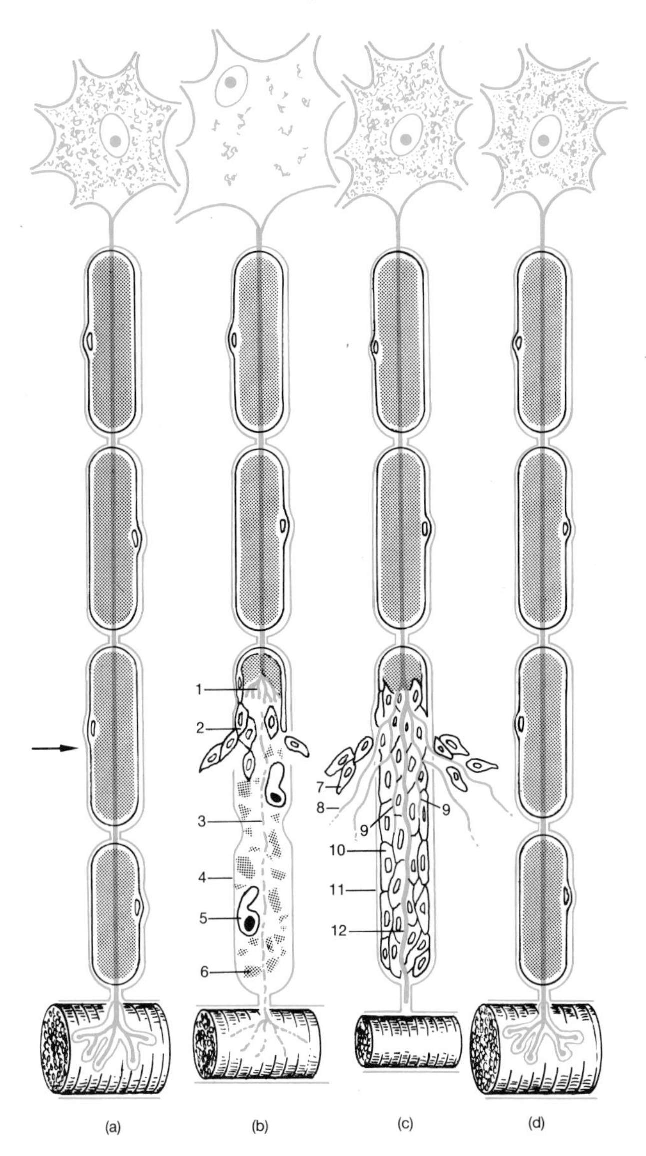

being removed by macrophages (5), but the basal lamina of the original Schwann cells remain as an intact tube (4). The muscle cell begins to atrophy.

(c) Three weeks after the cut. The proliferating Schwann cells (10) fill the tube of basal lamina (11). One axonal filament has wended its way to the end of the tube (12). Two supernumerary filaments are also within the tube (9). Other axonal filaments (8) and Schwann cells (7) have spread outside the tube. The cell body has returned to normal. The muscle cell shows marked disuse atrophy.

(d) Several months after the cut. The supernumerary filaments within the tube, and the filaments outside the tube, have gone. The axon has been re-myelinated and the motor end-plate re-established. The muscle cell has recovered.

some escape from the cut end of the fibre. This stage could be reached within a week of injury (Fig. 5.7(b)).

The distal stump experiences **descending degeneration** (Wallerian degeneration). Severe changes begin very quickly, so that by 24 hours the myelin sheath and axis cylinder are degenerate throughout the entire length of the fibre from the cut to its peripheral end. By the end of two weeks degeneration is maximal, the myelin having disintegrated into fatty droplets and the axis cylinder into small fragments; macrophages (derived from monocytes) phagocytose the debris. The **basal lamina**, however, remains intact as a continuous tube all the way to the axon terminal (Fig. 5.7(b)).

Within three weeks, **regeneration** is well on its way. Schwann cells from the proximal stump have filled the tube of basal lamina, and several filaments from the axonal leash have wended their way through the tube. Other filaments have wandered outside the tube (Fig. 5.7(c)). After several months (Fig. 5.7(d)), the wandering and supernumerary filaments have disappeared, and in the tube one surviving filament has been re-myelinated and reached the motor end-plate.

Certain conditions are needed for successful regeneration. The gap between the proximal and distal stumps must be small, and must not be filled by collagenous scar tissue. The two stumps must be correctly aligned. Of the various tubular structures that might guide the regenerating axon, the tube of **basal lamina** which belonged to the original nerve fibre is of critical importance. If the two stumps become misaligned, so that the regenerating axons of the proximal stump cannot find the basal lamina tubes of the distal stump, or if the distal stump is removed as in amputation of a limb, the filaments may continue to sprout from the proximal stump forming a painful mass known as an **amputation neuroma**.

Why do the regenerating axons of a cut nerve enter the tubes of basal lamina? First, because these tubes survive the cut. The perineurial sheath also survives the injury and is a sharply defined tube, but is obviously far too coarse to guide all the many regenerating axons enclosed within it to their individual targets. The endoneurium may also remain intact but is not organized into a private tubule for each nerve fibre. The Schwann cells distal to the injury are potential guidelines, but they rapidly disintegrate and are phagocytosed. In contrast, the basal lamina of the Schwann cells does not disintegrate, nor is it attacked by macrophages; therefore, delicate though it is, it offers every axon an accurate trail to its distant target. It has been observed that the basal lamina of muscle cells, endothelial cells, and pulmonary alveolar cells is not attacked by macrophages after degeneration of these cells, and that these cells also regenerate within their tubes of basal lamina. There is evidence that some substances found in basal laminae (a proteoglycan) might be associated with promoting the outgrowth of neurites. Possibly the distal nerve segments or cell debris may have a trophic influence on regenerating axons.

Because each proximal axonal stump sprouts many filaments, several may enter one tube of basal lamina. Only one of these filaments is finally selected and the

others disappear. Usually the survivor suits the sensory or motor function of the original fibre, but the selection is not entirely successful. If an afferent axon grows into a tube which ends in a motor end-plate, or vice versa, function will not be restored. In the human hand some sensory deficits always persist after severed nerves have been reconstructed.

5.7 Regeneration and plasticity in the neuraxis

Most neurons in the mammalian neuraxis survive for the life span of the individual, but with advancing age there may be a slow but steady loss. The olfactory neuroepithelial cell is a notable exception to the longevity of neurons, since it lasts only about a month and is then replaced by division of basal cells in the olfactory epithelium. The cell body of a mature neuron possesses no capacity whatever for reproducing itself, but axons in peripheral nerves do have good powers of regeneration (see Section 5.6). On the other hand, axons in the neuraxis have only a very limited capacity for regeneration. Thus if an axon in the brain or spinal cord is cut, the axonal stump nearer to the cell body sends out sprouts which penetrate the site of the lesion, but after a week or two they stop growing.

Various inconclusive explanations for this arrest have been suggested. Astrocytes may proliferate and fill the site of the lesion with a glial scar. Nerve growth factors derived from the blood may be unable to reach the sprouting axonal stumps, or may be blocked by antigrowth substances secreted by neuroglial cells. One strange aspect of this problem is that axonal regeneration does occur with precision in the neuraxis of fish and amphibia, but fails in that of reptiles, birds, and mammals.

The failure of axonal sprouting in the mammalian neuraxis does not, however, mean that **recovery** from a lesion cannot occur. In fact the central nervous system shows remarkable **plasticity** in its response to injury. Functional improvement begins during the first days or weeks, but this is due to resorption of oedema and blood, and to recovery of axons from anoxia. Then, in man, there begins a slow but progressive restoration of function (even after severe traumatic injuries, or vascular lesions such as a stroke), which is scarcely ever complete but can yet be so good that remaining deficits may escape even a sophisticated neurological examination. Several mechanisms appear to be involved. (i) Undamaged axons passing through the region of a lesion form collateral sprouts which spread out in the damaged area (**collateral sprouting**); these sprouts form new synapses at synaptic sites which have been vacated. (ii) Previously undisclosed pathways in the affected region may become functional (**unmasking**). (iii) The resumption of function may be a **learning process**, in which the brain modifies its neuronal circuitry by altering and adjusting synaptic connections, as may be assumed to occur during any kind of learning process in the immature individual; for this to happen patience is essential—learning to write is a slow process in both the child and the disabled adult. (iv) In man, **motivation** to recover is apparently of overwhelming importance. These mechanisms require months or years to occur, but in man the outlook is essentially hopeful. In veterinary practice, the prospect of progressive but very slow recovery must be weighed against the needs and abilities of the owner, and humanitarian considerations.

The reflex arc

5.8 Basic principles

The structural unit of the nervous system is the neuron, but the functional unit is the reflex arc. Somatic reflex arcs enable the animal to react to its external environment; autonomic reflex arcs regulate the internal environment by adjusting the functions of the circulatory organs and other viscera.

The simplest reflex arc consists of only two neurons, one afferent and the other efferent. Since these two neurons require only a single intervening synapse, this is known as a **monosynaptic reflex arc**. Monosynaptic arcs are somewhat unusual in mammals, but they do occur in the stretch reflex, or myotatic reflex, which controls the activity of skeletal muscle; the most familiar example of such a reflex is the patellar reflex. Usually, however, several **interneurons** lie between the afferent and the efferent neuron. (The diagrams in this book consistently show two interneurons, but the number is often greater than this.)

A simple reflex arc begins at the **receptor endings** of the **primary afferent axon**. As Fig. 5.1(b) shows, the receptor endings of the afferent neuron are continuous with receptor terminals, which in turn continue directly into the axon of the afferent neuron, the cell body being in a dorsal root ganglion. This location of the cell body applies to primary afferent neurons throughout the nervous system, with only two exceptions, namely the cell bodies of the mesencephalic trigeminal nucleus (see Section 6.6), and the first neuron of the olfactory pathway (see Section 8.5). After leaving the cell body the axon continues through the dorsal root into the neuraxis (central nervous system) where it ends in axon terminals. These form synaptic connections with a chain of **interneurons**, which in turn make synaptic connections with the efferent neuron. The **efferent (motor) neuron** (known as a **motoneuron** in a somatic reflex arc) has its cell body in the neuraxis, in the ventral horn in the case of the spinal cord. In a somatic reflex arc the axon of the motoneuron passes out through the ventral root into a peripheral nerve; its axon terminals act on its **effector organ**, which would be skeletal muscle fibres in this instance. (The effector organ is also known as 'the effector'.) In an autonomic reflex arc the 'efferent' neuron is the postganglionic neuron, with its cell body in an autonomic ganglion, the preganglionic neuron being really an interneuron. The autonomic effector organ may be cardiac muscle, smooth muscle, or glandular epithelium.

A reflex arc can be interrupted, under pathological conditions, at any point in its pathway. Thus there may be a defect in the afferent part of the pathway, or in the interneurons, or in the efferent part of the pathway. If the efferent part of the pathway is damaged, the interruption may occur at the motor cell body, in the peripheral nerve, at the effector ending, or in the effector organ itself, and it may be difficult to distinguish clinically between these various possible sites.

If an effector response of a reflex arc occurs only on the same side of the body as that which receives the stimulus the reflex is called an **ipsilateral reflex**; if the effector response is on the opposite side, it is a **contralateral reflex**. A spinal reflex arc in which all the neurons are confined to one segment of the spinal cord is known as an **intrasegmental reflex arc**; if the reflex involves neurons in several spinal segments it is described as an **intersegmental reflex arc**.

5.9 Decussation: the coiling reflex

One of the most remarkable characteristics of neurons within the neuraxis, particularly those with long axons, is the tendency for their fibres to **cross over** from one side to the other (**decussation**). This is a very primitive feature, since in all chordates from Amphioxus onwards there are ascending axons in the spinal cord which decussate in a commissure of white matter ventral to the central canal; in mammals, this region of the spinal cord is known as the **white commissure** (Fig. 12.3 and Section 10.13). In Amphioxus and the lower vertebrates there are also descending axons in the spinal cord which again decussate; in the higher vertebrates, including mammals, many of the descending tracts decussate. It is possible that decussation arose from the coiling reflex in early representatives of the chordate phylum. The coiling reflex is a defensive response which enables a very simple animal to curve its body away from a noxious stimulus (Fig. 5.8).

The coiling reflex (Fig. 5.8) is the most basic of all spinal reflexes, and is present in very primitive chordates, e.g. Amphioxus. These simple relatives of the vertebrates possess spinal chains of interconnected afferent and efferent neurons, but all of these neurons are entirely ipsilateral; the efferent neurons activate segmental myotomes on the one side of the body only. However, there are also interneurons in these animals, some at the cranial end of the spinal cord and others at the caudal end, each with a long axon (Fig. 5.8). The axon of such an interneuron at the **cranial** end of the cord decussates and **descends** the spinal cord; the axon of an interneuron at the **caudal** end decussates and **ascends** the spinal cord.

Each of these long interneurons (i) receives synapses from some adjacent

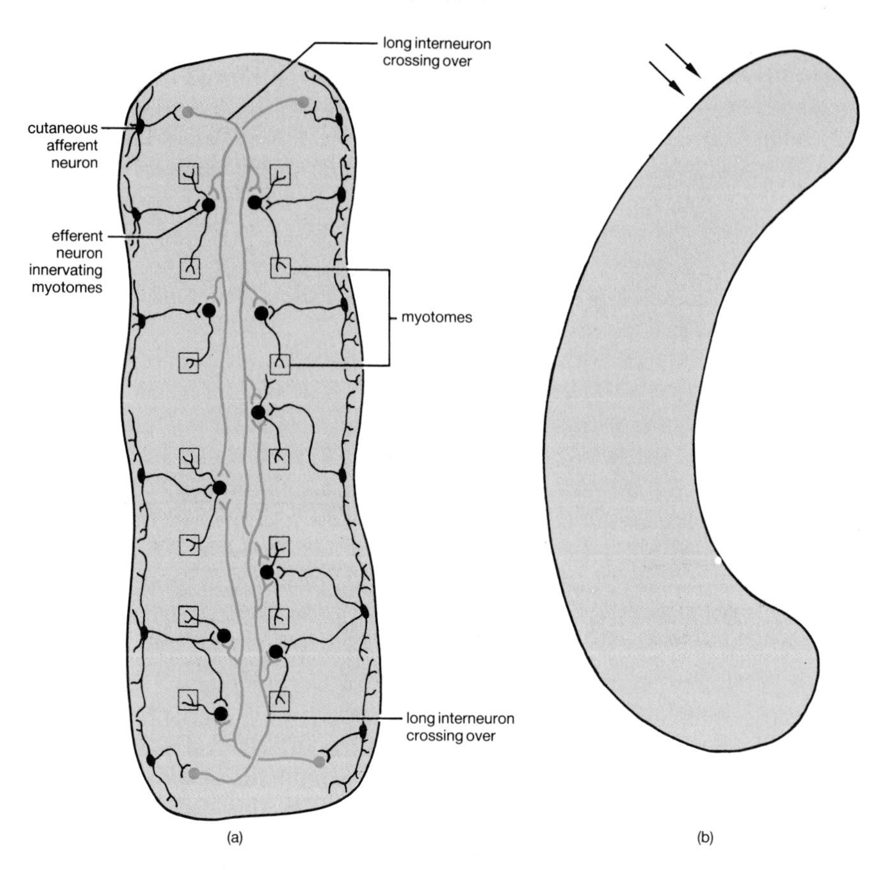

FIG. 5.8. Diagrams showing the structural basis of the coiling reflex in a primitive chordate animal. (a) Afferent cutaneous neurons make ipsilateral synapses with efferent neurons which innervate myotomes on the same side of the body. At each end of the body long interneurons (green) cross over and activate a series of efferent neurons and myotomes on the **opposite** side of the body. (b) A noxious stimulus received at the double arrows activates myotomes on the **contralateral** side, thus causing the body to coil away from the stimulus.

unilateral afferent neurons, and (ii) makes a series of synaptic contacts with the chain of unilateral efferent neurons of the **opposite** side of the body. Consequently, a noxious stimulus at either end of the body sets up the following sequence of neural and muscular activity; the stimulus excites the afferent neuron, the afferent neuron excites the interneuron, the interneuron excites the efferent neurons along the **opposite** side of the body, and the efferent neurons excite their myotomes; thus the myotomes contract in sequence along the side **opposite** to the stimulus. The result is that the body coils **away** from the stimulus. A similar defensive neuromuscular coiling reflex was probably formed by the decussation of the **optic nerves** in the most primitive vertebrates, causing the head to coil away defensively from a visual stimulus.

The decussating interneurons which mediate the coiling reflex in the most primitive vertebrates may represent the predecessors of the long ascending and descending axons which cross over in such an amazing way in the neuraxis of the highest vertebrates. As the evolution of the nervous sytem progressed, increasing numbers of these long decussating neurons could have developed. Ultimately this process could have led to the evolution of the decussating ascending and descending tracts which characterize the neuraxis of the highest vertebrates.

During the course of this evolutionary process, the defensive function of the coiling reflex must have been replaced in the higher vertebrates by the development of local withdrawal reflexes. These induce reflex flexion of the ipsilateral limb in response to a noxious stimulus.

The nerve impulse

Each neuron is either exclusively excitatory or exclusively inhibitory. All **primary afferent neurons**, i.e. those with their cell bodies in dorsal root ganglia, are excitatory. All motoneurons innervating **striated muscle** are also excitatory. On the other hand autonomic postganglionic motor neurons supplying smooth muscle, cardiac muscle, and glandular epithelium may be either excitatory or inhibitory. Interneurons within the neuraxis are again either excitatory or inhibitory.

5.10 Ion channels and gating mechanisms

When a volley of impulses arrives at a synaptic bulb on a resting interneuron or motoneuron, transmitter substance is released into the synaptic cleft. It is evident that the postsynaptic membrane must possess ion channels (pores) of two kinds. One allows Na^+ ions to pass from the extracellular fluid into the cytoplasm, thus causing an **excitatory postsynaptic potential**. The other type of channel allows K^+ ions to escape from the cell into the extracellular fluid; this gives rise to an **inhibitory postsynaptic potential**.

It is believed that a **gating mechanism** somehow enables each kind of channel to be unplugged by its appropriate transmitter substance. There are indications that the channel is composed, at least partly, of an integral protein macromolecule. The surface of this molecule possesses a biochemical receptor site, which recognizes and then binds the transmitter substance. As a result of the binding of a molecule of transmitter substance, the receptor protein molecule undergoes a conformational change which opens the channel.

Evidence from the ion selectivity of different channels has indicated that the sodium channel is a rectangular hole measuring about 0.3 to 0.5 nm. The potassium channel is narrower, about 0.3 nm in diameter. In aqueous solutions, water molecules become attached to the Na^+ and K^+ ions, causing them to become hydrated. Hydrated Na^+ ions are slightly larger than hydrated K^+ ions.

5.11 The membrane potential

In all cells an electrical potential exists whereby the inside is negative with respect to the outside: this is the **membrane potential (resting potential)**. Although it has been most intensively studied in nerve cells, the membrane potential is known to be a property of plant and animal cells generally. In nerve cells and muscle cells it is unstable, and capable of abrupt reversal as an **action potential**. Such unstable cells can be grouped together as **excitable cells**.

Sodium chloride is **outside** the cell, in the extracellular fluid, the Na ions being positively and the Cl ions negatively charged (Fig. 5.9(a)). Potassium, and largely unknown organic ions (P), are **inside** the cell, the K ions being positively and the organic ions negatively charged (Fig. 5.9(a)).

The Na^+, Cl^-, and K^+ ions can diffuse through the cell membrane by ion channels, but the organic anions are unable to diffuse because they are too large to go through the channels. Most of the Na^+ ions are kept **outside** the cell by the sodium–potassium pump, which trades Na^+ for K^+ ions, exchanging 3 Na^+ for 2 K^+. Most of the K^+ ions are kept **inside** the cell by an excess of negatively charged organic ions.

The essential cause of the membrane potential is the relative permeability of the cell membrane to the K^+ ions, and its relative impermeability to the Na^+ ions. Because the K cations can pass through the membrane relatively easily (from their high concentration within the cell), they leak rapidly from the cell leaving the organic anions behind. The leakage of K^+ ions continues until the tendency to diffuse is balanced by the electric field which is created by the residual organic anions. At the moment of equilibrium, the inside of a motoneuron is about 70 mV (thousandth of a volt) negative with respect to the outside. This difference of 70 MV is the membrane potential. Essentially it has been created by a 'potassium battery'.

5.12 The excitatory postsynaptic potential

Characteristics of the excitatory postsynaptic potential

The postsynaptic membrane lacks the ability to undergo the rapid alterations of inactivation and regeneration which produce the trains of action potentials typifying impulse activity. Instead the postsynaptic membrane responds to the release of E-substance by undergoing an **excitatory postsynaptic potential** (or **generator potential**). This is a depolarization which, in comparison with that of an action potential, is relatively small, relatively longlasting, decremental, and graded. It reduces the membrane potential from its resting value of 70 mV negative to a value nearer to the threshold of an action potential (i.e. towards 60 mV negative). In this content 'decremental' means a progressively decreasing intensity of depolarization, both with time and with distance from the synapse; 'graded' means that the depolarization can be varied in amplitude by interacting with (i) other excitatory postsynaptic potentials, thus allowing summation, and (ii) inhibitory postsynaptic potentials, thus allowing partial or complete cancellation of excitation.

Despite its decremental character, this graded depolarization, given adequate summation, may spread far enough over the surface of the nerve cell body to activate the specialised **'trigger region'**. This converts the sustained, decremental, depolarization into the explosive, non-decremental, action potentials that characterize impulse activity. The trigger region lies in the axon hillock and initial segment of the axon (see Section 5.2, **The myelin sheath**).

Thus the release of E-substance and the resulting excitatory postsynaptic potential resembles a stone thrown into a pond. A series of concentric ripples spread across the water. The ripples are sustained, but are decremental with time and distance.

Ionic basis of the excitatory postsynaptic potential

When E-substance is released into the synaptic cleft, ion channels are opened which allow the ingress of Na^+ ions through the postsynaptic membrane. This reduces the resting membrane potential below 70 mV negative.

It seems likely that the release of E-substance causes a non-specific increase in permeability to small cations (both Na^+ and K^+, and also to Cl^- ions), thus permitting these ions to flow along their concentration gradients. However, the concentration gradients are such that the predominant flow will be the ingress of Na^+ ions.

Summation

If a single excitatory impulse arrives at a neuron, the resultant excitatory postsynaptic potential will probably be too small to produce an action potential. The single excitatory impulse is then called a **subliminal impulse**. But the arrival of a rapid succession of subliminal excitatory impulses may lower the postsynaptic potential sufficiently to trigger an action potential. This is because the excitation accumulates, or summates. **Spatial summation** happens when the subliminal stimuli arrive at a neuron simultaneously, but along several different axon terminals. **Temporal summation** occurs when the subliminal stimuli arrive in close succession along a single axon terminal.

These characteristics can be illustrated by throwing stones into a pond. The ripples caused by a single stone may be too small to reach a floating trigger region on the other side of the pond; this would be a **subliminal stimulus**. A handful of similar stones, when flung simultaneously into many different parts of the pond, may together cause enough disturbance to rock the trigger mechanism into action; this would be **spatial summation**. Finally a series of identical stones thrown into the water at one place but in very rapid succession may again activate the trigger; this would be **temporal summation**.

Facilitation

The arrival of a continuous series of subliminal excitatory stimuli may be insufficient to reach the **threshold** of the trigger region of the axon, but may nevertheless place the cell body in a condition of facilitation. In this condition the postsynaptic potential could have been driven down from 70 mV and maintained at a level close to 60 mV negative in relation to the exterior of the cell. The arrival of one more volley of excitatory impulses could now, by further summation of excitatory postsynaptic potentials, lower the postsynaptic potential just a little more, so that the threshold of the trigger region is reached. On the other hand, if this neuron had not been already facilitated, the additional volley of excitatory stimuli alone would have been insufficient to induce an action potential. In the analogy of the pond, the surface is already in a condition of disturbance; one more stone, although in itself of only very small effect, may then rock the trigger to its threshold level.

5.13 The inhibitory postsynaptic potential

Characteristics of the inhibitory postsynaptic potential

The release of I-substance into the synaptic cleft causes an **inhibitory postsynaptic potential**. This increases the membrane potential towards 80 mV negative. The postsynaptic membrane is now hyperpolarized.

The inhibitory postsynaptic potential resembles the excitatory postsynaptic potential in being small, relatively longlasting, decremental, and graded. So long as the inhibitory postsynaptic potential is maintained by the arrival of volleys of inhibitory impulses, the neuron is held in a state of inhibition; during this phase of hyperpolarization, strong excitatory stimulation may be required to reduce the postsynaptic potential to 60 mV negative, the threshold of an action potential.

Ionic basis of the inhibitory postsynaptic potential

The I-substance released into the synaptic cleft increases the permeability of the postsynaptic membrane specifically to the smaller cations such as K^+. Thus a further egress of K^+ ions occurs through the postsynaptic membrane (Fig. 5.9(d)). On the other hand slightly larger cations such as the hydrated

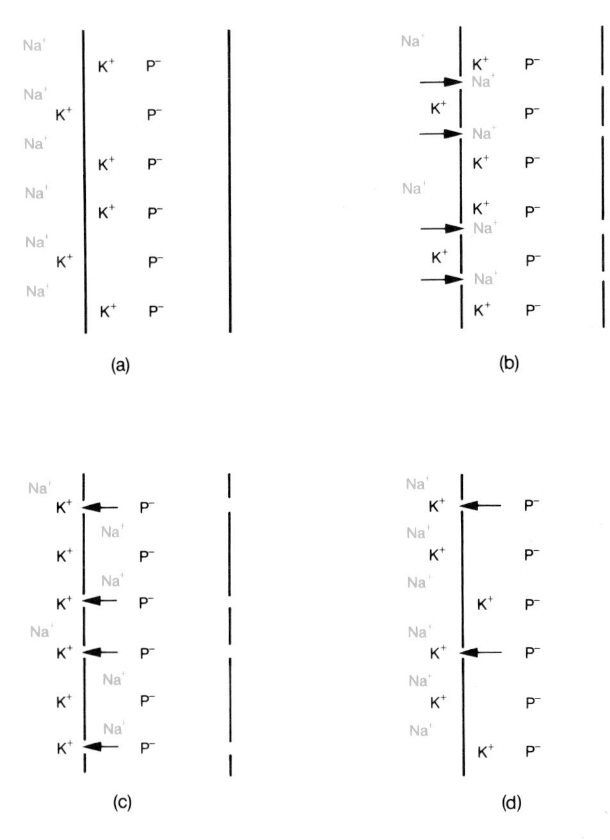

FIG. 5.9. Diagrammatic representation of the membrane potential. The diagram illustrates the ionic changes during an action potential, recovery after an action potential, and inhibition. The quantitative relationships of the ions are schematic only.

(a) The membrane potential (resting potential). K^+ ions leak out through the cell membrane leaving the organic anions P behind, until the tendency to diffuse is balanced by the electric field which is thus created. At equilibrium the inside of the cell is about 70 mV negative relative to the outside. This is the membrane potential.

(b) The action potential. Na^+ ions rush through ion channels in the cell membrane, making the inside positive, about +50 mV with respect to the outside. The total change in potential is therefore about 120 mV, and this change is the action potential.

(c) Recovery after an action potential. K^+ ions move out rapidly, restoring the resting potential. The Na^+ and K^+ ions are exchanged later during a slower recovery period.

(d) Inhibition. Inhibitory transmitter substance allows K^+ ions to escape. Thus the potential inside becomes more negative, reaching 75 to 80 mV negative. The cell membrane has become hyperpolarized.

Na^+, the diameter of which is greater than that of the hydrated K^+ ions, are unable to pass through.

There is evidence that inhibitory neurotransmitters (e.g. GABA) may act by increasing the permeability of the cell membrane to Cl^- ions, which are normally at a lower concentration inside a neuron than in the extracellular fluid. An inward flow of Cl^- ions hyperpolarizes a neuron.

Presynaptic inhibition

A neuron can be inhibited not only by postsynaptic inhibition as just described, but also be presynaptic inhibition. Presynaptic inhibition may depend on the existence of inhibitory axo–axonal synaptic bulbs (or more precisely, inhibitory termino–terminal synaptic bulbs); such inhibitory bulbs would end not on the dendrite or soma of an excitatory neuron, but on one of the synaptic bulbs of the excitatory neuron. The inhibitory fibre could then depolarize the excitatory synaptic bulb, thereby blocking its ability to release E-transmitter substance. The inhibitory fibre would therefore act by limiting the amount of transmitter substance released by the excitatory synaptic bulb.

More recently, however, this anatomical basis of presynaptic inhibition has been questioned and other more complex explanations have been offered.

Role of inhibitory neurons in the neuraxis

There are very large numbers of inhibitory neurons within the neuraxis. Some of them, such as the Renshaw cell which automatically inhibits a motoneuron in the ventral horn (Fig. 14.1), act individually. Others are grouped together to inhibit some particular activity, an example being the neurons of the pneumotaxic centre, which inhibit the inspiratory drive of neurons in the medullary respiratory centre in order to contribute to the rhythm of breathing. Yet other inhibitory neurons, including many in the cerebral cortex itself, have very widespread and diffuse effects on the neural regulation of locomotion and posture (see Section 13.3).

Reciprocal innervation

Inhibition plays an important part in the innervation of skeletal muscle, by means of reciprocal innervation. The principle of this mechanism is that, when the motoneurons innervating a muscle are reflexly stimulated by a volley of monosynaptic afferent impulses from a muscle stretch receptor (Fig. 10.1(a)), the motoneurons innervating the antagonistic muscles are simultaneously inhibited by the same afferent volley. Thus a monosynaptic afferent volley not only initiates muscular contraction but at the same time relaxes all the opposing musculature (see Section 10.5).

Disinhibition

In some regions of the neuraxis an inhibitory neuron makes synapses with a second inhibitory neuron. When these synapses release I-substance, the second inhibitory neuron is then prevented from inhibiting its own target neurons. This allows these target neurons to become dominated by excitatory stimuli. An indirect effect of this type has been called 'disinhibition'. The result of disinhibition is excitation. An example of disinhibition in the neuraxis is the inhibitory action of the lateral medullary reticular formation on the medial medullary reticular formation which is itself inhibitory; the net result is excitation of the ventral horn motoneurons associated with the medial medullary reticular formation (see Fig. 12.2 and Section 12.7).

5.14 The receptor potential

The receptor ending (i.e. the tip of a receptor terminal, which corresponds to the tip of a dendrite, see Section 5.1) has certain structural and functional properties which, not surprisingly, are essentially similar to those of the dendrite of an interneuron or motoneuron. In response to a suitable stimulus the receptor ending develops a **receptor** (or **generator**) **potential**; this resembles the excitatory postsynaptic potential of a dendrite, in being a relatively small, relatively longlasting, decremental, and graded depolarization (Fig. 5.12(a)). Moreover, like the dendrite, the receptor ending of an afferent axon (Fig. 5.12(b)) is unmyelinated (although sometimes it retains a thin Schwann cell covering). Furthermore, the receptor ending requires a trigger region to convert its initial decremental depolarization into impulse activity. In the interneuron or motoneuron this region is the axon hillock and initial segment of the axon: the analogous region on a receptor ending has not been firmly identified, but is probably the first node of Ranvier if the axon is myelinated, or the segment of the axon immediately central to the actual receptor region.

There is, however, one notable functional difference between the receptor ending and the dendrite of an interneuron or motoneuron: the receptor potential is always excitatory, never inhibitory, whereas the postsynaptic potential of the dendrite can be either excitatory or inhibitory.

5.15 The end-plate potential

The effector ending of an axon which innervates skeletal muscle makes a synaptic contact with its skeletal muscle fibre. The region of this synaptic contact is named a **motor end-plate** (Fig. 5.10). In a motor end-plate the cell membrane of the axonal effector ending is the presynaptic membrane; the sarcolemma of the muscle fibre (Fig. 5.11) is the postsynaptic membrane. Transmission from the axonal ending to the muscle fibre is achieved by the release of transmitter substance (acetylcholine). Just as in neuro–neuronal synapses, the postsynaptic membrane of the muscle fibre responds to the excitatory influence of the transmitter substance by undergoing a relatively small and longlasting but decremental depolarization, which in this instance is known as the **end-plate potential**. Furthermore, a trigger region is present on the sarcolemma of the muscle cell, presumably somewhere near the

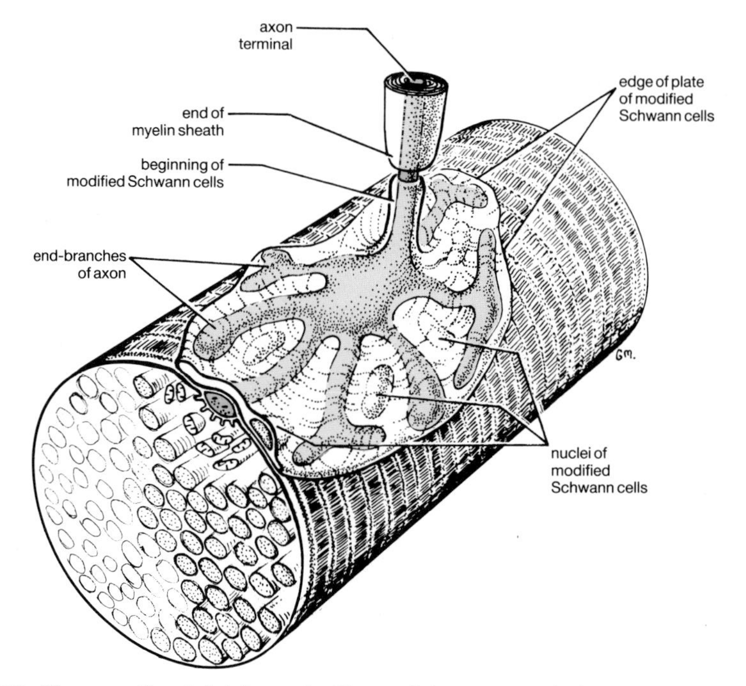

FIG. 5.10. Diagram of a skeletal muscle fibre and its motor end-plate. A muscle fibre has only one plate, oval in shape, about halfway along its length. The axonal terminal (green) abruptly loses its myelin sheath, and forms a number of end-branches which are covered by a thin flat plate of modified Schwann cells. Each end-branch of the axonal terminal forms a synaptic contact with the muscle fibre, one being shown in transverse section (see also Fig. 5.11).

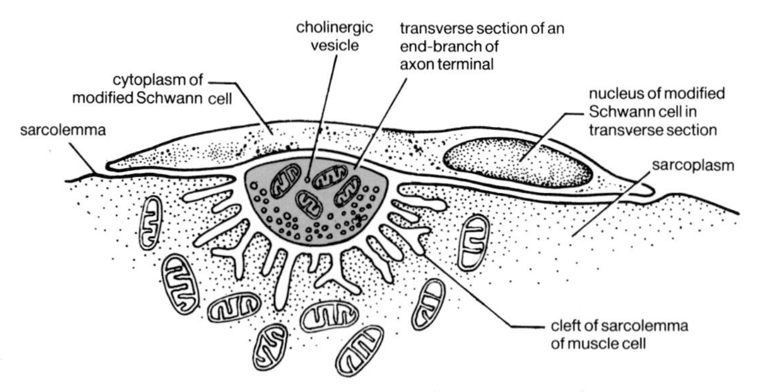

FIG. 5.11. Diagram of a synaptic contact within a motor end-plate. It shows an enlargement of the diagrammatic transverse section through the synaptic contact between an end-branch of the axonal terminal (green) and the underlying sarcolemma (see Fig. 5.10). The end-branch contains many cholinergic vesicles and numerous mitochondria. It lies in a groove on the surface of the muscle fibre, and is covered by a thin plate-like modified Schwann cell. The sarcolemma is furrowed by longitudinal clefts. The sarcoplasm under the clefts has many mitochondria and nuclei, but lacks myofibrils.

motor end-plate, which converts the end-plate potential into an action potential, resulting in contraction of the fibre.

However, the motor end-plate differs from a neuro–neuronal synapse in that the end-plate potential is always excitatory, never inhibitory. In this respect the end-plate potential resembles the receptor potential of an afferent receptor ending.

5.16 Summary of decremental potentials

There is much in common, both structurally and functionally, between (i) the dendrite of an interneuron or motoneuron and (ii) the receptor ending of a primary afferent neuron. Structurally, these components are really all dendrites, and therefore it is to be expected that they should share functional properties. Also belonging to this functional group are (iii) the cell membrane of a skeletal muscle fibre in the region of the motor end-plate, (iv) the cell membrane of a cardiac muscle fibre adjacent to its autonomic efferent axonal endings, and (v) the cell membrane of a smooth muscle cell in the region of its autonomic efferent axonal endings. Relatively small, longlasting, decremental, potentials characterize all five of these cell membranes.

Furthermore, all five of these structural entities require a specialized trigger region to convert the decremental potential into explosive, rapidly intermittent, non-decremental, action potentials. In the case of the interneuron and the motoneuron the trigger region is the axon hillock and initial segment: in the case of a myelinated afferent axonal ending it is probably the first node of Ranvier. In these instances, the trigger region therefore resides in an unmyelinated part of the axon. The trigger region of muscle cells has not been identified.

These then are similarities, structural and functional, between these five varieties of excitable cell. There are, however, some notable differences. The postsynaptic potential of the interneuron and motoneuron, and the end-plate potential of muscle cells, are all induced by a chemical transducer action, as is the receptor potential of a chemoreceptor afferent axon. On the other hand, the receptor potential of a mechanoreceptor is induced by mechanical distortion of the cell membrane (axolemma). Other types of receptor ending, e.g. thermal receptors, transduce yet other kinds of energy.

More important differences reside in the fact that the decremental potential may be excitatory only, or either excitatory or inhibitory. Thus, in the five structural entities which have just been listed, **either excitatory or inhibitory** potentials occur at (i) the postsynaptic membrane of an interneuron or motoneuron, and at the cell membrane of (iv) a cardiac muscle cell and (v) a smooth muscle cell. In contrast, purely **excitatory** potentials occur at (ii) the receptor ending of a primary afferent neuron and (iii) the cell membrane of a skeletal muscle cell in the region of the motor end-plate.

5.17 The action potential

Threshold

The cell body and dendrites of an interneuron or motoneuron are likely to be continually subjected to E-substance and I-substance released at excitatory and inhibitory synapses respectively. If inhibitory stimuli predominate, the neuron becomes hyperpolarized and is thence inhibited. If excitatory stimuli prevail, the postsynaptic potential may fall from a resting value of about 70 mV to about 60 mV negative. At this point the **threshold** of the trigger region is reached and an action potential results.

The role of the trigger region

The action potential is initiated by the 'trigger region' (**pacemaker region** or **regenerator region**) of a neuron or muscle cell. In interneurons and motoneurons the trigger region lies in the axon hillock and initial segment of the axon. Its position in axonal receptor endings and muscle cells is not clearly established, but in the former it is probably in the first (most peripheral) node of Ranvier when the afferent axon is myelinated.

In the trigger region, the relatively sustained but decremental excitatory postsynaptic potential is converted into the intermittent, non-decremental, action potentials which characterize impulse activity. Non-decremental means that the impulses produced in this way are conducted all the way to the axon terminals without any decrease in the degree of depolarization.

The ionic flux of the action potential

At the instant when the **threshold** of an action potential has been reached, the axolemma becomes highly permeable specifically to Na^+ ions, which then rush through (Fig. 5.9(b)). This makes the inside of the axon positive (about +50 mV) with respect to the outside. Thus the potential within the axon has abruptly altered from about −70 mV to about +50 mV, a total change of about 120 mV. This potential change is termed the **action potential** (or **spike potential**). Once begun, the action potential continues independently of the stimulus; thus a change in the intensity of the stimulus has no effect on the amplitude of the action potential. Furthermore an axon reacts to any stimulus either by an action potential or by remaining silent. Hence it follows the **all-or-none law**.

Nevertheless, the action potentials of an axon do not always have the same amplitude; various factors can change the energy reserves of an axon and thus alter the amplitude of its action potential. However, for any particular set of axonal conditions the axon will always respond maximally to a threshold stimulus.

The action potential passes along a non-myelinated axon like a series of small explosions along a train of gunpowder, each small explosion igniting the adjacent gunpowder. Thus the ionic flux occurring at a particular point on the axolemma disturbs the adjacent region of the axolemma, increasing its permeability to Na^+ ions. An inrush of Na^+ ions then occurs in this newly disturbed region. This in turn disturbs the next part of the axolemma. Thus the action potential moves along the axon in a self-propagating manner. In myelinated axons the action potential jumps from node of Ranvier to node of Ranvier. This leaping conduction is called **saltatory conduction**.

Recovery after an action potential

When the action potential ends, the axolemma ceases to be permeable to sodium ions and becomes permeable to potassium ions. Potassium ions quickly move out (Fig. 5.9(c)). The membrane potential therefore returns rapidly to its resting level. Thus the axon has lost some K^+ ions, but has made a **fast recovery** to its original electrical state and is ready for another impulse. The Na^+ and K^+ ions are **re-exchanged** later during a **slower recovery period**, by means of the Na/K pump. There are enough ions in one axon for a vast number of impulses; it is estimated that only one millionth of the potassium ions in an axon escape during a single action potential.

Refractory periods

While the axon is fully depolarized (i.e. during the action potential) it cannot be re-excited. This phase of depolarization is known as the **absolute refractory period**; it lasts about 0.5 ms (thousandths of a second).

During the recovery phase which follows an action potential, there is a brief overswing and the axon becomes temporarily hyperpolarized. This period is the **relative refractory period**. During this phase another action potential can be induced, though of less than normal amplitude; but this is possible only if the stimulus is considerably greater than the resting threshold value. Excitability is progressively restored during the relative refractory period.

The refractory characteristics of axons impose a limit on the number of impulses which an axon can transmit per second. If its absolute refractory period is 0.5 ms, an axon can be driven continuously at rates not exceeding 1000 impulses per second.

5.18 Concerning water closets

Summation, threshold, all-or-none characteristics, and refractoriness are illustrated by the flushing of a water closet. A single weak depression of the lever induces no more than a gurgle (a subliminal stimulus), but several similar depressions, rapidly repeated, may generate a flush and this illustrates **temporal summation** and **threshold**. Once the flush has been activated it follows the **all-or-none principle**. During the flush, and during the onset of refilling, further depressions of the lever cause no response; this is the **absolute refractory period**. Later in the refilling period a weak flush can be induced, but several relatively forceful depressions are required to do so; this is the **relative refractory period**.

5.19 Transducer mechanisms of receptors

The receptor ending is the site of the transducer mechanism which converts the energy of the stimulus into the energy of the nerve impulse. A mechanoreceptor converts kinetic energy into the nerve impulse, whereas a thermal receptor converts heat energy and a chemoreceptor converts chemical energy into the nerve impulse. Essentially a neuron which is responding to a transmitter substance is acting as a chemoreceptor. Some receptor pathways begin with a **receptor cell**. Examples include the neuro-epithelial cells of the spiral organ of Corti, the vestibular apparatus, and the retina. These cells act as the transducer, and release transmitter substance at synapses with their afferent (sensory) axons.

In the case of mechanoreceptor endings, the ending is distorted by the mechanical stimulus. Distortion stretches some part of the cell membrane (axolemma), thus causing the **receptor (generator) potential**. Possibly the stretching reduces the lateral attractive forces between the constituent molecules in this particular part of the cell membrane and thus reduces the resistance to the flow of ions across the membrane. Alternatively, the stretching may dilate pre-existing ion channels. The transducer process appears to be extremely focal; several different spots along the receptor membrane may be activated simultaneously, with spatial summation (Fig. 5.12(a)). The amplitude of the receptor potential is proportional to the total area of membrane which is activated.

Simple mechanoreceptor endings consist of a bare receptor ending (Fig. 5.12(b)), sometimes with a minimal covering of Schwann cell cytoplasm. Such receptor endings are distorted in direct proportion to the force of the mechanical stimulus which is applied to them: so long as they are distorted they develop a receptor potential. The amplitude of the receptor potential determines the frequency of the impulses produced by the trigger region. Thus the frequency of the impulse discharge is directly proportional to the degree of distortion. This is known as **frequency coding**. Such receptors are **slowly adapting** mechanoreceptors, as in the baroreceptor reflex from the carotid sinus or the Hering–Breuer respiratory stretch reflex.

Complex laminated mechanoreceptors consist of a central receptor ending, enclosed by concentric laminae which are slightly separated by extracellular fluid. The laminae are thin cell processes belonging to a pair of Schwann cells; the laminae interdigitate, much as the fingers of two hands can be interdigitated (Fig. 5.12(c), (d)). Outside the laminae there is a fluid-filled space. This in turn is surrounded by a ring-like series of slender capsular cells. The **Pacinian corpuscle** is an example.

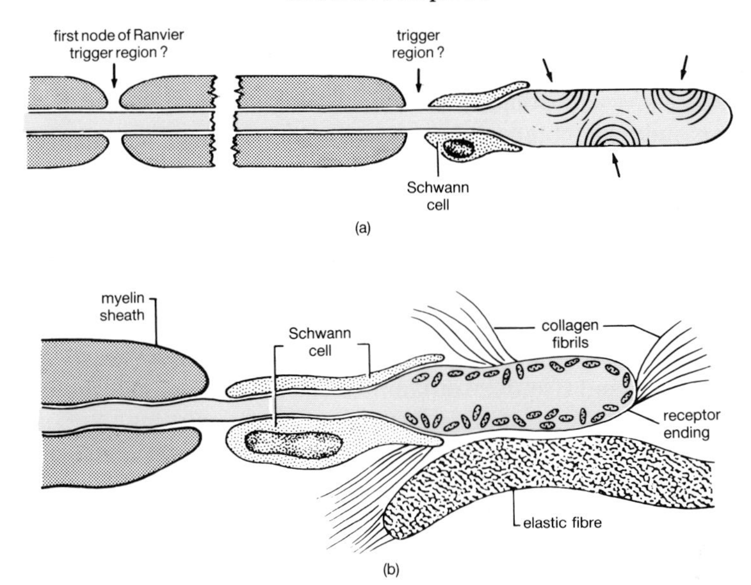

FIG. 5.12. **(a) Diagram of a bare receptor ending.** The ending could be a mechanoreceptor, chemoreceptor, or thermoreceptor. The arrows represent stimuli, causing decremental potentials (shown as waves). Summation of the decremental potentials causes a receptor potential, which is converted at the trigger region into an action potential. The site of the trigger region is uncertain.

(b) Diagram of a bare mechanoreceptor ending. The ending is enlarged, packed with mitochondria, and devoid of a Schwann cell covering. Its axonal membrane is close, or even attached, to collagen fibrils and an elastic fibre.

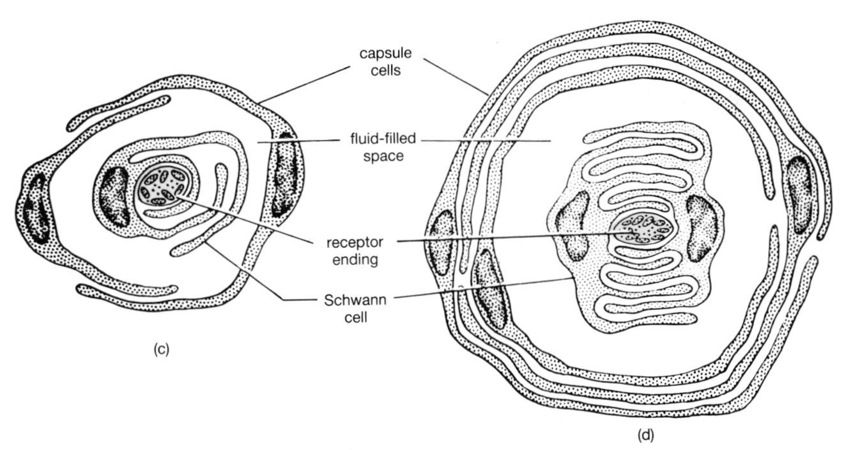

FIG. 5.12. **(c) Diagram of a very simple laminated mechanoreceptor.** The receptor ending is packed with mitochondria.

(d) Diagram showing the essential structure of a complex laminated mechanoreceptor. The central axonal receptor ending is enclosed by two interdigitating Schwann cells. Surrounding these cells is a fluid-filled space. Outside this is a capsule of ring-like cells. In the most advanced forms (such as the Pacinian corpuscle) the number of interdigitating lamellae belonging to the two Schwann cells is very much greater than shown here.

The whole structure, including the axonal ending and the laminae enclosing it, has an inherent elasticity which endows it with the tendency always to assume its resting, essentially circular shape. If the entire receptor complex is suddenly compressed externally (the dynamic phase of compression), a pressure wave is instantly transmitted through the fluids within. This distorts the receptor ending. The distortion generates a receptor potential, which in turn releases a train of action potentials. As in simple mechanoreceptor endings, the intensity of the receptor potential, and hence the frequency of the resulting action potentials, is determined by the area of the receptor ending which is distorted. If the external compression is sustained unchanged (the static phase of compression), the inherent elasticity of the receptor ending and its adjacent laminae causes these structures rapidly to resume their original shape; this they achieve by displacing the fluid of the fluid-filled space, even though the external capsule is still compressed into an oval shape. The receptor ending itself is now no longer distorted, and therefore the membrane potential quickly returns to its resting value and the receptor axon falls silent. It is therefore known as a **fast adapting** receptor. If the external compression is then suddenly released, the whole receptor complex springs back to its original circular shape. The returning movement of the external capsule and outermost laminae sets up another pressure wave in the fluids within, which distorts the receptor ending for the second time. This generates a second receptor potential, which releases another train of action potentials. However, the ending and its laminae quickly resume their resting shape, so the ending again reverts to its resting potential and the receptor axon once more falls silent.

This type of complex laminated mechanoreceptor therefore fires on the 'on' and the 'off' of compression, but is silent in between. Such receptors are capable of following rapid vibrations (up to 500 Hz). They respond to essentially the **rate** of pressure change, rather than to the magnitude of that pressure.

Neuroglia

In addition to the thousands of millions of neurons which make up the central nervous system, there are about five times as many supporting cells or neuroglial cells. There are three types of neuroglial cells, namely astrocytes, oligodendrocytes, and microglia. The first two are derived from the neurectoderm, but the microglia are mesodermal in origin. The neuroglial cells form the supporting framework of the central nervous system. They probably also play a part in the transport of gas, water, and electrolytes from the blood plasma to the nerve cells, and in the removal of metabolites. The majority of intracranial tumours arise from the neuroglia, forming different varieties of gliomas.

5.20 Astrocytes

Astrocytes form a continuous single cell layer immediately beneath the pia mater, cover nearly all (about 85 per cent) of the basal lamina of all blood vessels within the substance of the central nervous system by means of their perivascular feet (Fig. 5.13), and fuse with the ependymal cells. Thus they form almost a complete mosaic over the deep blood vessels. Because of the way in which their feet enclose the capillaries of the central nervous system, they may well contribute to the **blood–brain barrier**. Their normal function is probably to monitor and, by taking part in the transport of substances, to control the contents of the intercellular space. If neurons die within the neuraxis, the astrocytes multiply and tend to form so-called **glial scars** which may obstruct regenerating axons.

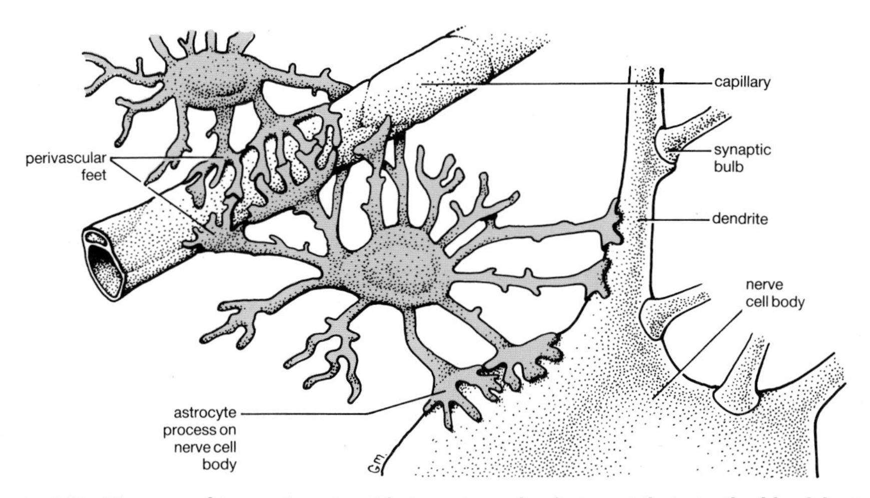

FIG. 5.13. Diagram of two astrocytes. Their perivascular feet contribute to the blood–brain barrier. In the diagram the spaces between the feet of these two astrocytes would be filled by the feet of other astrocytes. The perivascular feet, which thus cover the surface of the neuron and its processes, participate in the exchange of metabolites between the nerve cell, the blood, and the CSF.

5.21 Oligodendrocytes

Oligodendrocytes form and maintain myelin (Fig. 5.14) in the central nervous system, thus resembling the Schwann cell of the peripheral nervous system. However, in the central nervous system each oligodendrocyte encloses **several** axons, whereas the Schwann cell of the peripheral nervous system enwraps only one myelinated axon.

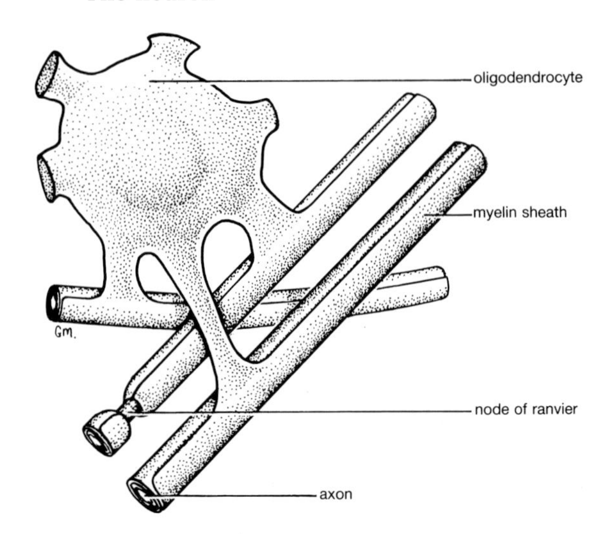

FIG. 5.14. Diagram of an oliogodendrocyte in the neuraxis. It forms a myelin sheath around three axons. Four of its cell processes are shown as transected stumps. One oligodendrocyte may myelinate up to 50 axons.

5.22 Microglia

As their name implies these are small cells. They function as macrophages and arise from monocytes of the bone marrow. They migrate into the neuraxis from the bone marrow.

6. Nuclei of the cranial nerves

General principles governing the architecture of the nuclei of the cranial nerves

At first sight the general architecture of the brainstem appears to be disconcertingly different from that of the spinal cord. However, on closer inspection it becomes apparent that the underlying organization of the brainstem shows a reassuring similarity to that of the spinal cord. Thus the dorsal and ventral horns of the spinal cord are directly continuous with the grey matter of the brainstem. Moreover, the four functional areas which typify the grey matter of the spinal cord can also be recognized in the grey matter of the brainstem. These four areas are the somatic afferent and visceral afferent components of the dorsal horn, and the visceral (autonomic) efferent and somatic efferent components forming, respectively, the lateral horn and ventral horn in the spinal cord. The visceral efferent component of the brain forms a pre- and a postganglionic relay as in the trunk.

Thus the basic design of the grey matter of the brainstem is essentially similar to that of the spinal cord. There are, however, three major differences in the anatomy of the brainstem which extensively modify its architecture as compared with that of the spinal cord, namely the shape and position of the central canal, fragmentation of the grey matter, and the presence of an additional neuronal component.

6.1 Shape and position of the central canal

In the brainstem the canal tends to be **dorsal** in position (Fig. 6.1). Furthermore it tends to be **wider** than in the spinal cord. These features cause the basic columns of grey matter of the brainstem to become V-shaped (rather than vertical), lying obliquely on either side of the widened central canal (Fig. 6.1).

6.2 Fragmentation of the basic columns of grey matter

In the spinal cord the horns of grey matter are continuous columns. Embryologically they are also continuous in the brainstem, but during development the 'horns' in the brainstem become separated from each other

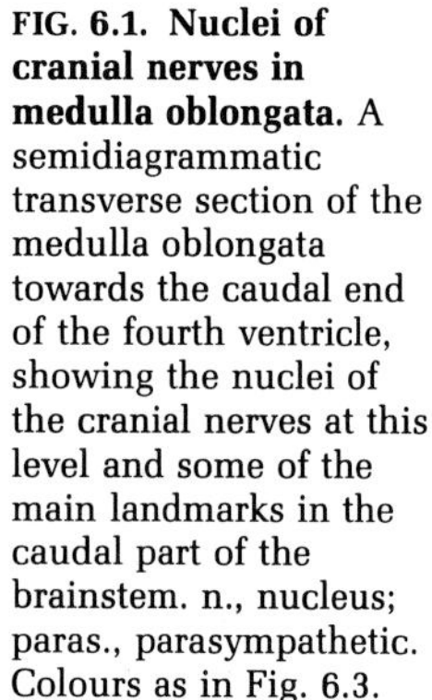

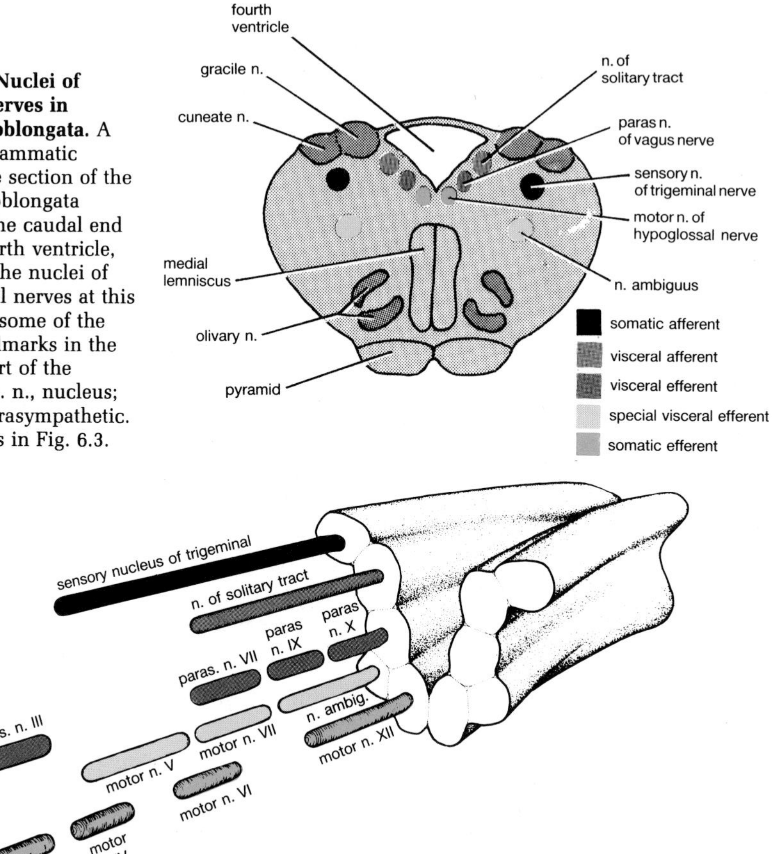

FIG. 6.1. Nuclei of cranial nerves in medulla oblongata. A semidiagrammatic transverse section of the medulla oblongata towards the caudal end of the fourth ventricle, showing the nuclei of the cranial nerves at this level and some of the main landmarks in the caudal part of the brainstem. n., nucleus; paras., parasympathetic. Colours as in Fig. 6.3.

FIG. 6.2. Lateral view diagram of the nuclei of the cranial nerves in the right side of the brainstem. The dorsal and ventral horns of the grey matter of the spinal cord, which are shown at the right-hand end of the diagram, become continuous rostrally with the nuclei of the cranial nerves. The nuclei tend to be arranged in a V-shape, because of the dorsal widening of the central canal in the caudal part of the brainstem. The sensory cranial nerve nuclei (the two, more dorsal, lines of nuclei) are continuous longitudinally. The motor cranial nerve nuclei (the three, more ventral, lines of nuclei) are broken up longitudinally into discontinuous islands of grey matter. The cranial nerve related to each nucleus (n.) is indicated by its number. Thus paras. n.III stands for parasympathetic nucleus of oculomotor nerve. ambig., ambiguus. Colours as in Fig. 6.3.

(probably because of the development of many intersecting bundles of axons). Thus the basic grey matter of the brainstem becomes broken **transversely** into four longitudinal columns, which are somatic afferent, visceral afferent, visceral efferent, and somatic efferent in function (Fig. 6.2). Moreover, the two efferent columns are also partly broken **longitudinally**, thus forming separate islands of grey matter along the brainstem which are now graced by the name of nuclei; however, these nuclei remain essentially in line with one another (Fig. 6.2). The two afferent columns (somatic and visceral afferent) do not break up longitudinally, but remain as two continuous longitudinal columns. They too become known as nuclei.

6.3 Development of an additional component, special visceral efferent

The lateral plate mesoderm of the embryonic head contributes to the striated muscle of the pharyngeal arches. This additional muscular component in the head is innervated by an additional neuronal component, known as the special visceral efferent neuronal component (also called branchial efferent). This arises as another efferent column of grey matter, between the visceral (autonomic) efferent and the somatic efferent columns (Fig. 6.3), but during development it gets squeezed out and comes to lie **ventrolateral** to the rest of the afferent and efferent nuclei (nucleus ambiguus in Fig. 6.1). Like the other efferent columns, however, it tends to break longitudinally into (three) isolated islands, and again these islands are now called nuclei. A special visceral efferent nucleus is allocated as follows to the cranial nerves which innervate the pharyngeal arches (Fig. 6.2): the motor nucleus of the trigeminal nerve supplying arch 1 (chewing muscles); the motor nucleus of the facial nerve supplying arch 2 (facial muscles); and one composite nucleus, the nucleus ambiguus, supplying arches 3, 4, and 6 (pharyngeal and laryngeal muscles, innervated by cranial nerves IX and X–XI).

It has long been accepted that the striated muscle of the pharyngeal arches is derived from the lateral plate mesoderm. It is now believed, however, that this muscle receives substantial additions from the cranial neural crest, but it is still by no means clear which of these sources actually predominates.

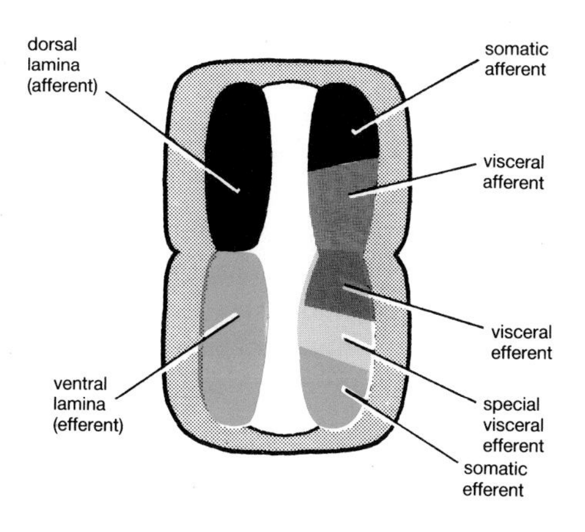

FIG. 6.3. The embryonic neural tube. The diagram shows the functional subdivisions of the developing grey matter in the brain.

6.4 The cranial nerves of the special senses

The olfactory, optic, and vestibulocochlear nerves (cranial nerves I, II, and VIII) are not segmental structures and are not governed by the architectural principles which apply to the other cranial nerves. The nerves of the special senses will be discussed in Chapter 8.

6.5 Summary of the architectural principles of the nuclei of the cranial nerves

(i) The nuclei of the cranial nerves lie in a V-shaped arrangement adjacent to the central canal, except for the special visceral efferent nuclei which are squeezed into a ventrolateral position.

(ii) The somatic afferent and visceral afferent nuclei are in line with the dorsal horn of the spinal cord; the visceral efferent nuclei are in line with the lateral horn; and the somatic efferent nuclei are in line with the ventral horn of the spinal cord.

(iii) The afferent nuclei are separated from each other transversely.

(iv) The efferent nuclei are separated transversely and tend also to be broken up longitudinally.

Names, topography, and functions of the cranial nerve nuclei

6.6 Somatic afferent nucleus

The somatic afferent region of the grey matter of the brainstem forms a single continuous and very elongated nucleus, the **sensory nucleus (nuclei) of the trigeminal nerve** (Figs. 6.1, 6.2), which is directly continuous with, and homologous to, the dorsal horn of the spinal cord. This nucleus receives the central projections of all of those primary afferent neurons which are involved in the somatic afferent pathways of the head (for example, from the cornea, Fig. 6.4). These include touch, pressure, pain, temperature, and the sense of joint movement, arising from the superficial and deep structures of the face. Four cranial nerves possess somatic afferent fibres, namely the fifth (trigeminal), seventh (facial), tenth (vagus), and probably also the ninth (glossopharyngeal) cranial nerves (Table 6.1). However, by far the largest component travels in the fifth cranial nerve, and it is because of this that the somatic afferent nucleus is named the sensory nucleus of the trigeminal nerve.

The sensory nucleus of the trigeminal nerve is in fact divided into three consecutive regions with different functions. The most caudal segment is called the nucleus of the

Table 6.1. Summary of the nuclei of the cranial nerves

Neurological component of nucleus	Name of nucleus	Cranial nerves associated with each nucleus	Main function of nucleus
Somatic afferent	Sensory nucleus of trigeminal nerve	V, VII, (IX), X	Afferents from ectoderm of oral cavity (V), and skin of head (V, also VII, IX, X). Also afferents of pain and kinaesthesia of the head.
Visceral afferent	Nucleus of solitary tract	VII, IX, X	Visceral afferents from pharynx (IX, X), larynx (X), and tongue (VII, IX, X); from cardiovascular receptor zones (IX, X); from respiratory and gastrointestinal tracts (X).
Visceral efferent	Parasympathetic nuclei of: oculomotor n. facial n. glossopharyngeal n. vagus n.	 III VII IX X	Efferents to iris and ciliary body (III), salivary glands (VII, IX), and thoracic and abdominal viscera (X).
Special visceral efferent	Motor of trigeminal n. Motor of facial n. Nucleus ambiguus	V VII IX together with X–XI complex	Efferents to muscles of chewing (V), to facial muscles (VII), and to pharyngeal and laryngeal muscles (IX, X–XI).
Somatic efferent	Motor of oculomotor n. Motor of trochlear n. Motor of abducent n. Motor of hypoglossal n.	III IV VI XII	Efferents to muscles of eyeball (III, IV, VI) and tongue (XII).

n., nerve.

spinal tract of the trigeminal nerve, the middle segment is called the principal (pontine) sensory trigeminal nucleus, and the most rostral segment is called the nucleus of the mesencephalic tract of the trigeminal nerve. The **nucleus of the spinal tract of the trigeminal nerve** is the cell station of the second neuron in the pathways carrying pain and temperature sensations from the face. It is therefore strictly homologous to the cells of the dorsal horn of the spinal cord which give rise to the spinothalamic tract. The **principal (pontine) sensory trigeminal nucleus** consists mainly of the cell bodies of the second neurons of the pathways of touch and pressure from the face. This nucleus is therefore strictly homologous to the cuneate and gracile nuclei (see Section 7.3). The separation of pain from touch in these two continuous nuclei enables the surgeon to relieve severe trigeminal neuralgia in human patients by eliminating the spinal trigeminal nucleus, while at the same time preserving intact the pathways of touch from the face (including the all important protective corneal reflex) by leaving the principal trigeminal nucleus intact. The **nucleus of the mesencephalic tract of the trigeminal nerve** apparently deals with all the incoming proprioceptive impulses from the muscle spindles of the jaw muscles and probably also those from the muscles of the eyeball. The cells of the mesencephalic trigeminal nucleus resemble histologically the pseudounipolar neurons of dorsal root ganglia; they may indeed be the cell bodies of **primary afferent neurons** which have arisen embryonically from the neural crest but have become embedded in the brainstem instead of forming a dorsal root ganglion. If so, these neurons of the mesencephalic trigeminal nucleus break the almost universal rule that **primary afferent neurons have their cell stations in dorsal root ganglia or in the homologous ganglia of cranial nerves:** the only other exception to this rule is the olfactory neuroepithelial cell (see Section 8.5).

6.7 Visceral afferent nucleus

The visceral (autonomic) afferent pathways from the head, and many from the trunk also, project into a single long nucleus, the **nucleus of the solitary tract** (Fig. 6.2). These afferent pathways arise in the embryonic head from the endodermal regions of the foregut, notably from the endodermal lining of the back of the tongue (the third embryonic pharyngeal arch) and from the endodermal lining of the pharynx and larynx (pharyngeal arches three, four, and six); these visceral afferent fibres travel in the ninth and tenth cranial nerves (Table 6.1). Others arise from the taste buds in the rostral two-thirds of the tongue and travel in the seventh cranial nerve (which is the nerve of the second embryonic pharyngeal arch, but invades the ectodermal lining of the front of the tongue belonging to the first arch).

Taste fibres are often called **special** visceral afferents as opposed to general visceral afferents, but all visceral afferents share the nucleus of the solitary tract.

Large numbers of other visceral afferent fibres arise from the trachea and lungs, and from the oesophagus and gastrointestinal tract, travelling in the vagus; these too project into the nucleus of the solitary tract. Yet others are cardiovascular afferents; these arise from the third and fourth embryonic aortic arches, constituting the carotid sinus nerve and aortic nerve, and travel finally in the ninth and tenth cranial nerves.

Thus the nucleus of the solitary tract receives projections from three cranial nerves, the facial, glossopharyngeal, and vagus nerves.

6.8 Visceral efferent nuclei

There are four visceral (autonomic) motor nuclei, lying in line with each other (Fig. 6.2). These are homologous to the lateral horn of the spinal cord, and contain neurons which resemble histologically those of the lateral horn. The most rostral of these nuclei is the **parasympathetic nucleus of the oculomotor nerve (Edinger–Westphal nucleus)**. This gives rise to the preganglionic autonomic axons of the third cranial nerve, which are responsible for regulating the muscles of the iris and ciliary body (Table

6.1). The next in the caudal direction is the **parasympathetic nucleus of the facial nerve** (rostral salivary or salivatory nucleus) supplying preganglionic fibres of the seventh cranial nerve to the mandibular and lingual salivary glands, lacrimal gland, and glands of the nasal mucosa. Caudal to this is the **parasympathetic nucleus of the glossopharyngeal nerve** (caudal salivary or salivatory nucleus), sending preganglionic fibres in the ninth cranial nerve for regulation of the parotid salivary gland (Fig. 6.4). The most caudal in this series of nuclei is the **parasympathetic nucleus of the vagus nerve** (dorsal motor nucleus of the vagus) (Fig. 6.1), which provides preganglionic vagal fibres to the lower respiratory and gastrointestinal tracts. These nuclei are often known as general visceral efferent, to distinguish them from the so-called special visceral efferent nuclei which are considered in the next paragraph.

6.9 Special visceral efferent nuclei

This is a series of three nuclei (Fig. 6.2), each giving rise to axons which innervate the striated muscles derived from the embryonic pharyngeal arches (Table 6.1). The most rostral is the **motor nucleus of the trigeminal nerve** (Vth motor nucleus), and this is responsible for the muscles of chewing, (which are derived from the first pharyngeal arch). The second is the **motor nucleus of the facial nerve** (VIIth motor nucleus) and innervates the facial muscles (all of which are derived from the second arch).

The third is an elongated nucleus, the **nucleus ambiguus** (Fig. 6.1), which innervates the muscle of the third arch (forming stylopharyngeus only) and the muscle of the fourth arch and sixth arches; the fourth and sixth arches form, respectively, the other pharyngeal muscles and the laryngeal muscles. Since stylopharyngeus is innervated by the ninth cranial nerve, and the rest of the pharyngeal muscles and the laryngeal muscles are innervated by the vagus, it follows that the nucleus ambiguus contributes to both the glossopharyngeal and vagus nerves.

It is probable that the chewing, facial, pharyngeal, and laryngeal muscles are not derived exclusively from the embryonic pharyngeal arches, but also receive contributions from the neural crest.

6.10 Somatic efferent nuclei

There are four somatic efferent nuclei, arranged in a line (Fig. 6.2). These innervate the muscles derived from the myotomes of the somites of the head, and therefore their axons travel in the four cranial nerves which are homologous to the ventral roots of the spinal nerves. The nuclei themselves are homologous to the ventral horn of the spinal cord, and consist of neurons which resemble histologically those of the ventral horn.

The most rostral is the **motor nucleus of the oculomotor nerve** (Figs. 6.2, 6.4). This nucleus controls the orbital muscles which arise from the first somite of the head, namely the ventral oblique muscle, the recti muscles except the lateral one, and the levator palpebrae superioris muscle (Table 6.1).

Next in line is the **motor nucleus of the trochlear nerve** (Figs. 6.2, 6.4), supplying the only muscle which arises from the second somite of the head, namely the dorsal oblique muscle. Since this cranial nerve is homologous to a ventral spinal root its axons ought to emerge from the **ventral** aspect of the brainstem, but in fact they spring from the **dorsal** surface. The explanation is that the nucleus is ventral and in line with the other somatic efferent nuclei (Fig. 6.2), but its axons pass dorsally from the nucleus and then **cross** the midline **dorsal** to the rostral part of the fourth ventricle. The axons continue laterally and emerge from the **dorsal** surface on the **opposite** side from which they started. The crossing over of all its axons and their dorsal emergence are together unique among the cranial nerves.

Some oculomotor fibres also cross over before emerging.

The third nucleus of the somatic efferent series is the **motor nucleus of the abducent nerve** (Figs. 6.2, 6.4). This nerve emerges ventrally, as it should, and innervates the lateral rectus muscle, and also the retractor bulbi muscles in those mammals which have them including the domestic species but not man.

The most caudal of the somatic efferent nuclei is the **motor nucleus of the hypoglossal nerve** (Figs. 6.1, 6.2, 6.4). In many mammals this elongated nucleus continues directly into the ventral horn of the spinal cord, and its cells are of the same type as the motoneurons of the ventral horn. Its fibres supply both the intrinsic and the extrinsic tongue muscles.

Reflex arcs of the nuclei of the cranial nerves

Since the nuclei of the cranial nerves (except those of the special senses) are homologous to the grey matter of the spinal cord, it is to be expected that they should participate in reflex arcs resembling those of the spinal cord. As in the spinal nerves, the **primary afferent neurons** have their cell bodies in 'dorsal root ganglia', i.e. in the trigeminal ganglion, geniculate ganglion (of the facial nerve), the ganglion (proximal or distal) of the glossopharyngeal nerve, or the ganglion (proximal or distal) of the vagus nerve. The central axons of the primary afferent neurons project into either the somatic afferent nucleus or the visceral afferent nucleus depending on whether the reflex arc is somatic or visceral, and form synapses there with an interneuron. The interneuron projects into one or other of the three types of

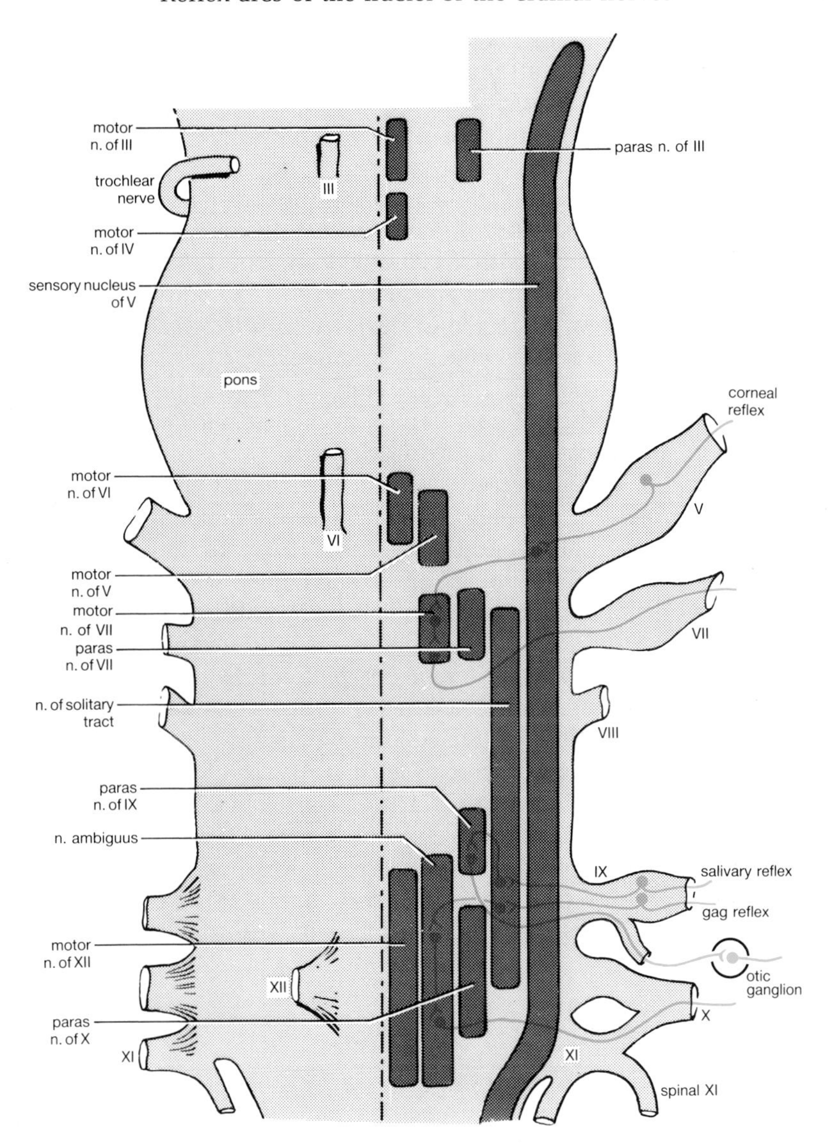

FIG. 6.4. Ventral view diagram of the cranial nerve nuclei and some of their reflex pathways. The nuclei (n.) of the cranial nerves are seen as though the left half of the brainstem were transparent. The neuronal pathways of the corneal reflex, a salivary reflex, and the gag (retching) reflex are traced through the relevant nuclei on the right side of the diagram. As shown, the corneal reflex starts from receptor endings in the cornea, and ends as motor fibres to muscles closing the eyelids. The salivary reflex begins with taste receptors on the back of the tongue, and terminates in postganglionic endings in the parotid salivary gland. The gag reflex is initiated by mechanoreceptor endings in the pharyngeal wall, and ends in motor fibres to muscles closing the pharynx. A nucleus (n.) is named according to the cranial nerve with which it is associated. Thus paras. n. of III indicates parasympathetic nucleus of the oculomotor nerve. The cranial nerves themselves are labelled by roman numerals.

efferent nuclei, depending on whether the effector pathway is somatic efferent, visceral efferent, or special visceral efferent.

Examples of such reflex arcs are shown in Fig. 6.4. The upper part of the right side of the diagram shows the neuronal pathway of the **corneal reflex**. The primary afferent neuron arises from the cornea, travels in the trigeminal nerve, and has its cell body in the trigeminal ganglion. Its central axonal continuation projects into the sensory nucleus of the trigeminal nerve, and synapses there with an interneuron. The interneuron projects into the motor nucleus of the facial nerve, where it synapses with a special visceral efferent neuron. The axon of this neuron travels in the facial nerve to the orbicularis oculi muscle, which closes the eyelids.

The lower middle part of Fig. 6.4 shows the neuronal pathway of a **salivary reflex**. A primary afferent neuron from taste buds at the back of the tongue has its cell body in the glossopharyngeal ganglion, and projects to the nucleus of the solitary tract (the visceral afferent nucleus). An interneuron here projects to a preganglionic neuron in the parasympathetic nucleus of the glossopharyngeal nerve. The axon of this preganglionic neuron passes to the otic ganglion where it synapses with a postganglionic neuron. The postganglionic neuron projects to the parotid salivary gland.

At the bottom of Fig. 6.4 the neuronal connections of the **gag reflex** are shown. In this reflex, stimulation of the pharyngeal wall elicits retching movements of the pharyngeal muscles. The primary afferent neuron travels in the glossopharyngeal nerve, the cell body being in the glossopharyngeal ganglion; the central projection is to an interneuron in the nucleus of the solitary tract. The interneuron projects to a special visceral efferent neuron in the nucleus ambiguus. The axon of this neuron leaves the brainstem in the cranial component of the spinal accessory (XIth) nerve, transfers to the vagus (Xth nerve), and passes through the pharyngeal rami of the vagus to innervate muscles which constrict the pharynx (e.g. the cricopharyngeus muscle).

Significance of the nuclei of the cranial nerves in clinical neurology

Any of the nuclei of the cranial nerves may be damaged by lesions of the brainstem. The signs are predictable from a knowledge of the functional responsibilities of these cranial nerves. For example, bilateral destruction of the Vth motor nucleus of the trigeminal nerve will cause the lower jaw to droop and prevent chewing. The reflexes which can be tested to examine the cranial nerves clinically are outlined in Chapter 19. Plainly the ability to recognize an isolated central lesion of a cranial nerve is clinically useful in itself. The main value, however, of being able to identify a lesion of a cranial nerve nucleus is that this reveals the **level** of the lesion in the brainstem. Most lesions of the brainstem interrupt some or other of the general motor or sensory pathways, and may thus cause more or less widespread disturbances in the head, trunk, and limbs. Signs of a lesion in a cranial nerve nucleus may enable the level of such a lesion in the brainstem to be more or less precisely identified.

7. Medial lemniscal system

A very wide range of afferent information is collected by a vast array of receptors in the periphery of the body and transmitted to the cerebral cortex where it is consciously perceived. It is projected by two main systems of spinal tracts, namely the gracile–cuneate and spinothalamic systems; these eventually converge into a single great pathway, the medial lemniscal system. There is also a third major spinal pathway for afferent information which is to some extent consciously perceived, i.e. the ascending reticular formation, but this remains separate from the medial lemniscal system and is considered in Chapter 9.

7.1 Conscious sensory modalities

At least 12 conscious modalities (sensations) can be recognized. They can be divided into four groups according to their mode of central transmission. (i) The first group comprises the modalities of **touch**, **pressure**, and **joint proprioception**. These three modalities are transmitted in the cuneate and gracile fascicles. The external energy that gives rise to these modalities is always purely mechanical, the receptor endings being low-threshold mechanoreceptors. (ii) The modalities of **pinprick pain**, **heat**, and **cold** form the second group. In man these modalities are transmitted in the spino-thalamic tract. In the domestic mammals their pathways are uncertain. (iii) The third group contains the modality of **true pain** only. In man, this modality is transmitted through the ascending reticular formation. In the domestic species, the precise pathway of this modality is unclear, but is probably again the ascending reticular formation. (iv) The final group consists of the modalities of the special senses, namely **vision, hearing, balance, taste**, and **olfaction**. The central pathways of the special senses are considered in Chapter 8.

The sense of **kinaesthesia** is the awareness of the precise position and movements of the parts of the body and especially the limbs. This sense obtrudes scarcely at all on an individual's consciousness, but it enables the limbs to be moved over uneven terrain without stumbling, or to carry out such useful functions as eating chocolates in total darkness which requires that the hand must 'know' the position of both the box and the mouth. Other names for this modality are 'proprioception' and 'joint sensibility'. It has been widely believed that the modality of kinaesthesia is mediated mainly or solely by receptors associated with joints, i.e. joint proprioceptors, but it is now known that receptors associated with muscles, i.e. muscle proprioceptors, also contribute and are probably at least as important as joint proprioceptors. This chapter considers joint proprioceptors and their pathways. The contribution of muscle proprioceptors to kinaesthesia is mentioned briefly, but the pathways of muscle proprioceptors are outlined in Chapter 9.

7.2 Peripheral receptors of touch, pressure, and joint proprioception

Anatomists in the nineteenth century believed that each sensory modality must be related to one particular structural type of receptor. Later on this concept of receptor specificity was strongly challenged, but the majority of receptors are once again believed to be so selective that under natural conditions they transmit information of one type only. The receptors of touch, pressure, and joint proprioception are mechanoreceptors transmitting general somatic afferent impulses from the skin and joints. Structurally they are of three types, i.e. free nerve endings, neurite–receptor cell complexes, and complex laminated receptors.

Free nerve endings

Myelinated, and more often unmyelinated, nerve fibres form abundant aborizations in the dermis and the stratum germinativum of the epidermis. These terminate in bare receptor endings (Fig. 5.12(b)), having lost their Schwann cell covering. Presumably these are slowly adapting receptors (see Section 5.19). Free nerve endings belonging mainly to unmyelinated fibres also occur in joint capsules and ligaments where, like all joint receptors, they are evidently tension receptors.

Neurite-receptor cell complexes

These consist of **granular cells (Merkel cells)** in which each cell is intimately associated with an axonal ending from a myelinated fibre. The cell is said to be 'granular' because it contains numerous dense-cored vesicles. The cell and its axonal ending together constitute a 'neurite-receptor cell complex'. Such complexes occur immediately beneath the epidermis, forming very slight but perceptible swellings on the surface of the skin (in the cat). The Merkel cell may be the primary receptor, or it may simply modulate the receptor characteristics of its associated axonal ending. Physiologically, however, the axonal ending behaves as a slowly adapting mechanoreceptor, discharging continuously when the skin is distorted and falling silent when the skin is not deformed.

In the Nomina Histologica the Merkel cell is now known as epithelioidocytus tactus.

Complex laminated receptors (encapsulated receptors)

In principle these have the laminated structure shown in Fig. 5.12(c), (d) and occur in a continuum of sizes and degrees of complexity. The largest and most complex, the **Pacinian corpuscle**, is visible to the naked eye (about 2 mm long). This type is fast adapting (see Section 5.19), and fires on the 'on' and the 'off'. It is capable of following fast vibrations. Smaller varieties (e.g. so-called **Ruffini endings**) are slowly adapting despite their laminated structure, the laminae evidently being unable to filter out the effect of sustained distortion. The largest types occur in subcutaneous tissues and many other tissues. They also occur in joint capsules and ligaments, as do many of the smaller varieties; these are all tension receptors transmitting information about the tension in the capsule and ligaments.

7.3 Pathways of touch, pressure, and joint proprioception

Neuron 1

The cell location of the primary afferent neuron (Fig. 7.1) is in a dorsal root ganglion (or, in the head, an equivalent ganglion of a cranial nerve). The axon enters the dorsal horn. Here it divides, forming short branches and a long collateral.

The **short branches** make synapses with (i) interneurons, some forming part of reflex arcs; and with (ii) the ascending reticular formation, except for the primary afferents of joint proprioception, which are atypical in **not** projecting into the ascending reticular formation (see Section 9.2). These short branches are phylogenetically primitive.

A single **long collateral** ascends in the dorsal funiculus, in the **gracile** or **cuneate fascicle**. This ends by making synapses in the gracile nucleus or (medial) cuneate nucleus. Phylogenetically, these long collaterals are an advanced refinement occurring only in the highest vertebrates (reptiles, birds, and mammals).

In the gracile and cuneate fascicles the axons are arranged in an orderly fashion (see the inset in Fig. 7.1). The axons relating to the hindlimb pass as near to the midline as possible, before turning cranially to ascend within the gracile fascicle. The axons coming in subsequently from the trunk and forelimb also pass as far towards the midline as they can and then ascend, those from the forelimb and neck travelling in the cuneate fascicle. Thus the gracile and cuneate fascicles represent a map of the body, the hindlimbs being near the midline and the neck being placed laterally. This is known as **somatotopic localization** (see Section 7.10). The gracile and cuneate fascicles are exceptional in that the hindlimbs are represented **medially**: in the many other examples of somatotopic localization in the spinal cord and brainstem which will be mentioned later, the hindlimbs are placed **laterally**.

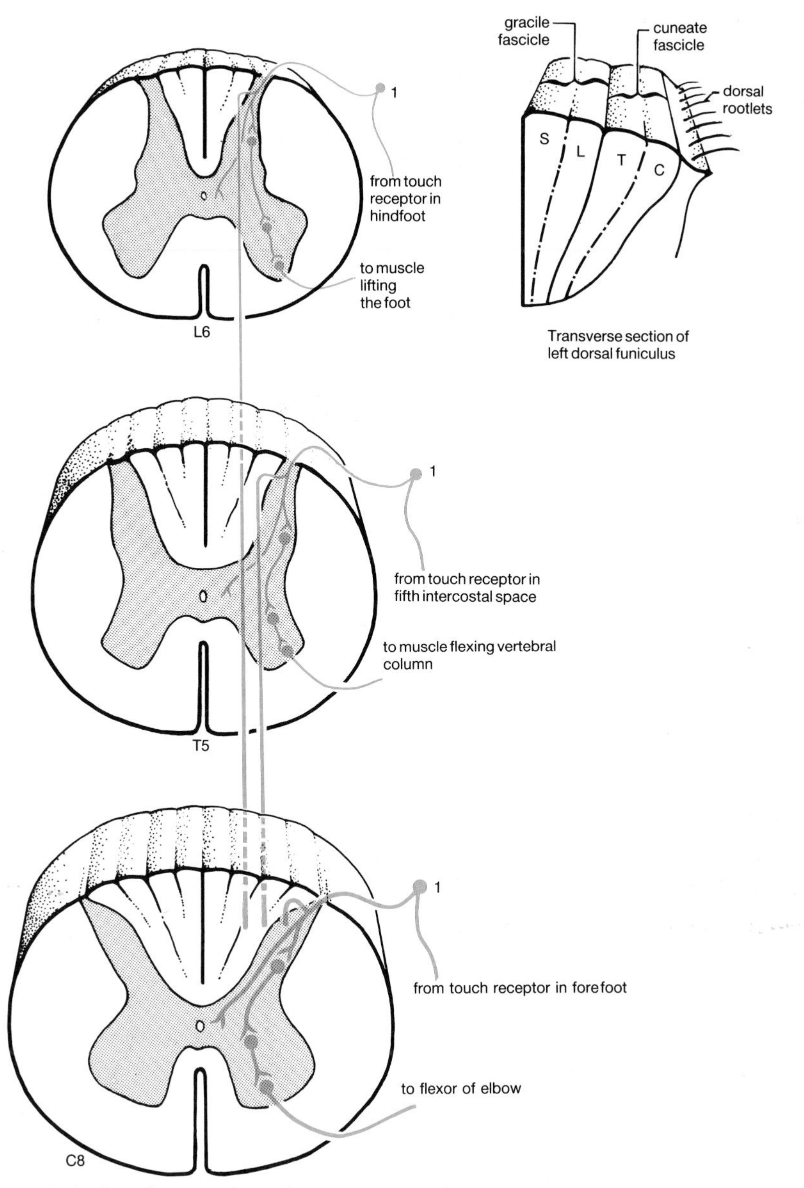

FIG. 7.1. Spinal pathways of touch, pressure, and joint proprioception. These pathways ascend in the gracile and cuneate fascicles of the dorsal funiculus. Each fibre goes across the funiculus as far as possible towards the mid-line; this causes the fibres to be arranged somatopically, the caudal segments of the body being medial in the funiculus, and the cranial segments lateral. This is shown in the inset diagram, where S indicates sacral segments, L lumbar, T thoracic, and C cervical. The first neuron (1) in this pathway enters the dorsal horn and forms three varieties of collateral: (a) a long collateral forming the gracile or cuneate fascicle; (b) a short collateral shown contributing to a reflex arc; and (c) another short collateral projecting into the reticular formation at the centre of the grey matter. See Fig. 7.3 for the continuation of these pathways to the cerebral cortex.

This picture is complicated by the probability that numbers of joint proprioceptive pathways from the hindlimb transfer from the gracile fascicle to the dorsal spinocerebellar tract, in the first lumbar segments of the spinal cord.

For details of this transfer see Section 7.4, **Effects of lesions in the dorsal funiculus.**

Neuron 2

The cell location is in the **gracile nucleus** or (medial) **cuneate nucleus** in the medulla oblongata (Fig. 7.3). The axon **crosses over** to the opposite side of the neuraxis (decussates) in the deep arcuate fibres, and continues rostrally in the **medial lemniscus**. Strictly, the cell location of the second neuron of the forelimb pathway is in the **medial** cuneate nucleus; the **lateral** cuneate nucleus belongs to the spinocerebellar pathway (see Section 9.4).

In the pathways of touch, pressure, and joint proprioception arising from the **head**, the synapse between neuron 1 and neuron 2 occurs in the various parts of the sensory nucleus of the trigeminal nerve: the axons of neuron 2 cross over and join the medial lemniscus, which now has four components namely head, forelimb, trunk, and hindlimb (Fig. 7.3).

The second neuron ends by making a synapse in the thalamic nuclei.

Neuron 3

The cell location (Fig. 7.3) is in the **ventral group of thalamic nuclei**. The axon projects to the **primary somatic sensory area** of the cerebral cortex. In the human cerebral cortex the body is here represented upside-down; the head is ventral to the body, but the right way up. The area of cerebral cortex allocated to each part of the body corresponds with the richness of the sensory nerve supply of the peripheral structures concerned, and not with the actual size of those structures. These topographical characteristics can be represented by a **sensory 'homunculus'**, which closely resembles the motor homunculus in Fig. 11.3. Thus in man the hand has a very much bigger cortical representation than the whole of the trunk put together. The positions of the primary somatic sensory area in the brains of cat and man are indicated in Figs. 8.2 and 8.3. The **sensory 'felunculus'** in Fig. 8.2(b) shows that in the primary somatic sensory area of the cat the body and also the head are upside down, and that the head and forelimb have much the greatest area of representation.

The cell location of neuron 3 is in the ventrocaudal nucleus of the ventral group of thalamic nuclei (see Section 17.17).

7.4 Effects of lesions in the dorsal funiculus

A unilateral lesion confined to the dorsal funiculus gives ipsilateral deficits of cutaneous sensation and joint proprioception. The tactile placing response and proprioceptive positioning tests are inadequately performed or fail, and the posture and gait may be unsteady, the limbs being irregularly abducted or even adducted during locomotion.

It was previously believed that the dorsal funiculi are mainly or even solely responsible for transmitting the modality of joint proprioception. This view seems to have been partly based on proprioceptive deficits in cases of **tabes dorsalis** in man, in which degeneration of the dorsal funiculus is a common feature (though the dorsal roots are also involved). However, it is now clear that the pathways of joint proprioception must be more complex than hitherto supposed. For instance, experimental destruction of the dorsal funiculi in the cervical region in monkeys gives severe ataxia of the forelimbs but less pronounced ataxia of the hindlimbs. In the dog, experimental bilateral section of the dorsal funiculi at C4 results in a high stepping gait of the forelimbs but not of the hindlimbs. Similar results have been obtained in cats and monkeys.

The explanation of these different joint proprioceptive deficits in the forelimbs, as compared to the hindlimbs, following bilateral cervical lesions of the dorsal funiculi appears to be that the great majority of the proprioceptive fibres from the joints of the hindlimb (90 per cent from the stifle joint of the cat) leave the gracile fascicle in the first lumbar segments. They then synapse in the dorsal horn (in the **thoracic nucleus**) with neurons, the fibres of which ascend in the dorsal spinocerebellar tract. These fibres end ipsilaterally near the gracile nucleus in a special nucleus known as **nucleus Z**, which may perhaps be regarded as a displaced fragment of the gracile nucleus; the neurons in nucleus Z form fibres which travel in the deep arcuate fibres to join the medial lemniscus, and thus the signals in this pathway reach the cerebral cortex. In contrast, the joint proprioceptive fibres from the forelimb do remain in the cuneate fascicle. These observations apply at least to the cat and monkey. (For further details of nucleus Z see Section 9.5.)

If, as seems likely, a similar transfer of joint proprioceptive pathways from the gracile fascicle to the dorsal spinocerebellar tract also occurs in the dog and horse, it may clarify the joint proprioceptive deficits observed in **wobbler dogs** and **horses** (see Section 4.11, 4.12). These deficits severely affect the hindlimbs, which tend to be erratically abducted (or in the dog, crossed over) during walking, with marked swaying of the hindquarters; joint proprioceptive positioning tests on the hindlimbs show obvious abnormalities. In contrast, the actions of the forelimbs during walking may seem normal, and a joint proprioceptive deficit in these limbs may only become apparent during careful proprioceptive testing. The main lesion lies typically in segments C6 or C7 of the spinal cord in the dog and C3 to C5 in the horse, and affects both the grey and the white matter. In wobbler dogs cranial to C6 and C7 there is some degeneration of fibres in the dorsal funiculus, but there is also degeneration in the dorsolateral region of the lateral funiculus. In wobbler horses, cranial to the main lesion degenerate fibres are again found laterally and dorsolaterally in the lateral funiculus, but degeneration in the dorsal funiculus is mild or in some cases absent. In both species, the degenerate fibres in the dorsolateral region of the lateral funiculus lie in the region of the dorsal spinocerebellar tract, where most of the proprioceptive pathways from the joints of the hindlimb might be located. Thus it is conceivable that joint proprioceptive pathways from the hindlimbs may be more extensively disrupted than those from the forelimbs, especially in the horse where degeneration in the dorsal funiculus is relatively mild or even absent.

A further complication in the interpretation of clinical deficits of joint sensibility (kinaesthesia) is that **muscle proprioceptors** (i.e. muscle spindles and Golgi tendon organs) are now known to contribute to the perception of limb position. From the turn of the centry to the 1960s abundant experimental researches had seemed to indicate that these muscle receptors were unable to record the absolute position of joints. More recently, however, many other experiments have clearly demonstrated that muscle proprioceptors do indeed make a major contribution to the awareness of the position and movement of joints, along with classical joint proprioceptors. If this awareness were provided by joint proprioceptors alone, the surgical replacement of joints in man would be followed by a loss of such awareness, but in fact these operations produce no serious kinaesthetic deficits. The pathways of muscle proprioception from the hindlimb and forelimb are summarized in Section 9.5

Since afferent fibres from the muscle spindles of the hindlimb ascend in the dorsolateral region of the lateral funiculus (in the dorsal spinocerebellar tract) they are likely to be involved in the kinaesthetic deficits associated with the lesions in wobbler dogs and horses. In other words, the proprioceptive deficits of the hindlimb that are so characteristic of wobblers of both species may be partly due to the interruption of the hindlimb muscle proprioceptor fibres in the dorsal spinocerebellar tracts, as well as to the interruption of hindlimb joint proprioceptive pathways which have transferred from the gracile fascicle to the region of the dorsal spinocerebellar tract of the lateral funiculus.

In summary, two broad functional anatomical generalizations can now be stated about kinaesthesia and its pathways: (i) all afferent impulses from joints are not projected exclusively through the dorsal funiculi, since some or many of those from the joints of the hindlimb are projected through the lateral funiculi; (ii) the sense of kinaesthesia is derived both from joint proprioceptors and from muscle proprioceptors.

7.5 Peripheral receptors of pain

Pain receptors in somatic structures probably consist of fibres with free branching endings. In man they give rise to two main types of pathway, those of pinprick pain and those of true pain.

Pinprick pain

The sensation of pinprick pain such as that caused by a superficial penetration with a fine sharp needle, is transmitted from the periphery by thin (2–6 μm diameter) **myelinated** fibres (which are classified as A-III (or A-delta) fibres, see Section 10.1). These sensations are conducted relatively fast, are accurately localized, and do not outlast the provoking stimulus; not everyone calls this pain. Within the human spinal cord such sensations are carried to the brain by the specific pathway of the **spinothalamic tract**. (For the definition of specific and non-specific pathways see Section 5.5.)

True pain

The perception of true pain, such as that induced by pushing a large blunt needle deep into the skin, consists of much more severe sensations ranging from itch to downright agony. It is carried from the periphery by much thinner (about 0.3–1.3 μm) **unmyelinated** (C) fibres. These sensations appear to be carried to the cerebral cortex by the non-specific pathway of the **ascending reticular formation** (see section 9.9).

7.6 Spinothalamic tract of man

This is the specific pathway for pinprick pain and also for temperature sensation. Some touch fibres also travel in the ventral part of the spino-thalamic tract.

Neuron 1

The cell location (Fig. 7.2) is in a dorsal root ganglion (or, in the head, in an equivalent ganglion of a cranial nerve, Fig. 7.3). The axon enters the dorsal horn (or the equivalent cranial nerve nucleus in the head). Here it gives off: (i) many short branches, some making synapses with interconnecting neurons (often forming parts of reflex arcs), and others synapsing with the reticular formation (see Section 9.8); (ii) a short branch making a synapse in the dorsal horn (or equivalent cranial nerve nucleus, in the head) with neuron 2.

Neuron 2

In the spinal cord, the cell location of **neuron 2** is in the dorsal horn (Fig. 7.2). In **man**, the axon decussates and ascends in the spinothalamic tract on the opposite side of the spinal cord. At the level of the pons, this tract blends with the **medial lemniscus**. Here the medial lemniscus, which in the medulla oblongata is vertical in position like a book-end, has fallen on its face (Fig. 7.3, middle diagram). Rostral to this level the pathways of pinprick pain and temperature share the medial lemniscus with the pathways of touch, pressure, and kinaesthesia, forming the **medial lemniscal system**.

In similar pathways of temperature and pinprick pain arising from the tissues of the **head** in man, the cell location of neuron 2 lies in the **sensory nucleus of the Vth nerve** (Fig. 7.3); the axon of neuron 2 crosses over and enters the medial lemniscus, which is lying on its face (Fig. 7.3).

Strictly, the cell location of neuron 2 in somatic pain pathways from the head lies in the **nucleus of the spinal tract of the trigeminal nerve** (see Section 6.6). For visceral pain afferents of the head (e.g, pain afferents from the larynx travelling in the vagus, or pain afferents from the pharynx travelling in the IXth nerve), the cell station will be either the **nucleus of the solitary tract**, or more probably the **nucleus of the spinal tract of the trigeminal nerve**.

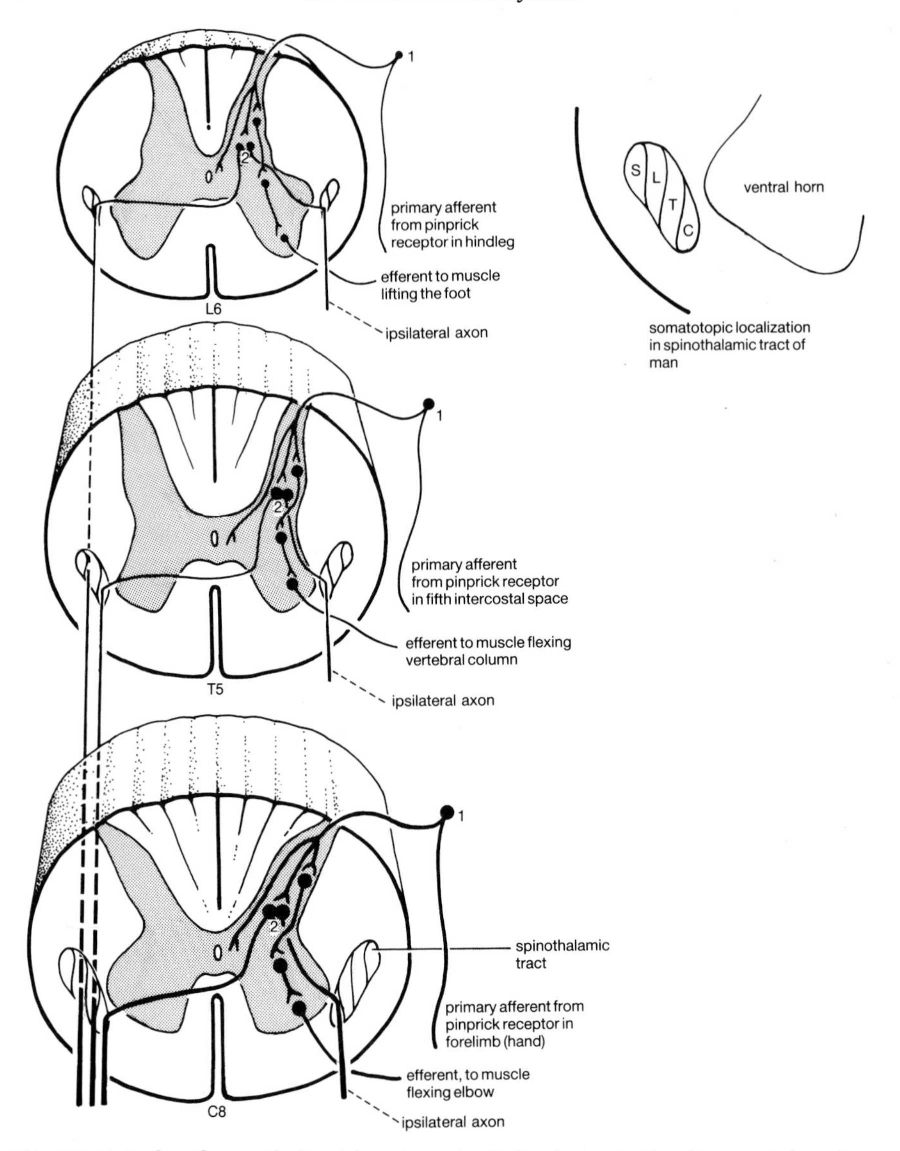

FIG. 7.2. Spinal pathway of pinprick pain: spinothalamic tract. The diagram is based on the well-established spinothalamic tract of man. 1 and 2 represent the first two neurons in this pathway. It is a characteristic of the human spinothalamic tract that the second neuron decussates and ascends on the contralateral side; the axons go across as far as they can and assemble themselves somatotopically, with the sacral fibres (S) in the most lateral position and the cervical fibres (C) most medial (see inset, in which L is lumbar and T thoracic). The arrangement of these spinothalamic pathways in the domestic mammals is not clear. There is evidence suggesting that in these species: (a) the axons of the second neuron, 2, enter the lateral column on **both** sides of the spinal cord, and to illustrate this an additional ipsilateral axon is shown on the right side of the three transverse sections, passing from L6 towards T5 and so on; and (b) these ascending axons often re-enter the spinal grey matter, and synapse with additional neurons whose axons then rejoin the spinothalamic tract of the same or opposite side. Consequently, the spinothalamic tract of the domestic mammals may be more bilateral, diffuse, and multisynaptic than that of man. See Fig. 7.3 for the continuation of these pathways to the cerebral cortex.

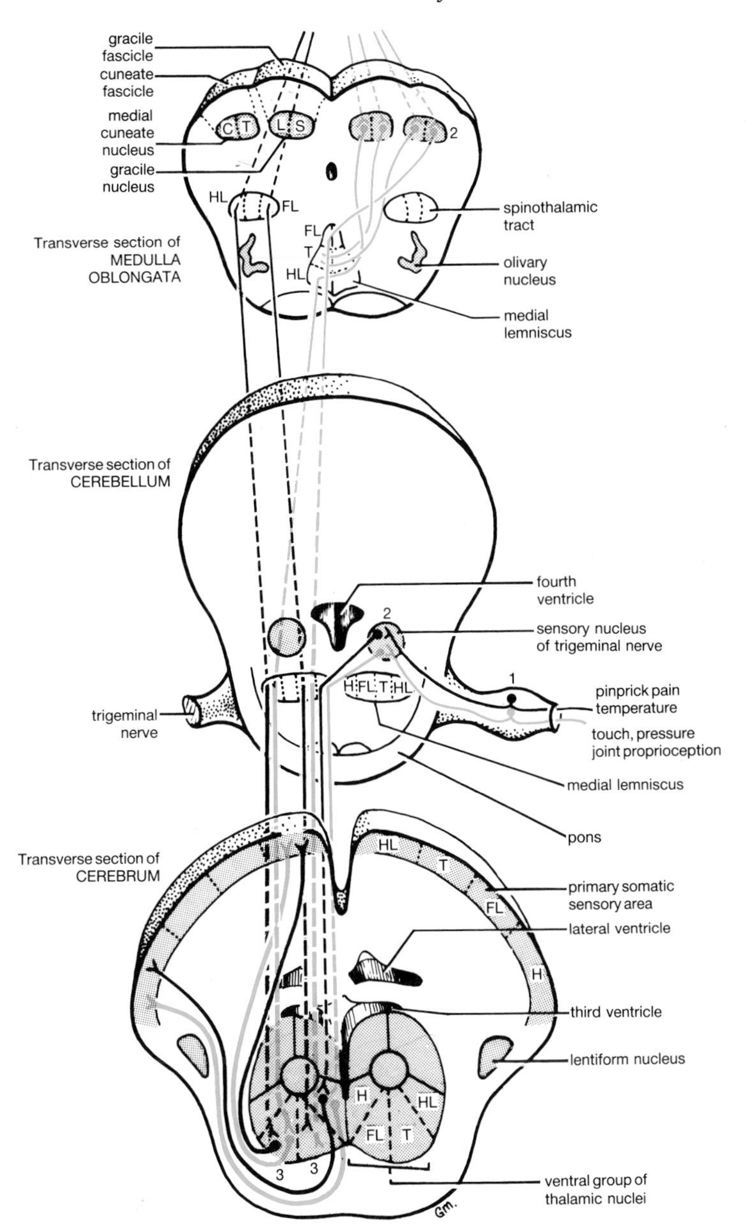

FIG. 7.3. Pathways of the medial lemniscal system within the brain. The spinal pathways of touch, pressure, and joint proprioception are shown in green. The second neuron (2) lies in the cuneate or gracile nucleus, and projects through the medial lemniscus to the third and final neuron (3) in the thalamus. The spinothalamic pathways (black) join the medial lemniscus and project to neuron 3 in the thalamus. The third neuron projects to the primary somatic sensory area of the neocortex. The input from the head is shown entering through the trigeminal nerve, and then joining the medial lemniscus. In the medulla, the medial lemnisci are vertical, each facing its partner like two book-ends pressed face to face. In the pons the two book-ends have fallen on their faces; this makes the hindlimb pathways lateral and forelimbs medial. The medial lemniscus is now positioned to absorb the spinothalamic tract. Point-to-point localization occurs throughout. H, head; FL, forelimb; T, trunk; HL, hindlimb.

Neuron 3

The cell location is in the **ventral group of thalamic nuclei** (Fig. 7.3). The axon projects to the primary somatic sensory area of the cerebral cortex.

Lateral and ventral spinothalamic tracts are sometimes mentioned in man, but there appears to be no anatomical justification for these subdivisions.

7.7 Spinothalamic pathways in domestic mammals

The axons of neuron 2 seem to be distributed to the lateral funiculus on **both** sides of the spinal cord (Fig. 7.2), thus forming a **bilateral** pathway. The ascending axons may re-enter the grey matter and make synapses with new neurons; the axons of these latter neurons return to the spinothalamic tract on the same or the other side of the spinal cord. The pathway in the domestic mammals is therefore both **bilateral** and **multisynaptic**. Some of its axons may mingle with those of the adjacent **propriospinal system** (Fig. 12.3), and therefore it is difficult or impossible to say whether the pathway of pinprick pain is truly 'spinothalamic' or partly propriospinal and partly spinothalamic.

7.8 Spinocervical tract (spinocervicothalamic tract)

Occurring in carnivores among the domestic species, this tract arises mainly from cutaneous receptors of touch and pressure, and some cutaneous receptors conveying pinprick pain. It eventually joins the medial lemniscus, thus forming part of the medial lemniscal system. In the spinal cord it lies in the dorsolateral part of the lateral funiculus (Fig. 12.3).

The cell location of neuron 1 is in a dorsal root ganglion. Neuron 2 has its cell location in the dorsal horn, and its axon ascends on the same side in the spinocervical tract lying in the lateral funiculus (Fig. 12.3), finally making synapses ipsilaterally with neuron 3 in the **lateral cervical nucleus** at the level of C1 and C2 (Fig. 21.1(a)). The axon of neuron 3 decussates and joins the medial lemniscus, forming synapses with neuron 4 in the ventral group of thalamic nuclei. Thus this pathway has a minimum of *four* neurons of which the *third* decussates. Evidently the modalities transmitted by this tract in the cat quite closely resemble those transmitted by the human spinothalamic tract. In man the spinocervical tract is small or even absent. There may be an inverse relationship between the degree of development of the spinocervical and spinothalamic tracts in different species.

7.9 Species variations in the medial lemniscal system

Cuneate and gracile fascicles

These appear to be well developed in the domestic mammals generally, possibly because of their role in kinaesthesia and the importance of this in locomotion.

Spinothalamic pathway

In man the spinothalamic tract is a specific pathway transmitting the modalities of pinprick pain, and heat and cold. A unilateral spinothalamic tract (comprising a direct chain of three neurons as in man) appears not to exist as a specific pinprick pain pathway in the domestic mammals. There is evidence that in these species the spinothalamic pathway may be bilateral and multisynaptic.

Because of the uncertainty about its anatomy and the modalities it transmits, it is difficult (if not impossible) to exploit the spinothalamic tract for diagnostic purposes in the domestic mammals.

7.10 Somatotopic localization

The gracile and cuneate fascicles (and the spinothalamic tract of man) show somatotopic localization throughout; so also do the medial lemniscus, the thalamic nuclei, and the primary sensory area of the cerebral cortex itself. Somatotopic localization means that the axons and nerve cells in a pathway are arranged in constant positions in relation to each other, these positions coresponding to those parts of the limbs and trunk which the axons represent. Strictly the term 'somatopic' excludes the head, since 'soma' means body (but it can be used to cover both the head and body). An alternative term, 'point-to-point localization', does include the head. Somatotopic or point-to-point localization also occurs in the somatic motor systems, e.g. in the lateral corticospinal and rubrospinal tracts (Fig. 12.3).

7.11 Blending of tracts in the spinal cord

Almost all the ascending and descending tracts undergo some mixing in the spinal cord; they are not separated into well-defined bundles. The only general exceptions are the gracile and cuneate fascicles, but in several of the domestic species, e.g. the sheep, even the gracile and cuneate fascicles are blended with other tracts, since a part of the corticospinal pathway travels in the dorsal funiculus (see Section 11.4). For a simplified summary of the positions of the tracts see Fig. 12.3.

7.12 Brainstem

The brainstem is the mid-line core of the cranial neuraxis, continuing the spinal cord into the cranial cavity. It consists of all of the brain except the cerebral hemispheres and cerebellum (Fig. 7.4). It therefore comprises the medulla oblongata, pons, and cerebellar peduncles of the **hindbrain**; the whole of the **midbrain** (mesencephalon); and the thalamus, geniculate bodies, hypothalamus, and pineal body, together with all other parts of the **diencephalon**.

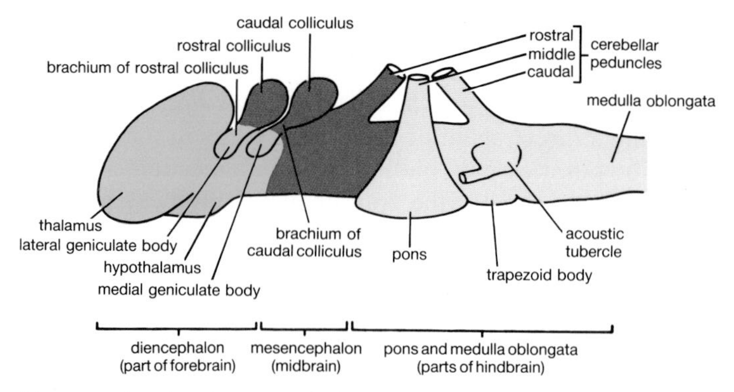

FIG. 7.4. **Diagram showing the principal components of the brainstem.** The forebrain, midbrain, and hindbrain components are shown in three different colours.

The diencephalon is one of the two great subdivisions of the **forebrain** or **prosencephalon**, the other subdivision being the cerebral hemispheres or **telencephalon**. The **hindbrain** is the **rhombencephalon**; its two components are the **myelencephalon** (medulla oblongata) and **metencephalon** (pons and cerebellum). In the early embryo there are three primary brain vesicles, the prosencephalon, mesencephalon, and rhombencephalon. Soon the prosencephalon forms two enlargements, i.e. the telencephalon and diencephalon; at the same time the rhombencephalon also develops two swellings, i.e. the metencephalon and the myelencephalon.

7.13 Summary of the medial lemniscus system

(i) The conscious modalities (apart from the special senses) are: touch, pressure, joint proprioception, pinprick pain, heat, cold, and true pain.

(ii) The impulses of touch, pressure, and joint proprioception are transmitted through the cuneate and gracile fascicles. The impulses of pinprick pain, and heat and cold, pass through the spinothalamic tract in man. The impulses of true pain are conducted by the ascending reticular formation of man, and probably of the domestic mammals also, but this is **not** part of the medial lemniscal system.

(iii) Carnivores also have a well-developed spinocervical tract, consisting of touch and pressure fibres which eventually join the medial lemniscal system.

The spinocervical tract has at least four neurons of which the third decussates, thus differing from the rest of the medial lemniscal system. In addition to touch and pressure, it may carry other modalities which in man are transmitted by the spinothalamic tract, notably pinprick pain.

(iv) In the medial lemniscal system three neurons transmit the conscious modalities (those listed in item 1 above, except true pain) from the periphery of the body to the cerebral cortex.

(v) The first neuron is always in a spinal dorsal root ganglion, or in the equivalent 'dorsal root' ganglion of a cranial nerve.

(vi) The second neuron typically decussates.

(vii) All paths converge into the medial lemniscus eventually, thus forming the medial lemniscal system.

(viii) All paths finally relay through the thalamus, which contains the cell station of neuron 3. This neuron is in the ventral group of thalamic nuclei.

(ix) The end of the medial lemniscal system is the primary somatic sensory area of the cerebral cortex.

(x) The medial lemniscal pathways show point-to-point localization including the cortex itself.

(xi) In domestic mammals the cuneate and gracile fascicles are well developed. There appears to be an ill-defined spinothalamic tract which is multisynaptic, and tends to be bilateral rather than unilateral; the modalities which it transmits are uncertain.

8. Special senses

8.1 Vision

Neuron 1

The first neuron in the pathway of vision is the bipolar cell of the retina (Fig. 8.1). This receives impulses from the neuroepithelial cells of the retina, i.e. the rods and cones.

Neuron 2

The second neuron is the ganglion cell of the retina (Fig. 8.1). Its axon lies in the **optic nerve**, and continues through the optic chiasma and **optic tract**. Its terminal filaments make synapses with neuron 3 in the **lateral geniculate nucleus** (within the lateral geniculate body).

Other terminal filaments of neuron 2 do not reach conscious levels of perception but form the afferent components of visual reflexes. They enter the left (or right) **rostral colliculus** (part of the **tectum** of the midbrain) (Fig. 8.4) making synapses there with neurons of four possible pathways:

Pathways co-ordinating eyeball movements

In each rostral colliculus there are neurons which project to the motor nuclei of the IIIrd, IVth, and VIth cranial nerves of both the left **and** right side (i.e. the left rostral colliculus acts not only on the motor nuclei of the left side but also on those of the right side). These pathways co-ordinate the movements of the eyeball in response to visual stimuli, such as reflexly turning the eyes towards a flash of light or a moving object. This co-ordination ensures that the object observed is kept focused on the area centralis (macula lutea or yellow spot) of the retina. (For **voluntary** eye movements see Section 11.1.)

Pathways constricting the pupil

There are also neurons in each rostral colliculus which project to both the left and the right **parasympathetic nuclei of the oculomotor nerve** (Edinger–Westphal nuclei); these enable **both** pupils to constrict in response to a light stimulus in **either** the left **or** right eye alone. Thus the left rostral colliculus controls both the left and the right parasympathetic nuclei of the oculomotor nerve. The constriction which occurs in the eye receiving the light stimulus is known as the **direct response**: the constriction in the other eye is called the **consensual response**.

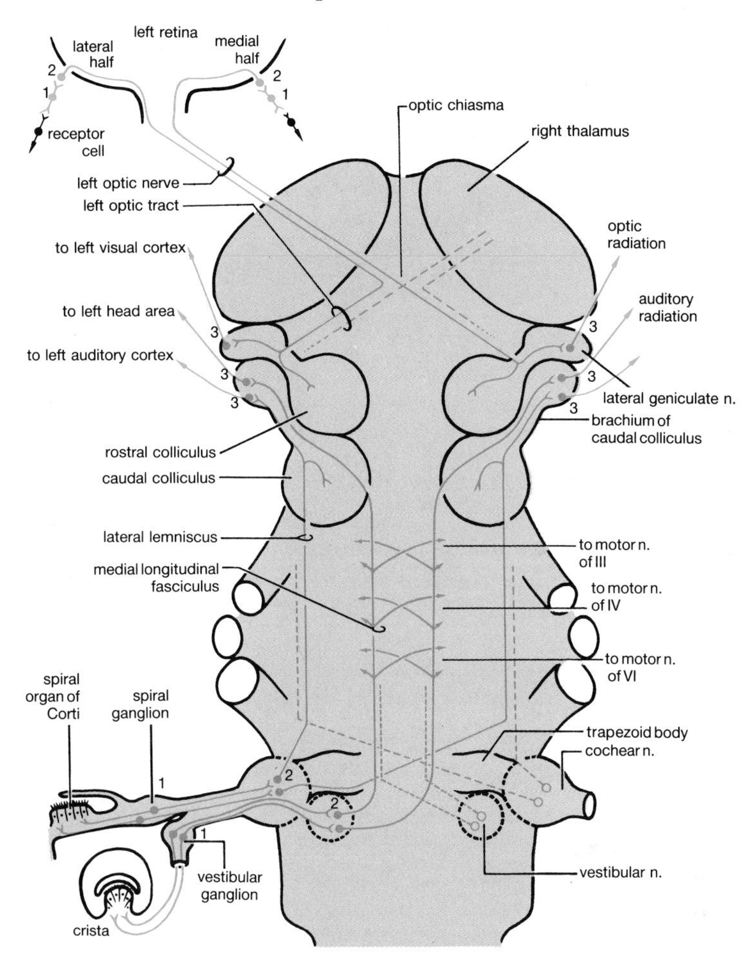

FIG. 8.1. **Dorsal view diagram showing visual, auditory, and vestibular projections to the cerebral cortex.** The visual pathways apply to an animal with overlapping visual fields, such as a cat. 1, 2, and 3 are the first, second, and third neurons in the visual, auditory, and vestibular pathways. The projections to the colliculi are shown in simplified form, the visual projections to the rostral colliculi being **separate** retinal axons, not collaterals as shown. The auditory simplifications are noted in Section 8.2. n., nucleus. The cranial nerves are indicated by roman numerals.

Pathways controlling turning of head and neck

Third, there are neurons in each rostral colliculus the axons of which decussate and descend through the brain and into the cervical spinal cord, forming the **tectospinal tract** (see Section 12.13). These give reflex turning of the head and neck towards a sudden source of light or a movement. In this instance each rostral colliculus controls only the muscles on the contralateral side of the neck.

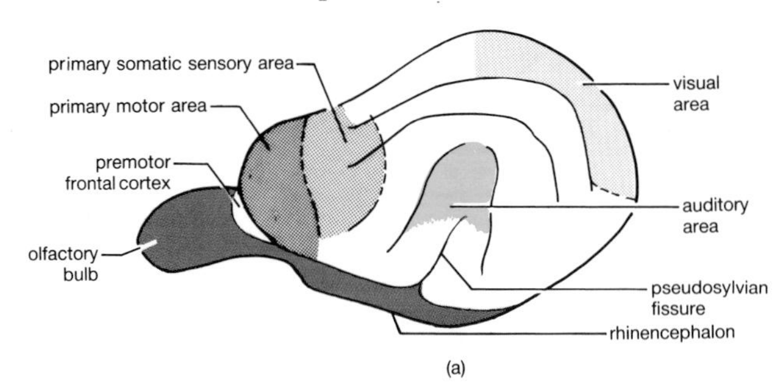

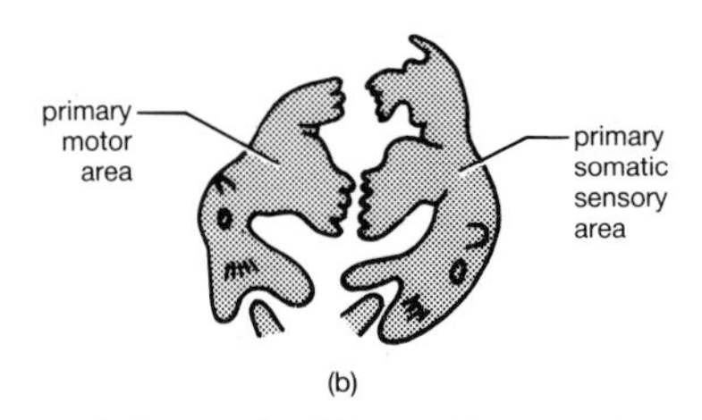

FIG. 8.2. **(a) Left lateral view of the cerebral hemisphere of the cat.** The diagram shows the rhinencephalon and the four projection areas, i.e. the primary motor area, primary somatic sensory area, visual area, and auditory area. The primary motor and primary somatic sensory areas spread on to the medial surface of the hemisphere. The projection areas of the cat are not related to particular fissures on the cortex. The visual area lies on the occipital lobe and extends on to its medial aspect. The auditory area is situated on the temporal lobe. For the lobes of the cerebral cortex in the carnivore see Fig. 17.2(b). (Based on Campbell (1905) *Histological studies on the location of cerebral function,* Cambridge University Press.)

(b) Lateral view diagram of the left primary motor and primary somatic sensory areas of the cat. The two areas are drawn to the same scale, and in the same relative position as in (a); they spread on to the medial surface of the hemisphere. They can be expressed as a motor and a sensory 'felunculus', in which the areas of cortex allocated to the parts of the head and body are indicated by the sizes of the components of the 'felunculi'. (Based on Woolsey (1958), in *Biological and biochemical bases of behaviour* (ed. H. F. Marlow and C. N. Woolsey), by courtesy of the University of Wisconsin Press.)

Ascending reticular formation

Fourth, some visual neurons in the rostral colliculi project into the ascending reticular formation (see Section 9.8, and Fig. 8.4).

Strictly, the terminal filaments of neuron 2 which enter the rostral colliculi belong to **separate** retinal axons and are not collaterals as shown in Figs. 8.1 and 8.4. About 80 per cent of the axons in the optic tract end in the lateral geniculate nucleus; the other 20 per cent end in the rostral colliculus.

In the **optic chiasma** (Fig. 8.1) there is partial decussation of the axons of neuron 2. Decussation is **almost complete** in mammals with eyes which are directed laterally, e.g. rodents. In the horse, in which the visual fields of the

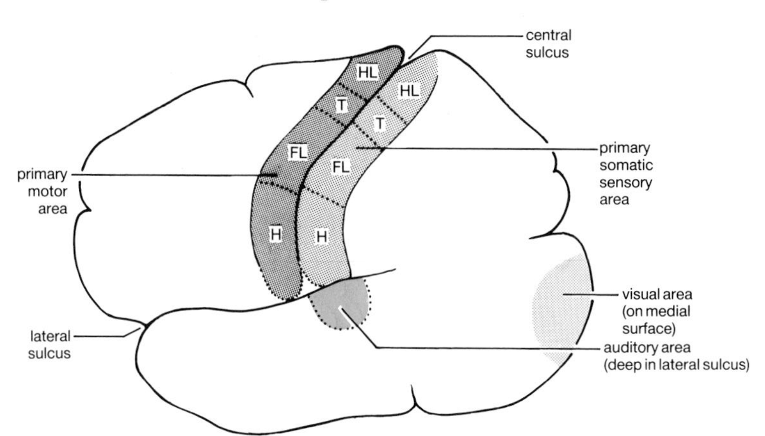

FIG. 8.3. Left lateral view of the cerebral hemisphere of man. The diagram shows the four projection areas, i.e. the primary motor area, the primary somatic sensory area, the visual area, and the auditory area, which are related to sulci as shown. The approximate sizes of the motor and somatic sensory areas are indicated as: H, head; FL, forelimb; T, trunk; HL, hindlimb. The details of these areas resemble Penfield's motor homunculus (Fig. 11.3).

two eyes overlap only slightly, about 80 to 90 per cent of the fibres decussate. In the dog and cat the proportion of fibres which decussate is smaller, about 75 and 66 per cent, respectively. Those that cross over arise from the nasal (medial) part of the retina, and the remainder come from the temporal (lateral) part, essentially as in Fig. 8.1. In man, all the axons from the nasal half of the retina decussate, while none of those from the temporal half crosses over. Thus each eye in man is represented about equally on both the left and right cerebral cortex; this may be associated with development of stereoscopic vision. The reduced degree of decussation in the dog and cat, and in particular in man, is a specialization of the highest mammals.

The visual pathways were probably among the first to undergo decussation during vertebrate evolution (see Section 5.9).

Note that the lateral part of the retina in man and carnivores sees mainly the nasal visual field; the medial half of the retina sees the temporal visual field.

Neuron 3

The cell location is in the **lateral geniculate nucleus**, within the **lateral geniculate body** (Fig. 8.1). The axon projects to the **visual area** of the cerebral cortex (Figs. 8.2(a), 8.3) in a band of fibres called the **optic radiation**. The visual area lies on the dorsocaudal and medial aspect of the occipital lobe (Fig. 8.2(a)).

The lateral geniculate nucleus may be regarded as a part of the **ventral group of thalamic nuclei**. The geniculate nuclei and thalamus belong to the **diencephalon** (Fig. 7.4).

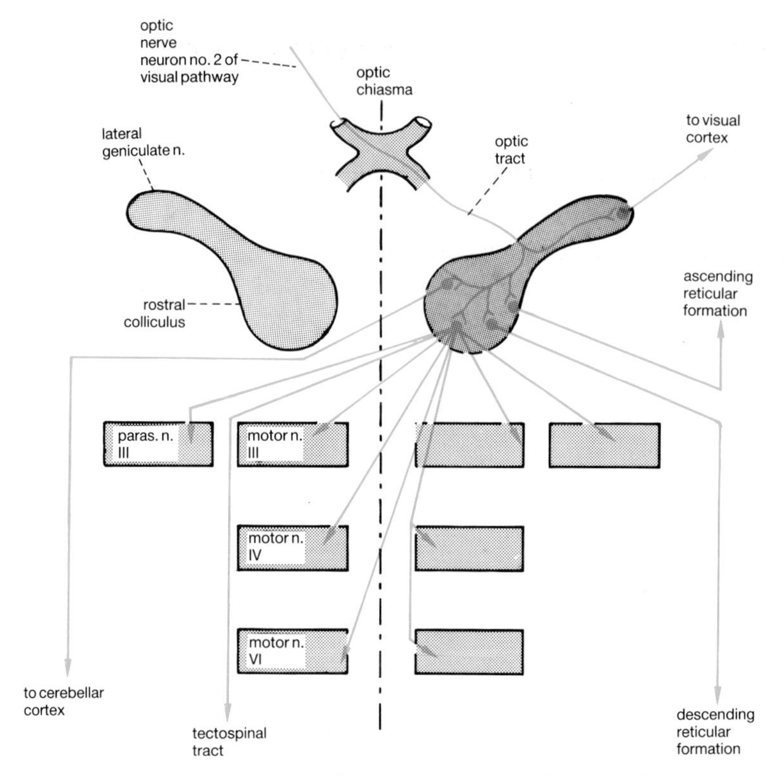

FIG. 8.4. **Diagram summarizing visual projections from the rostral colliculus (optic tectum).** The rostral colliculus on the right side is shown projecting to the motor nuclei of the oculomotor, trochlear, and abducent nerves (motor n. III, motor n, IV, and motor n. VI, respectively) on **both** sides of the brainstem; these projections control conjugate movements of **both** eyeballs towards a source of light. The rostral colliculus on the right side also projects to the parasympathetic nuclei (paras. n. III) of the oculomotor nerve (Edinger–Westphal nuclei) on **both** sides of the brainstem; these projections cause light striking one eye to constrict the pupil in **both** eyes. Furthermore the rostral colliculus projects into the tecto-spinal tract; this projection is contralateral only and causes turning of the head and neck towards a source of light. The projection from the rostral colliculus into the descending reticular formation initiates **dilation** of the pupil by sympathetic pathways passing through the cervical sympathetic trunk. The ascending reticular formation receives a visual input from the rostral colliculus, which contributes to the alerting function of the ascending reticular formation. Visual impulses are also projected from the rostral colliculus (tectum) to the contralateral cerebellar cortex, thus aiding the maintenance of posture. The left and right parasympathetic nuclei of the oculomotor nerve are not widely separated as indicated by the diagram but lie very close to each other adjacent to the midline.

For visual defects see Sections 17.14 and 19.2, and Cases 7, 9, 10, and 20 in Chapter 19.

8.2 Hearing

Neuron 1

The cell location is in the modiolus of the inner ear, i.e. the first neuron in the pathway of hearing is the cell of the spiral ganglion of the VIIIth nerve (Fig. 8.1). This neuron receives impulses from the neuroepithelial cells in the spiral organ (of Corti).

The axon runs in the VIIIth nerve, and enters the brainstem. It ends in synapses, in the cochlear nuclei, with neuron 2.

Neuron 2

The cell location is in the cochlear nuclei (Fig. 8.1).

Many of the axons decussate at once, forming a transversely orientated mound, the **trapezoid body**. Each axon then turns rostrally into the (contralateral) **lateral lemniscus** and ends in a synapse with neuron 3, in the contralateral **medial geniculate nucleus**. Other axons do not decussate but project to the ipsilateral medial geniculate nucleus, ascending in the (ipsilateral) lateral lemniscus. Thus each ear is projected to both the left and right side of the forebrain.

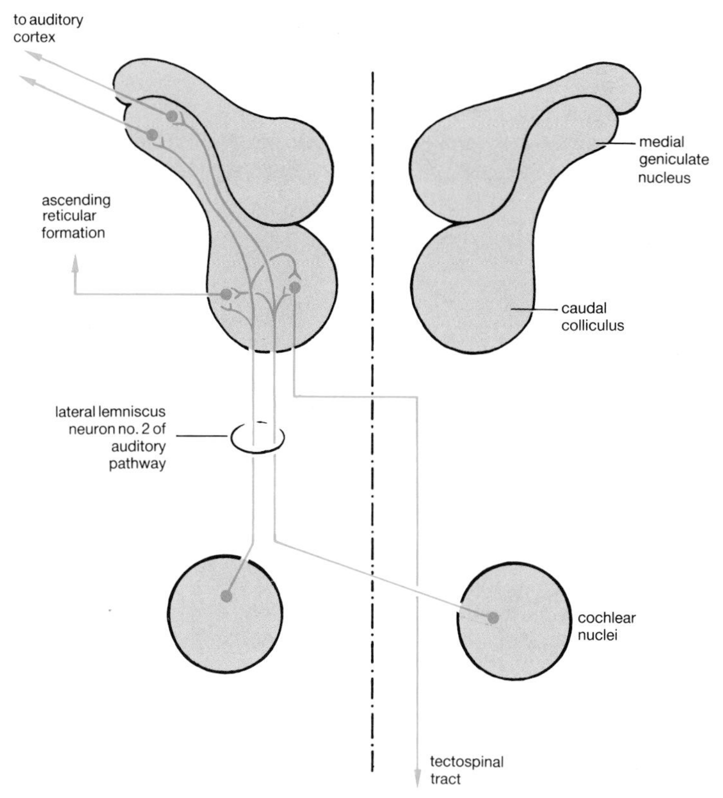

FIG. 8.5. **Diagram summarizing auditory projections to and from the caudal colliculus (auditory tectum).** The auditory stimulus which activates the tectospinal tract causes reflex turning of the head towards a source of sound. The tectospinal tract decussates. The ascending reticular formation receives an alerting input from the auditory pathways.

The axons of neuron 2 also give collaterals to motor neurons in the caudal colliculus which form the **tectospinal tract** (see Section 12.13 and Fig. 8.5). These collaterals form the afferent component of auditory reflexes, e.g. head-turning in response to a sudden sound; they fail to reach conscious levels of awareness, thus resembling the retinal pathways which initiate visual reflexes.

Some axons of neuron 2 are believed to make synapses with a tertiary neuron in the dorsal and ventral **nuclei of the trapezoid body** (previously known as the **superior olivary nucleus**). The axons of these tertiary neurons project via the lateral lemniscus to the caudal colliculus, where they synapse with neurons projecting to the medial geniculate body. From there, the neuronal pathway continues to the auditory cortex.

Neuron 3

The cell location is the **medial geniculate nucleus**, within the **medial geniculate body** (Fig. 8.1). The axon projects, as the **auditory radiation**, to the auditory area of the cerebral cortex (Figs. 8.2(a), 8.3). The medial geniculate nucleus belongs to the **ventral group of thalamic nuclei** (see Section 17.17).

This is a simplification. Usually four neurons are involved, No. 3 being in the caudal colliculus and No. 4 in the medial geniculate nucleus. Only a minority of paths go direct from the cochlear nucleus to the medial geniculate nucleus, and have only the three neurons.

The auditory pathway projects into the ascending reticular formation (see Section 9.8).

8.3 Balance

Nerve pathways to cerebral cortex

Neuron 1

The cell location is in the vestibular ganglion of the VIIIth nerve (Fig. 8.1). The neurons of the vestibular ganglion receive impulses from five sites: the **crista of the ampulla** of each of the three semicircular ducts, and the **macula of the utricle** and the **macula of the saccule**.

The **cristae** record **movement** of the head. Their neuroepithelial cells are stimulated by the drag of the gelatinous cupula which is caused by the movement of the endolymph as the head turns. The **maculae** record the **position** of the head. Gravity acts on the statoconial membrane of each macula, thus pulling on the microvilli and cilium of the neuroepithelial cells. This fires the neuroepithelial cells, which adapt only very slowly. Any tilting of the head changes the pattern of firing of the receptor cells.

The macula of the utricle is parallel to the ground with its ‘hairs’ pointing dorsally, and the macula of the saccule is vertical with its ‘hairs’ pointing laterally.

The axon of neuron 1 projects into the vestibular nuclei of the same side.

Neuron 2

The cell location is in the **vestibular nuclei** (Fig. 8.1), which (as expected) lie near the cochlear nuclei in the medulla oblongata. Some of the axons from these neurons ascend on the same side, while others decussate and then ascend on the other side (note the similarity to the second neuron of the pathway of hearing). They project to the **medial geniculate nucleus** (again like the hearing pathway). They give off numerous side-branches to the motor nuclei of cranial nerves III, IV and VI, and thereby strongly influence the movements of the eyes; it is these projections to the ocular nuclei which produce the **nystagmus** that characteristically follows rotation of the head (postrotatory nystagmus). These ascending fibres travel in the **medial longitudinal fasciculus (bundle)**.

Probably many of the fibres in this fasciculus do not reach the medial geniculate nucleus but project directly to the motor nuclei of the ocular nerves. The fasciculus also contains many descending vestibular fibres which continue into the spinal cord.

Neuron 3

The cell location is in the **medial geniculate nucleus**, within the **medial geniculate body** (Fig. 8.1). The axon projects to the cerebral cortex, probably close to the face region of the primary somatic sensory area.

Note that although the cortex receives balance projections, it is not responsible for the postural control which maintains balance; this control is maintained by the hindbrain.

Other connections of the vestibular nuclei

Two other major projections of the vestibular nuclei are described under the extrapyramidal motor system (see Chapter 12) and its feedback circuits (see Chapter 13). These are:

(i) The **vestibulospinal tract**. This descends ipsilaterally and projects mainly on alpha neurons (Fig. 12.2).

(ii) The vestibular nuclei project direct to the cortex of the ipsilateral flocculonodular lobe of the **cerebellum** (Figs. 13.1, 15.3). The return pathway from a cerebellar nucleus to the vestibular nuclei is also ipsilateral; this is a massive projection, giving the cerebellum a strong influence over the activity of the vestibular nuclei. These to-and-fro pathways between the vestibular nuclei and the cerebellum travel in the **caudal cerebellar peduncle** (Fig. 7.4 and see Section 15.3).

There are also:

(iii) **Numerous projections into the ascending and descending reticular formation**. Some of these descending reticular projections are involved in the vomiting and cardiovascular reactions which may occur in vestibular disturbances.

There are two vestibulospinal pathways of unequal size. The larger one is the vestibulospinal tract. The smaller one forms part of the medial longitudinal fasciculus (Fig. 12.3) (see Section 12.12).

Vestibular disease

Disorders of the vestibular system are relatively common, both in man and in the domestic mammals. The lesion may be peripheral (i.e. in the vestibular nerve or in the vestibular apparatus itself), or it may be central (i.e. within the neuraxis).

In both man and the domestic mammals **nystagmus** is often one of the clinical signs of vestibular disease. Nystagmus is an involuntary rhythmic oscillation of the eyeball, usually from side to side (horizontal) and nearly always affecting both eyes equally. Typically it has a slow phase in one direction and a quick phase in the other direction.

Nystagmus can be induced in normal individuals. For example it occurs normally after rotation, and is then called **postrotational nystagmus**. A similar kind of nystagmus occurs after watching the passing scenery from a railway carriage. If nystagmus occurs when the head is still and there is no rotation or movement of the surroundings, it is called **spontaneous nystagmus**.

Spontaneous nystagmus is usually pathological in origin. The slow phase represents an abnormal dragging away of the eyeball from the normal axis of vision; the fast phase represents a quick recovery to the normal axis of vision. The slow phase therefore tends to indicate the side of the lesion (i.e. if the slow phase is to the left, the lesion is likely to be on the left side). Despite this relationship, it is customary to describe nystagmus clinically in terms of the fast phase. If the nystagmus only occurs when the head is placed in an unusual position, e.g. by turning the head laterally or dorsally, it is known as **positional nystagmus**.

In man, spontaneous nystagmus is the commonest single sign of vestibular disease. Peripheral vestibular disease in man, but not central vestibular disease, is often accompanied by subjective symptoms, notably vertigo and nausea.

In the **domestic mammals**, spontaneous nystagmus quite often occurs in vestibular disease, but the most characteristic sign is **tilting of the head**, the face being rotated downwards on the side of the lesion. There is likely to be reduced tone in the limbs on the affected side, and a tendency to **fall** or even to **roll** over and over **towards** the side of the lesion. The animal may walk in a **circle** (typically towards the side of the lesion), but this is not a very common sign of vestibular disease. In some cases vomiting occurs. After a time (sometimes after only a few days) the spontaneous nystagmus disappears, probably through compensation within the nuclei of the motor nerves of the eyeball. However, positional nystagmus may then be induced, even when the spontaneous nystagmus has been gone for many weeks. Postrotational nystagmus is lost in vestibular disease. **Disturbance of the cerebellum** (see Section 15.14) may produce somewhat similar signs, including spontaneous nystagmus and a general insecurity of posture with, at least in man, a tendency to fall towards the side of the lesion. However, in cerebellar disease the head is usually **not tilted**, and the signs tend to be **worse during movement**. The nystagmus of cerebellar disease is really an **intention tremor**.

In **head tilt** the head rotates about its long axis, the tip of the nose remaining more or less in its normal position, apart from being rotated. In **head aversion**, as in cerebral lesions (see Section 17.14), the tip of the nose turns to one side and may go up or down as well. Sometimes, the limbs on the side of the lesion show hypertonus, presumably because an **irritative** lesion is affecting the vestibular nuclei; in this type of case the animal tends to fall on the side **opposite** to the lesion. In peripheral vestibular lesions the spontaneous nystagmus is said to be typically either horizontal or rotatory. In central vestibular lesions it is apparently horizontal, rotatory, or vertical; a consistent vertical nystagmus is believed to indicate a central vestibular lesion.

The difficulties of distinguishing between vestibular and cerebellar disease are accentuated by the extensive two-way exchanges of projections between the vestibular nuclei and the cerebellum. A lesion affecting these projections can give signs which could, therefore, be partly vestibular and partly cerebellar.

Complications arising from a combination of cerebellar and vestibular disturbance appear to be shown in the so-called **paradoxical vestibular syndrome** in dogs. The lesion (typically a choroid plexus papilloma) affects the cerebellar peduncles and may also involve the vestibulocerebellum (Fig. 15.3). The head is tilted down on the side **opposite** to the lesion. There is evidence from experimental lesions of the cerebellar peduncles and vestibular nuclei that section of the **juxtarestiform body** (see Section 21.26), dorsal to rather than within the medulla oblongata, can cause the head to be tilted down on the side opposite to the lesion (i.e. a paradoxical head tilt). The juxtarestiform body carries afferent and efferent cerebellar–vestibular fibres. Experimental ablation of one flocculus and the nodulus can also result in tilting of the head down on the opposite side to the floccular lesion.

8.4 Taste

Neuron 1

The cell location is in the 'dorsal root' ganglia of the VIIth or IXth cranial nerves for **taste buds** on the rostral two-thirds or caudal two-thirds of the tongue, respectively, or in the 'dorsal root' ganglion of the Xth nerve for taste buds near the epiglottis. The axon runs in the facial (VIIth), glosso-pharyngeal (IXth), or vagus (Xth) nerve.

Neuron 2
The cell location is in the **nucleus of the solitary tract**.

Neuron 3
The cell location is (probably) in the **ventral group of thalamic nuclei**. The axon projects to the cerebral cortex, near the 'face' part of the primary sensory area.

Taste is the least sophisticated of the special senses. Man can detect only four basic sensations—sour, salt, sweet, bitter. These appear to be responses to specific types of ions, dissolved in the liquids of the mouth. Much of what we interpret as taste is really smelling of the contents of the oral cavity; a cold reduces the sense of 'taste'. There is evidence that in other mammals the taste buds can be specifically sensitive to other substances; for instance, the rat can taste water.

8.5 Olfaction proper: the sense of smell

Neuron 1
This is the **olfactory neuroepithelial cell** itself (Fig. 8.6) and the axon is the central process of the neuroepithelial cell. The cell lies in the olfactory epithelium on the ethmoturbinate bones. These nerve cells break the almost universal rule that primary afferent neurons have their cell bodies in a

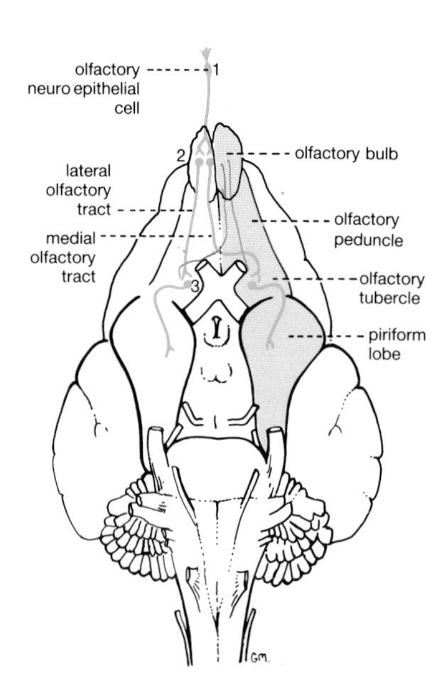

FIG. 8.6. Ventral view diagram of olfactory pathways. The olfactory pathways are shown arising from a neuroepithelial cell on the right side of the olfactory epithelium. Nos. 1 to 3 indicate the three neurons in the olfactory chain. The left rhinencephalon is shown in green.

dorsal root ganglion or in the equivalent ganglion of a cranial nerve; the neurons of the mesencephalic trigeminal nucleus are the only other exception (see Section 6.6). They also break another almost universal rule, namely that neurons are permanent cells throughout the life of the animal: olfactory cells last only about a month and are then replaced by the division of basal cells in the olfactory epithelium. The axon passes through the cribriform plate of the ethmoid and ends in synapses with neuron 2 in the **olfactory bulb**. Because of the continuous replacement of the cell body new axons are always present in the olfactory nerves.

Neuron 2

The cell location is in the **olfactory bulb** (Fig. 8.6). The axon passes caudally in the **olfactory peduncle**, which divides into the medial and lateral **olfactory tracts** (visible on the ventral surface of the forebrain). The axons in the lateral tract of the same side end in the olfactory tubercle. Some of the axons in the medial tract decussate, and project to the olfactory tubercle on the contralateral side (Fig. 8.6).

Some olfactory pathways project from the olfactory bulb through the medial olfactory tract to the **septal nuclei**. The next neurons in this chain mainly project from the septal nuclei into the **medial forebrain bundle** and end in the **hypothalamus**. Others, however, continue caudally in the medial forebrain bundle to reach the reticular formation of the brainstem; from here, further projections make synapses in the **parasympathetic motor nuclei** of the facial, glossopharyngeal, and vagus nerve, thus completing pathways by which olfactory stimuli induce **salivation** and **gastric secretion**.

Neuron 3

The cell location is in the **olfactory tubercle** (Fig. 8.6). The axon projects to the olfactory region of the cerebral cortex, i.e. to the **piriform lobe**.

Thus there are again three neurons, but there is no relay in the thalamus. This is the only sensory pathway which projects to the cerebral cortex without a relay in the diencephalon. This is because the thalamus was evolved after the rhinencephalon.

8.6 The rhinencephalon

The rhinencephalon (Gk. *rhis*, nose; Gk. *enkephalos*. brain), or smell-brain, is phylogenetically the oldest part of the forebrain. It lies on the ventral aspect of the cerebral hemisphere (Figs. 8.2(a), 8.6)). In primitive vertebrates it is concerned only with olfaction. This function is still retained by the olfactory components of the rhinencephalon, and remains of great importance in many mammals. Other parts of the rhinencephalon have subsequently been adapted to carry out functions associated with emotional responses; these are the limbic components of the rhinencephalon.

Olfactory components of the rhinencephalon

As would be expected from the anatomy of the olfactory pathway, the following areas give rise to a sense of smell when stimulated: olfactory bulb, peduncle, and tracts; olfactory tubercle; and piriform lobe.

Limbic components of the rhinencephalon

Large areas of the rhinencephalon are apparently only remotely, if at all, associated with olfaction. These include in particular the hippocampus and structures associated with it. They also include the amygdaloid body, the septal nuclei, and the habenular nuclei.

(i) The **hippocampus** is an area of primitive motor cortex, very ancient phylogenetically (the archicortex, see Chapter 18), which has been rolled into the floor of the lateral ventricle (Figs. 21.15(c), 21.20). Topographically, the rostral end of the hippocampus is directly continuous with the piriform lobe. The hippocampus seems to be involved in the control of the motor expression of **emotion** and **rage**, in the process of **learning** and **memory**, and in **instinct** (see Section 16.2).

(ii) The **amygdaloid body** is buried deeply in the piriform lobe. It consists of a complex of nuclei. The afferent connections of these nuclei are numerous and complicated, and their origins include the olfactory bulb, temporal cortex, thalamus, hypothalamus, and the nucleus of the solitary tract. The efferent projections are equally complicated. One large group of efferent fibres passes in the **stria terminalis**, and many of these end in the hypothalamus and septal nuclei; some of these connections of the amygdaloid body are shown schematically in Fig. 21.18. Other efferent projections from the amygdaloid nuclei pass to the thalamus, neocortex, and allocortex. Some projections also extend to the nucleus; these may perhaps explain the somatic movements, such as turning movements of the head, which sometimes occur during experimental stimulation of the amygdaloid body and may have encouraged the frequent inclusion of it in the basal nuclei. In general, however, the amygdaloid body appears to be involved in a very wide range of autonomic, endocrine, and somatic motor responses associated with **emotion**, including those of respiration, circulation, dilation of the iris, gastric motility, micturition, piloerection, and the somatic expressions of fear, rage, and aggression.

(iii) The **septal nuclei** form a small area of grey matter in the **septum pellucidum** (Fig. 21.14(b)). Their main input is from olfactory regions of the rhinencephalon, but they probably contribute little if anything to the sense of smell. They project to the **habenular nuclei** via the **habenular stria** (see Sections 16.4, 16.5). They also have afferent and efferent connections with the hippocampus by means of the **precommissural fornix**, and to a lesser extent connections with the amygdaloid body and hypothalamus (Fig. 21.18). They appear to be involved in behavioural reactions to external stimuli, particularly those of a sexual and aggressive nature.

(iv) The **habenular nuclei** lie in the roof of the third ventricle (Fig. 21.14(b)). They receive olfactory impulses via the septal nuclei, and then project to the hindbrain reticular formation and the parasympathetic motor nuclei of cranial nerves in the medulla oblongata. By these pathways the habenular nuclei participate in autonomic responses to olfactory and emotional stimuli (see Section 16.5).

8.7 Concept of the limbic system

Essentially the 'limbic system' is meant to include those parts of the forebrain which are concerned with visceral processes, especially those associated with the display and control of emotion, and hence also memory, behaviour, and personality. Unfortunately, as researches into fibre connections and behavioural studies advance, the limbic system seems to be well on the way to including all the main regions of the brain. Consequently the term is losing its value, and (like some other time-honoured concepts such as the pyramidal and extrapyramidal systems) may eventually have to be abandoned.

As would be expected from the steadily increasing range of its frontiers, there is no generally accepted definition of what the limbic system includes. The term is derived from the latin *limbus*, a border. The 'limbic' concept was expressed originally (in the late-nineteenth century) as the **limbic lobe**, a ring of grey matter on the medial aspect of each hemisphere, at the **border** between the central region of the forebrain and the peripheral hemisphere, encircling the interventricular foramen. The main constituents of the limbic lobe were considered to be the hippocampus, parahippocampal gyrus, and cingulate gyrus.

Gradually this notion became expanded into the concept of the **limbic system**, which again is characterized by the ring-like patterns of its constituents (Fig. 21.18). According to fairly general usage, the term now includes the following components. In the cerebral hemisphere: (a) the hippocampus; (b) the parahippocampal gyrus; (c) the cingulate gyrus; (d) the amygdaloid body; (e) the septal nuclei; (f) the fimbria and fornix; (g) the cingulum; (h) the stria terminalis; and (i) the medial forebrain bundle. In the diencephalon: (i) the habenular nuclei; (ii) the rostral thalamic nuclei; (iii) the hypothalamus; and (iv) the mamillary body. In the midbrain: the reticular formation. It will be observed that many of these structures are part of the rhinencephalon (a, b, d, e, f, and h), but all the others are not. Moreover, the truly olfactory parts of the rhinencephalon are not parts of the limbic system. Therefore, the term limbic system is in no way synonymous with rhinencephalon.

8.8 Thalamus and geniculate nuclei

The geniculate nuclei may be regarded as part of the ventral group of thalamic nuclei. The thalamus and geniculate nuclei are the last relay station of those afferent pathways that project to the cerebral cortex. No sensory pathways can reach the cerebral cortex without making a final relay with a neuron in the thalamic or geniculate nuclei, with the solitary exception of the olfactory pathway, which is more ancient phylogenetically than the thalamus. The thalamus is discussed more fully in Sections 17.17–17.22.

8.9 Summary of conscious sensory systems

(i) There are 12 conscious modalities, forming four distinct groups: group 1—touch, pressure, and joint proprioception; group 2—pinprick pain, heat, and cold; group 3—true pain; group 4—vision, hearing, balance, taste, and olfaction.

(ii) The pathways of the modalities in groups 1 and 2 converge on and share the medial lemniscus. Consequently they are often known as the **medial lemniscal system**.

(iii) The pathways conducting the modalities of groups 1, 2, and 4 are generally built on a chain of three neurons. The first neuron is typically in a dorsal root ganglion (spinal or cranial). The second neuron decussates (except for some fibres in the visual, hearing, and balance pathways). The third neuron (except in the olfactory pathway) lies in the ventral group of thalamic nuclei. It projects to the cerebral cortex, either to the primary somatic sensory area, or to the auditory or visual area.

(iv) The pathways conducting the modalities in groups 1 and 2 are arranged point-to-point in the neuraxis.

(v) The specific systems conducting the modalities in groups 1, 2, and 4 are also connected to the non-specific neuronal network of the ascending reticular formation. An exception is the pathway of joint proprioception which does **not** project into the ascending reticular formation.

(vi) The pathways of the modalities in groups 1, 2, and 4 have the triple function of inducing specific stimulation of the cerebral cortex, activating local reflexes, and altering the non-specific ascending reticular formation.

(vii) The modality of true pain (group 3) is conducted not by the medial lemniscal system, but by the ascending reticular formation.

(viii) The olfactory pathway is very primitive and departs from some of the principles followed by all the other systems. Thus the primary afferent 'neuron' is the neuroepithelial cell itself and this does **not** have its cell station in a dorsal root ganglion; second, the pathway bypasses the thalamus.

9. Spinocerebellar pathways and ascending reticular formation

There are two great ascending afferent systems which remain largely outside the realm of conscious perception—the spinocerebellar pathways and the ascending reticular formation. The spinocerebellar pathways transmit information from muscle proprioceptors about the activity of all skeletal muscles to the cerebellar cortex. When it was discovered at the end of the nineteenth century that muscle spindles are supplied by large afferent fibres, it was at first assumed that this afferent information would be utilized by the central nervous system to establish conscious awareness of the position and movement of the limbs (i.e. kinaesthesia). During the next 60 to 70 years this concept was almost totally abandoned in favour of the view that the activity of muscle afferents is not transmitted to the cerebral cortex, and that conscious awareness of the position and movement of the limbs is projected to the cerebral cortex almost exclusively by joint proprioceptors via the cuneate and gracile fascicles. However, since the 1960s it has become clear that muscle afferents are indeed an important source (together with the classical joint afferents) of the information on which the conscious perception of limb position and movement is based. Thus the cerebral cortex remains unaware of whether one particular skeletal muscle or another is in action, but it can translate detailed muscular activity into a generalized perception of the movement and position of the limbs.

The pathways of the ascending reticular formation do reach the cerebral cortex, but they consist of diffuse networks of primitive neurons which have only a limited capacity for projecting specific sensory information.

Peripheral receptors

9.1 Spinocerebellar pathways

Impulses originate in muscle proprioceptors (muscle spindles and Golgi tendon organs). These lead to the dorsal and ventral spinocerebellar tracts from the hindlimb and trunk, and to the spinocuneocerebellar pathway and cranial spinocerebellar tract from the forelimb. They transmit the modality of muscle proprioception only. The structure and function of muscle spindles and Golgi tendon organs is discussed in Sections 10.1–10.5.

9.2 Ascending reticular formation

Every type of afferent pathway, except muscle proprioception and joint proprioception, sends abundant connections into the ascending reticular formation. Hence the afferent modalities of touch, pressure, pinprick pain, heat, cold, and true pain project into it; so also do the five modalities of the special senses (vision, hearing, balance, taste, and olfaction), which project into the ascending reticular formation of the brainstem.

Spinocerebellar pathways

9.3 Hindlimb

Dorsal spinocerebellar tract

This tract is also called the **direct spinocerebellar tract**. The tract is as direct as it can be, since (i) it does not decussate, (ii) it lies close to the dorsal horn, and (iii) it enters the cerebellum by the nearest peduncle.

Neuron 1

The receptor is a muscle spindle. The cell location of the first neuron is in the dorsal root ganglion of a spinal nerve serving the hindlimb (Fig. 9.1). Its central axonal process projects into the dorsal horn.

Some Golgi tendon organs also contribute to the dorsal spinocerebellar tract.

Neuron 2

The cell location is in the base of the dorsal horn of (roughly) the same segment as neuron 1 (Fig. 9.1). The axon does **not** decussate. It ascends in the dorsal spinocerebellar tract in the dorsolateral region of the lateral funiculus (Fig. 12.3), enters the cerebellum by the **caudal cerebellar peduncle**, and ends somatotopically on the cerebellar cortex. These fibres are of very large diameter (about 20 μm) and fast conducting (about 120 m/s).

The aggegation of cell bodies in the base of the dorsal horn constitutes the **thoracic nucleus**, which was previously known as the dorsal nucleus of Clarke, or **Clarke's column**.

Ventral spinocerebellar tract

The ventral tract is also known as the **indirect spinocerebellar tract**. This is reasonable since its axons (i) decussate twice, (ii) lie relatively far away from

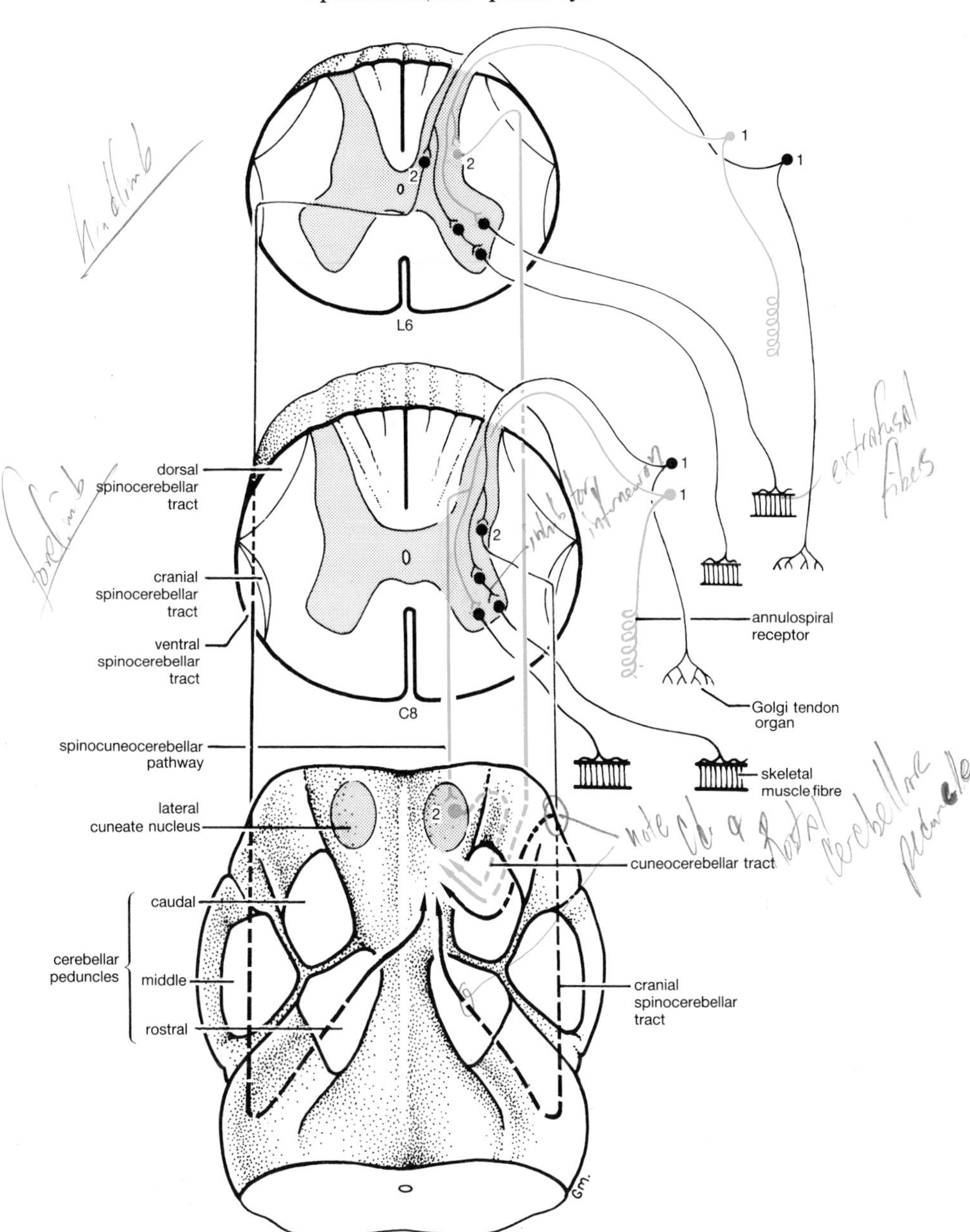

FIG. 9.1. Dorsal view diagram of spinocerebellar pathways. The spinocerebellar pathways project from annulospiral receptors (green) and Golgi tendon organs (black) to the cerebellar cortex. The muscles of the hindlimb are served by the dorsal and ventral spinocerebellar tracts; the muscles of the forelimb are served by the spinocuneocerebellar pathway and the cranial spinocerebellar tract. All four pathways have only two neurons (1 and 2). Three of these pathways remain ipsilateral; the fourth (ventral spinocerebellar) decussates twice, and is therefore effectively ipsilateral.

the dorsal horn, and (iii) make a long journey rostrally to enter the cerebellum by the most rostral of the cerebellar peduncles.

Some of the axons of this tract do not decussate a second time, but end contralaterally in the cerebellar cortex.

Neuron 1
The receptor is a Golgi tendon organ. The cell location is in the dorsal root ganglion of a spinal nerve serving the hindlimb (Fig. 9.1).

Neuron 2
The cell location is in the base of the dorsal horn (Fig. 9.1). The axon decussates immediately. Then it ascends rostrally in the ventral spinocerebellar tract, in the ventrolateral region of the lateral funiculus (Fig. 12.3), to the level of the **rostral cerebellar peduncle**. Here it turns caudally into the rostral cerebellar peduncle, and enters the cerebellum. In the cerebellum itself it **decussates for the second time**. The axon therefore ends, in the cerebellar cortex, on the same side of the body as that from which the pathway started.

9.4 Forelimb

Spinocuneocerebellar pathway

This pathway is equivalent to the dorsal spinocerebellar tract of the hindlimb.

Neuron 1
The receptor is a muscle spindle. The cell location is in the dorsal root ganglion of a spinal nerve serving the forelimb (Fig. 9.1). The axon enters the ipsilateral cuneate fascicle of the dorsal funiculus (Fig. 12.3), and ascends to the lateral cuneate nucleus.

Some Golgi tendon organs may be involved.

Neuron 2
The cell location is in the **lateral cuneate nucleus** of the medulla oblongata. The axon travels rostrally in the **cuneocerebellar tract**, entering the cerebellum via the ipsilateral caudal cerebellar peduncle (Fig. 9.1), and ends somatotopically on the ipsilateral cerebellar cortex.

The lateral cuneate nucleus may be regarded as homologous to the thoracic nucleus, i.e. Clarke's column, of the spinal cord.

Thus this pathway resembles the dorsal spinocerebellar tract in (i) arising (primarily) from muscle spindles, (ii) being ipsilateral throughout, (iii) being composed of two neurons, and (iv) using the caudal cerebellar peduncle.

Cranial spinocerebellar tract

This tract is equivalent to the ventral spinocerebellar tract of the hindlimb.

Neuron 1

The receptor is a Golgi tendon organ. The cell location is in the dorsal root ganglion of a spinal nerve serving the forelimb (Fig. 9.1).

Neuron 2

The cell body is in the base of the dorsal horn (from C1 to T1). The axon enters the ipsilateral cranial spinocerebellar tract, which lies on the medial aspect of the ventral spinocerebellar tract (Fig. 12.3). The axon enters the cerebellum through both the rostral and the caudal cerebellar peduncle, and ends somatotopically on the ipsilateral cerebellar cortex.

Thus the cranial spinocerebellar tract resembles the ventral spinocerebellar tract in (i) the receptor origin, (ii) its site in the spinal cord, and (iii) the number of neurons (two). However, the cranial tract differs from the ventral tract in (i) being ipsilateral throughout, and (ii) entering the cerebellum by the caudal cerebellar peduncle as well as the rostral peduncle.

9.5 Projections of spinocerebellar pathways to the cerebral cortex

As mentioned in the introductory paragraph of this chapter, information from muscle proprioceptors is not projected exclusively to the cerebellar cortex, but also indirectly reaches the cerebral cortex. In the cerebral cortex it makes a major contribution to the general perception of the movement and position of the limbs (the sense of **kinaesthesia**). The pathway of muscle proprioceptor information to the cerebral cortex is reasonably clear for the hindlimb. Axons in the dorsal spinocerebellar tract form collaterals, as the tract enters the medulla oblongata. Here the collaterals and the tract diverge: the collaterals remain in the medulla, whereas the tract enters the caudal cerebellar peduncle. The collaterals now make synapses with neurons in a displaced fragment of the gracile nucleus (*known as **nucleus Z***). The axons of these neurons join the medial lemniscus, and project via the ventral thalamic nuclei to the primary somatic sensory area of the cerebral cortex. In principle, therefore, these muscle proprioceptor pathways have dual projections: (i) directly to the cerebellar cortex, where they remain below conscious levels: and (ii) indirectly, via the medial lemniscal system, to the cerebral cortex where they contribute to the conscious sense of kinaesthesia.

The displaced fragment of the gracile nucleus is known as **nucleus Z**. The pathway from muscle proprioceptors of the hindlimb to the cerebral cortex therefore consists of four neurons: neuron 1, in a dorsal root ganglion; neuron 2, in the base of the dorsal horn (the thoracic nucleus); neuron 3, in nucleus Z; and neuron 4, in the ventral thalamic nuclei.

A pathway from the muscle proprioceptors of the forelimb to the conscious cerebral cortex is also present. It probably consists of collaterals from the spinocuneocerebellar pathway, which synapse with neurons in **nucleus X**. This nucleus lies near the lateral cuneate nucleus. Rostral to this point the pathway is the same as from nucleus Z.

As for the hindlimb pathway, there are four neurons in the pathway of forelimb muscle proprioceptors to the cerebral cortex.

9.6 Functions of the spinocerebellar pathways

The spinocerebellar pathways inform the cerebellum about the activities of all the skeletal musculature throughout the whole body. This enables the cerebellum to adjust and perfect this muscular activity. These mechanisms are discussed in Sections 15.9–15.12.

The spinocerebellar pathways also inform the cerebral cortex about the movement and position of the limbs (kinaesthesia).

9.7 Species variations

The dorsal and ventral spinocerebellar tracts are well developed in ungulates. This is predictable from the importance of locomotion in these species. The cranial spinocerebellar tract and spinocuneocerebellar pathways are known to occur in the cat.

Ascending reticular formation

9.8 Organization

Constituent neurons

The reticular formation consists of an ascending and a descending system, and is composed of a network of neurons forming the most primitive part of the neuraxis. It corresponds to the simple nerve networks of invertebrates and primitive vertebrates. Being primitive, its neurons make synaptic contacts with many other widely separated neurons.

Input

The ascending reticular system receives projections from the pathways of touch, pressure, pain, heat, and cold, at every segment of the spinal cord (Fig. 9.2) and brainstem (Fig. 9.3). In the brainstem it also receives

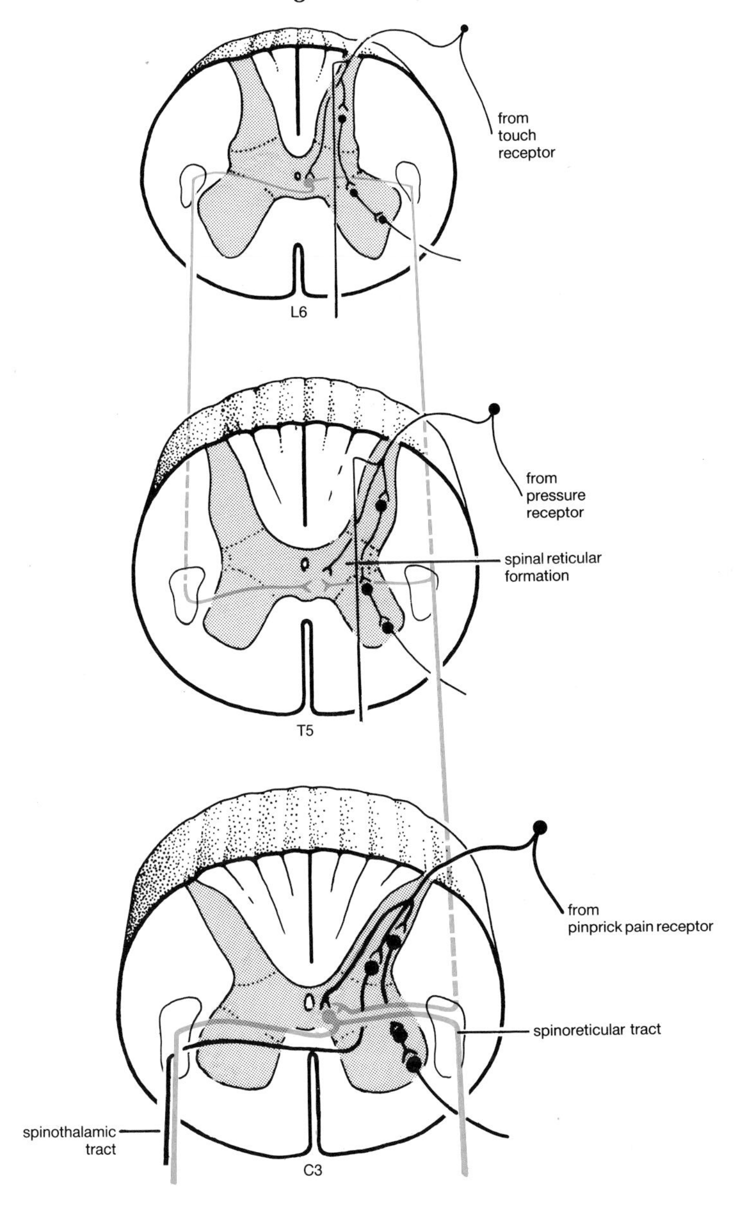

FIG. 9.2. **Dorsal view diagram of the ascending reticular formation: spinal part.** All the afferent modalities, **except** joint propriception and muscle proprioception, project into the ascending reticular formation. The neurons of the ascending reticular formation are shown in green. They have long axons which form the spinoreticular tract; they also have many short collaterals (only a few being shown here) which project to **both** sides of the reticular formation. Essentially this is therefore a **midline** network of **primitive** neurons. This diagram is continued by Fig 9.3, which begins by repeating the cross-section of the spinal cord at C3 and continues the pathway as far as the thalamus.

projections from the special senses (Fig. 9.3). Thus it receives an input from all the afferent modalities, except those of muscle and joint proprioception.

In the spinal cord

Within the spinal cord the system is known as the ascending spinal reticular formation, and includes spinoreticular fibres which form a somewhat indistinct **spinoreticular tract** (Figs. 9.2, 12.3).

In the brain

The system continues rostrally through the brainstem and into the thalamus (Fig. 9.3). Abundant reticular formation exists throughout the whole length of the brainstem, and is named according to its topographical position: hence (i) the medullary reticular formation; (ii) the pontine reticular formation; (iii) the midbrain reticular formation under the colliculi; and (iv) the thalamic reticular formation. It receives afferent projections from the special senses, via the rostral colliculus for the visual input and via the caudal colliculus for the auditory input (Fig. 9.3), and from the head generally (except for muscle and joint proprioceptors).

Projection to the cerebral cortex

This pathway takes the form of a diffuse projection from the thalamic reticular formation to all parts of the cortex (Fig. 9.3). The group of thalamic nuclei which is concerned with this diffuse projection is the **central group** (see Section 17.19).

The central group of thalamic nuclei is generally known as the **intralaminar nuclei**.

Site of cell location of reticular formation throughout the CNS

Throughout the spinal cord and brainstem the cell bodies of the reticular formation tend to lie in the grey matter which surrounds the central canal, i.e. in the spinal, medullary, etc. reticular formation (Figs. 9.2, 9.3). Their ascending axons give off numbers of collaterals which pass from side to side of the spinal cord and brainstem (Figs. 9.2, 9.3), and therefore essentially the ascending reticular formation is a **midline** system.

Descending reticular formation

The descending system of the reticular formation (see Section 12.2) threads its way through the ascending reticular system, the two systems being intimately intermingled and exchanging many synaptic contacts. The functions of these interrelationships are still obscure, but it is beginning to become clear that the descending system must be able to block selectively certain channels within the ascending system; this may explain how specific pain stimuli can be transmitted through the diffuse non-specific networks of

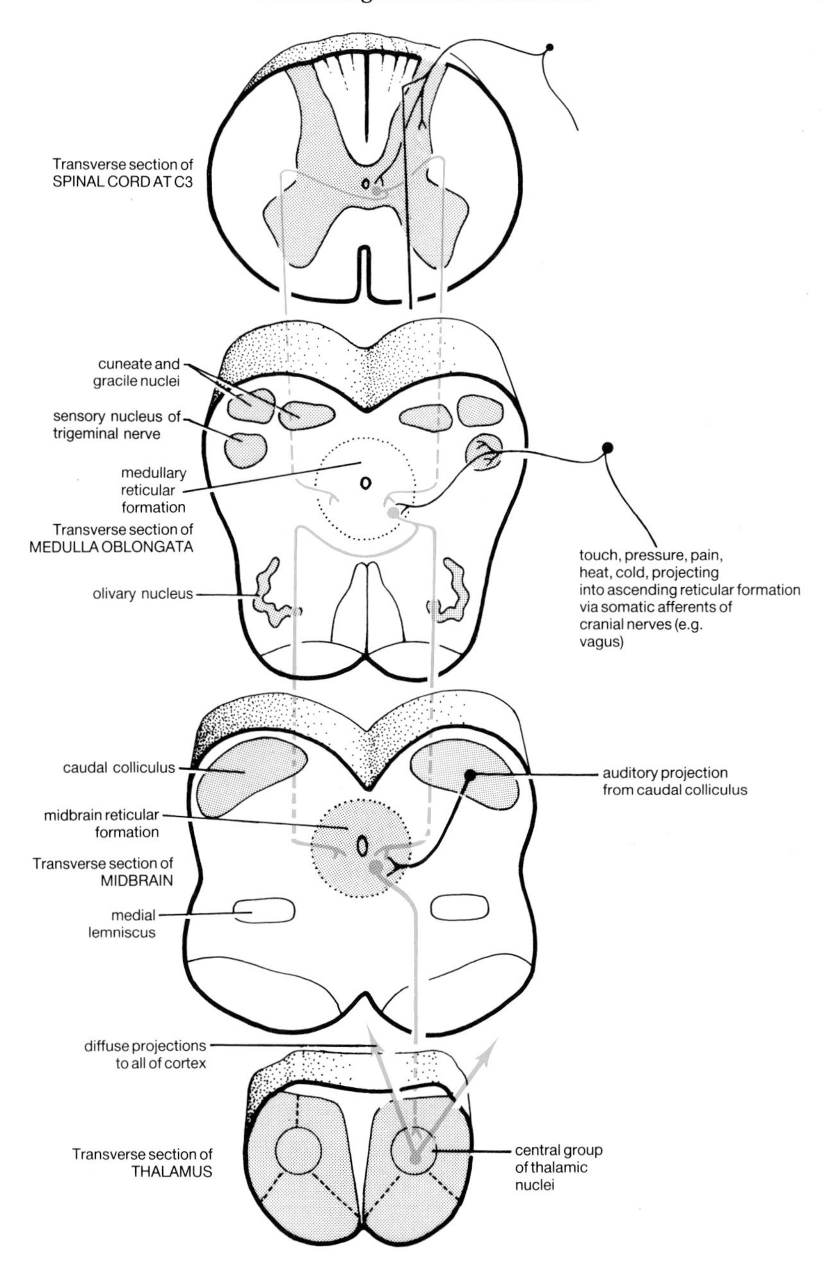

FIG. 9.3. Dorsal view diagram of the ascending reticular formation; cranial part. All the special senses project into the ascending reticular formation. So do the afferent modalities, arising in the head, of touch, pain, etc., but **not** joint proprioception and muscle proprioception. The ascending reticular neurons shown (green) in the diagram are much simplified; they have many collaterals to both sides of the brainstem. See also Fig. 9.2, which shows the spinal part of this pathway.

the ascending reticular formation and be to some extent localized by the cerebral cortex.

Phylogeny

The specific sensory systems of the spinal cord discussed in Chapter 7 have all evolved from the primitive non-specific neuronal network of the reticular formation. This process must have started early in the evolution of all but the lowest vertebrates. It appears to have happened through a progressive differentiation of the non-specific network enabling the peripheral sensory systems **to employ some particular parts of the network more than others**. By this process, specific neuronal pathways gradually crystallized out from the non-specific primitive neuronal net of the spinal cord. In mammals the development of these specific neuronal pathways (that is the medial lemniscal system, together with the pathways of the special senses) enables a much more detailed analysis of the external environment than is possible in submammalian forms, which depend largely or entirely on a non-specific network. However, the specific neuronal pathways of mammals are not independent of the non-specific neuronal network; indeed all the specific pathways (except those of joint and muscle proprioception) project into the reticular formation. Thus the specific and the non-specific pathways are welded into a single integrated system for the complex analysis of the signals coming in from the environment. It is mainly this **sensory analyser system**, rather than the increasing arsenal of motor mechanisms, which gives mammals their neuronal superiority over non-mammalian forms.

The immense importance of the non-specific pathways is shown by the fact that even in man more than 50 per cent of all neurons in the spinal cord and brain may belong to the ascending or the descending reticular formation.

9.9 Functions of the ascending reticular formation

Arousal

The ascending reticular formation arouses the whole cerebral cortex, thereby enabling specific stimuli (touch, temperature, special senses, etc.) to be perceived by the primary sensory area, visual area, etc. It is essentially facilitatory in action.

Because of this arousal function, the ascending reticular system has been called the **reticular activating system**.

Transmission of true pain

The more severe sensations of true pain, ranging from itching to agony, are probably transmitted mainly by the ascending reticular formation by means of the spinoreticular tract; this would be one activity of the ascending reticular formation which definitely does reach the conscious levels of the cerebral cortex. (Pinprick pain, in man anyway, is transmitted by the spino-thalamic tract (see Section 7.6).) The sense of true pain (as opposed to pinprick pain) is carried from the periphery by thin unmyelinated nerve fibres (about 0.3–1.3 μm in diameter). These sensations are slowly conducted, poorly localized, and outlast the provoking stimulus. They are believed to be induced by the release of a histamine-like substance from damaged tissues and detected therefore by chemoreceptor endings; the essentially chemical nature of the stimulus probably explains its longlasting character. The ascending pathways in the spinal cord appear to be based on the spinoreticular tract, in the ventral part of the lateral funiculus (Fig. 12.3), although the adjacent **propriospinal fibres** may also be involved (see Section 10.13). In the brainstem this non-specific pathway further ascends through the medullary, pontine, and midbrain reticular formation, and the central group of thalamic nuclei, and finally forms widespread projections over the cerebral cortex (Fig. 9.3).

The thin, unmyelinated, high-threshold fibres which transmit true pain from the periphery are 'C' fibres (see Section 10.1). They project into a special region of the dorsal horn known as the **substantia gelatinosa**, running continuously throughout the length of the spinal cord; the substantia gelatinosa consists of densely packed neurons with short axons, which are diffusely interconnected on the ipsilateral side by axons forming the **dorsolateral tract** (Fig. 12.3), previously known as **Lissauer's tract**. The pathway of true pain continues as projections from the substantia gelatinosa to the reticular formation, and thence up the spinoreticular tract (Fig. 9.4).

The substantia gelatinosa forms an important part of the **gate mechanism of** (true) **pain** (Fig. 9.4). The thin 'C' fibres excite pain pathways through the substantia gelatinosa, i.e. they **open** the gate to pain. If the gate is opened, volleys of pain impulses may be transmitted by extensive lengths of the substantia gelatinosa, thence into the ascending reticular formation, and finally to the cerebral cortex; such impulses can give the sensations of severe pain from what seem to be widespread regions of the body, the sensations being diffuse and longlasting in contrast to pinprick pain. Research has so far identified five mechanisms which can **close the gate to pain**.

(i) The substantia gelatinosa receives collaterals from the large-diameter cutaneous mechanoreceptor axons which form the gracile and cuneate fascicles. These collaterals make synapses with the terminals of the 'C' fibres (Fig. 9.4), subjecting them to **presynaptic inhibition**. The inhibitory influence of these fibres probably explains why mother says that rubbing the site of the injury makes the pain better.

(ii) The incoming axons of pinprick pain form collaterals which project upon and excite short inhibitory interneurons. These interneurons are **enkephalinergic** (Fig. 9.4). They inhibit the neurons which project pain impulses from the substantia gelatinosa to the reticular formation. These projections of pinprick pain fibres are

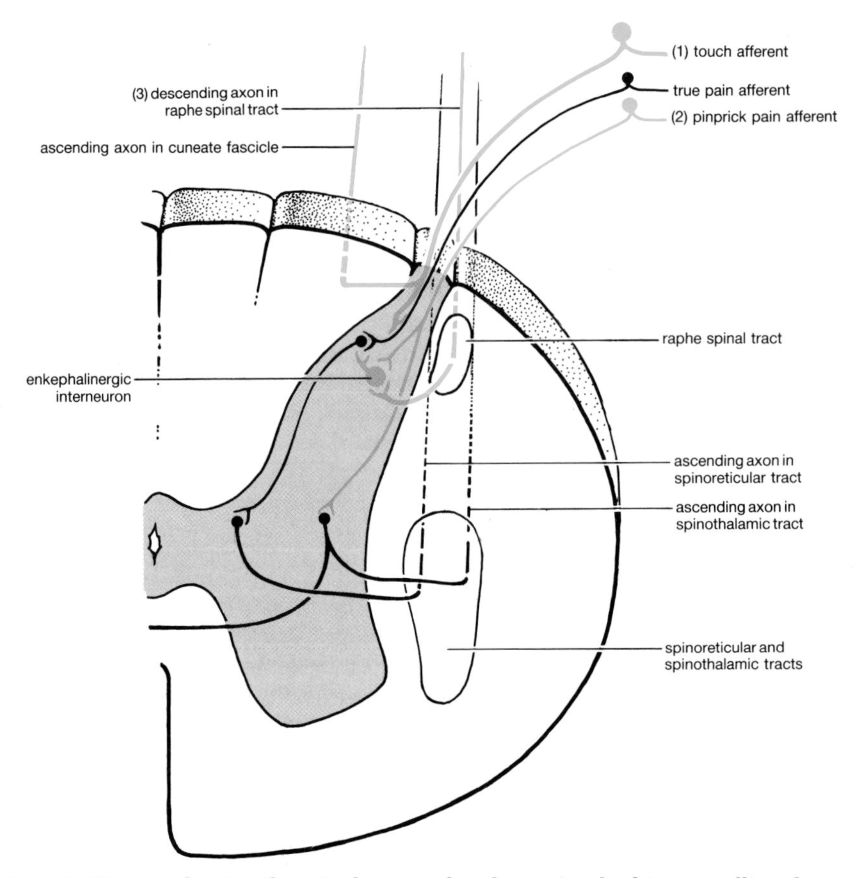

FIG. 9.4. Diagram showing the spinal neuronal pathways involved in controlling the gate mechanism of pain. The pathway of true pain begins with an unmyelinated 'C' fibre, which projects on a neuron in the substantia gelatinosa near the tip of the dorsal horn. This neuron in turn projects on a neuron in the spinal reticular formation, which sends its axon along the spinoreticular tract to the brain and transmits the conscious sensation of true pain. Three neuronal systems (green) inhibit the pain pathway. (1) The axon of a cutaneous mechanoreceptor of touch or pressure contacts the central terminal of the 'C' fibre presynaptically, subjecting it to presynaptic inhibition. (2) A pinprick myelinated fibre projects upon and excites an enkephalinergic interneuron, which inhibits the pain neuron in the substantia gelatinosa. (3) A descending fibre in the raphe spinal tract (part of the descending reticulospinal system) also excites the inhibitory enkephalinergic interneuron.

one of the two mechanisms which explain how **acupuncture** relieves pain; in this instance the acupuncture needle should be inserted segmentally, for example into the sensory field of the sciatic nerve to control sciatic pain in man (often effectively).

(iii) The downgoing **raphe spinal tract** (one of the reticulospinal pathways) peels off axons which end in **serotoninergic terminals**; again these project upon and excite the inhibitory enkephalinergic neurons in the substantia gelatinosa (Fig. 9.4). This reticulospinal pathway may be activated initially by the ascending reticular formation, which projects into the central grey matter (periaqueductal grey) of the midbrain reticular formation and acts there by enkephalinergic synapses. The **periaqueductal grey matter** then drives the **nucleus raphe magnus** (within the raphe nuclei of the medullary reticular formation). The neurons of the nucleus raphe magnus form the **serotoninergic** fibres which project down the spinal cord as the raphe spinal tract (Fig. 12.3). These circuits appear to be the second mechanism by which **acupuncture** acts; in this instance the needles are not inserted segmentally, but (for example) may be placed between the toes in order to relieve the pain of migraine in man (in general, less effectively). Alternatively, the initial ascending pathway of this circuit may lie in axons of the spinothalamic tract, some of which are known to project into the midbrain reticular formation.

(iv) It is a common experience that a person totally committed to intense physical activity, such as a soldier in battle, may not notice a severe injury. This is believed to be due to a massive release of endorphins from the hypothalamus.

(v) **Hypnosis** can be highly effective in relieving pain in man. The mechanism is not understood, but is known not to depend on enkephalinergic pathways.

Of these five mechanisms for controlling the gate to pain, (ii), (iii), and (iv) are all known to involve **enkephalinergic pathways**, because all three of them can be reversed by **naloxone**. The other two mechanisms do not require enkephalinergic activity, since they are not affected by naloxone.

In man, it has been found that otherwise intractable pain can be controlled by electrical stimulation of either the periaqueductal grey or the nucleus raphe magnus; the former stimulation site is used clinically. The analgesic effect is antagonized by naloxone. The raphe spinal tract must be intact. It has been suggested that analgesia induced by **morphine** may act through the periaqueductal grey, the raphe nuclei, and the dorsal horn, all of these being sites where the drug can mimic the action of enkephalins.

Recent evidence has suggested that the use of the **twitch** in horses may be equivalent to acupuncture. The twitch is a stout stick about 50 cm long, with a loop of rope at one end. The loop is twisted round the horse's upper lip. The horse then becomes quiet and appears sufficiently sedated to allow minor painful procedures to be carried out. The application of the twitch causes the plasma concentration of endorphin to increase; naloxone blocks the sedating actions of the twitch.

Visceral pain

In man, and probably in the domestic species also, the great majority of the **peripheral pathways** of visceral pain travels in sympathetic nerves. Territorially these sympathetic afferent pain pathways resemble the sympathetic efferent pathways. Thus if the sympathetic nerves relating to segments T6 to T10 carry **efferent** pathways **to** the stomach, they also carry **afferent** pain pathways **from** the stomach. The incoming axons reach the dorsal root ganglia by passing through the white rami communicantes and entering the dorsal roots of the spinal nerve. The cell location of these primary afferent neurons is, of course, in the dorsal root ganglia of the thoracolumbar segments.

The only exceptions to the rule that the peripheral pain pathways from the viscera travel in sympathetic nerves are (i) the pelvic viscera, which send their pain pathways through the **pelvic** and **pudendal nerves** and thence through the dorsal roots of sacral spinal nerves, and (ii) the respiratory tract and oesophagus which send theirs through the **vagus**. Pain axons in the pelvic and pudendal nerves therefore have their cell location in the dorsal root ganglia of the sacral segments; those of the vagus have their cell stations in the distal ganglion of the vagus.

The vagus has two sensory ganglia, the distal (nodose) and the proximal (jugular) ganglia, but the proximal ganglion is **somatic** afferent in function.

In **man**, the **central pathways** of these pain sensations from the viscera are again basically of two types, as for somatic structures. Some impulses travel through the **spinothalamic tract**, essentially via three neurons as before; these probably enable the cerebral cortex to localize approximately the source of the pain. The axons are arranged somatotopically alongside the axons carrying somatic pain. Other impulses reach the cerebral cortex via **spinoreticular fibres**; these probably convey a sensation of more severe pain from generalized visceral areas. In the **domestic mammals** the central pathways of visceral pain remain obscure, but it seems likely that the primitive neuronal networks of the ascending reticular formation would mediate this function.

Sleep

Sleep is probably due to temporary inhibition of the ascending reticular formation; awakening is caused by the arrival of volleys of stimuli entering the reticular formation from the pathways carrying the classic sensory modalities (touch, pain, sound, etc.). There is, however, experimental evidence for a **sleep centre** in the diencephalon. This is believed to project inhibitory pathways to the ascending reticular formation, thus damping down its facilitatory action on the cerebral cortex, but the mechanism is not yet understood (see Section 16.3). Lesions in this 'sleep centre', which is in the rostral hypothalamus, produce insomnia. The caudal hypothalamus seems to contain a **'waking centre'**, since lesions here cause sleepiness.

Lesions in the caudal part of the hypothalamus in man sometimes produce repeated episodes of overpowering sleepiness during daytime, the condition being known as **narcolepsy**. Similar clinical observations have been reported in Shetland ponies and in dogs and cats. In certain breeds of dog (including the Doberman Pinscher) it appears to have a genetic basis. Often no actual lesions can be found. Each episode may last for several minutes, and resembles rapid eye movement sleep (paradoxical sleep); apart from the eye movements, there is no somatic motor activity (in contrast to epileptic seizures).

Effects of destruction: general anaesthesia

Destruction of the ascending reticular formation leads to **coma**, even when all the other main ascending tracts are intact. **General anaesthetics** probably selectively depress the ascending reticular formation by blocking transmission at its synapses.

Summary of spinocerebellar pathways and ascending reticular formation

(i) Two separate systems remain largely outside the realm of consciousness, namely the muscle proprioceptive (spinocerebellar) system, and the ascending reticular formation.

(ii) There are four spinocerebellar pathways. Two serve the hindlimb, i.e. the dorsal and ventral spinocerebellar tracts. Two serve the forelimb, i.e. the spinocuneocerebellar pathway and the cranial spinocerebellar tract.

(iii) The dorsal spinocerebellar tract and the spinocuneocerebellar pathway project information from muscle spindles. The ventral spinocerebellar tract and the cranial spinocerebellar tract project information from Golgi tendon organs.

(iv) There are only two neurons in each of the spinocerebellar pathways.

(v) The spinocerebellar pathways do not decussate, except for the ventral spinocerebellar tract which decussates twice and this in effect is the same as not decussating at all.

(vi) The spinocerebellar pathways project directly to the cerebellar cortex through the cerebellar peduncles, and indirectly to the cerebral cortex by collaterals which join the medial lemniscal system in the medulla oblongata.

(vii) The spinocerebellar system informs the cerebellum about the state of contraction of the skeletal musculature, and informs the cerebral cortex about the movement and position of the limbs.

(viii) Ungulates have a well-developed spinocerebellar system, related to the importance of locomotion.

(ix) In the ascending reticular formation there are chains of several neurons, corresponding to the primitive neuronal networks of lower vertebrates.

(x) The neurons of the ascending reticular formation have many transverse collaterals, thus amounting to a midline system.

(xi) The ascending reticular formation receives an input from all the afferent pathways arising from the spinal nerves, from the segmental cranial nerves, and from the cranial nerves of the special senses—except for the modalities of joint proprioception and muscle proprioception.

(xii) The ascending reticular formation projects diffusely all over the cerebral cortex, via the central (intralaminar) group of thalamic nuclei.

(xiii) The ascending reticular formation arouses the whole cerebral cortex. Therefore it is sometimes called the reticular activating system. It alerts the cortex and enables the cortex to perceive the specific modalities.

(xiv) In spite of its generally non-specific structure and function, the ascending reticular formation has some capacity for transmitting specific modalities, notably the sensation of true pain.

(xv) The ascending reticular formation combines with the medial lemniscal system and the special senses to analyse afferent signals originating from the internal and external environment.

(xiv) The reticular formation (its ascending and descending systems together) is the most primitive part of the neuraxis, and the largest single component of the neuraxis even in the highest mammals.

10. Somatic motor systems: general principles

10.1 Motor neurons in the ventral horn of the spinal cord

There are two types of somatic efferent neurons in the ventral horn. (i) The **skeletomotor neuron** innervates the ordinary skeletal muscle fibre, i.e. the extrafusal muscle fibre (Fig. 10.1(a)). It is often known as a **motoneuron**. (ii) The **fusimotor neuron** innervates the intrafusal muscle fibre (Fig. 10.1(a)), which forms part of a muscle spindle (Fig. 10.1(b)).

The skeletomotor neuron has a myelinated axon of large diameter (up to 20 μm) and is therefore fast conducting. It is classed as an A alpha neuron

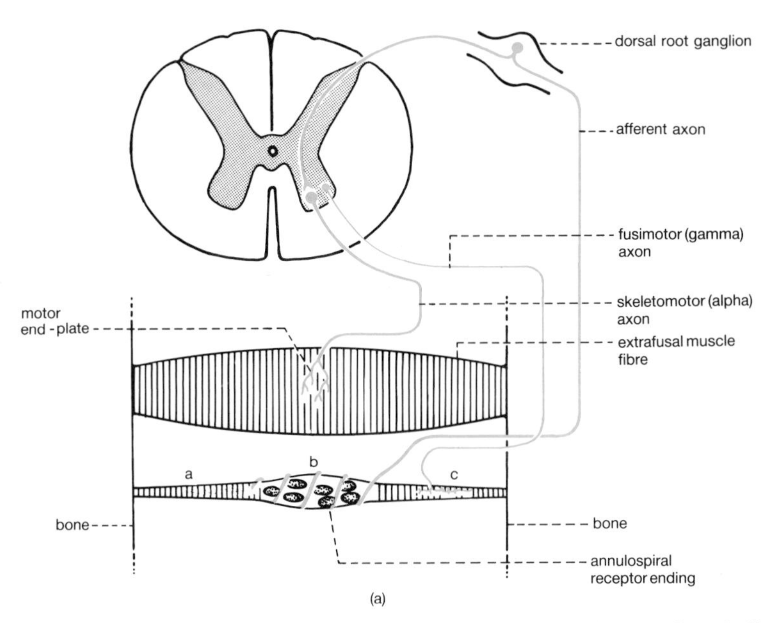

FIG. 10.1. (a) Diagram of the basic components and innervation of a muscle spindle. The intrafusal muscle fibre (either a nuclear bag or nuclear chain fibre, Fig. 10.1(b)) comprises two contractile regions, a and c, joined by a stretchable midregion, b. The annulospiral receptor ending is wound round b. The central continuation of the annulospiral receptor ending is an axon of large diameter. In the ventral horn it makes a monosynaptic reflex arc with a skeletomotor (alpha) motoneuron which also has a thick axon. The contractile regions of the intrafusal muscle fibre are innervated by a fusimotor (gamma) neuron, which has a thin axon and terminates in a single motor end-plate.

because of the thickness of its axon. The fusimotor neuron has a thinner myelinated axon (2–8 μm) which is slower in conduction. It is classed as an A gamma neuron. Because of this classification the skeletomotor neuron is often called the **alpha neuron**, and the fusimotor the **gamma neuron**.

Of the total motor outflow from the ventral horn, about one-third of the axons are from fusimotor neurons and two-thirds from alpha neurons.

The category of A neurons includes all myelinated fibres, except preganglionic autonomic efferent neurons which are category B. Category C includes all unmyelinated fibres. The efferent A fibres are subdivided into A-alpha, A-beta, and A-gamma fibres. The alpha fibres are about 17 μm in diameter, and supply extrafusal muscle fibres only; the beta fibres give a branch to a muscle spindle and then innervate 'slow twitch' extrafusal muscle fibres; the gamma fibres are about 2–8 μm in diameter and supply intrafusal muscle fibres only. The afferent A fibres are subdivided into groups I, II, and III. Group I fibres are up to 20 μm in diameter, and arise from annulospiral receptors and Golgi tendon organs; group II fibres are between 5 and 15 μm in diameter, and arise mainly from cutaneous mechanoreceptors; group III fibres range between 1 and 7 μm, and include pinprick pain receptors from various tissues of the body, sensory endings from walls of blood vessels, and endings from the follicles of down hairs.

10.2 Structure of the muscle spindle

A muscle spindle (fusus neuromuscularis) consists of two components: (i) a small striated muscle fibre, the **intrafusal muscle fibre**, together with the **fusimotor (gamma) neuron** which innervates the intrafusal muscle fibre; the middle section of the intrafusal muscle fibre is stretchable; (ii) the **annulospiral receptor** consisting of an afferent axon, the ending of which is wound round the stretchable middle region of the intrafusal muscle fibre (Fig. 10.1(a)). A fibrous capsule encloses the spindle and indirectly attaches to bone (Fig. 10.1(b)).

The muscle spindle detects stretch, by responding to an increase in muscle length. Its motor innervation maintains the length of the stretchable midregion close to the threshold of its annulospiral receptor. Further elongation of the midregion causes the receptor to fire, the rate of firing being related to the increase in length.

The afferent axon of the annulospiral receptor has its cell station in a dorsal root ganglion. The axon is of **large** diameter (10–20 μm), and makes a **monosynaptic** connection with the skeletomotor neuron of the ventral horn (Fig. 10.1(a)). Consequently, the conduction velocity of this reflex arc is rapid.

Each spindle consists of from 4 to 20 intrafusal fibres, depending on the species. There are two types of intrafusal fibre. In the **nuclear bag fibre** (Fig. 10.1(b)) the nuclei are clustered in the middle of the fibre, which is slightly expanded and lacks obvious striations; the **nuclear chain fibre** is much shorter and more slender, its

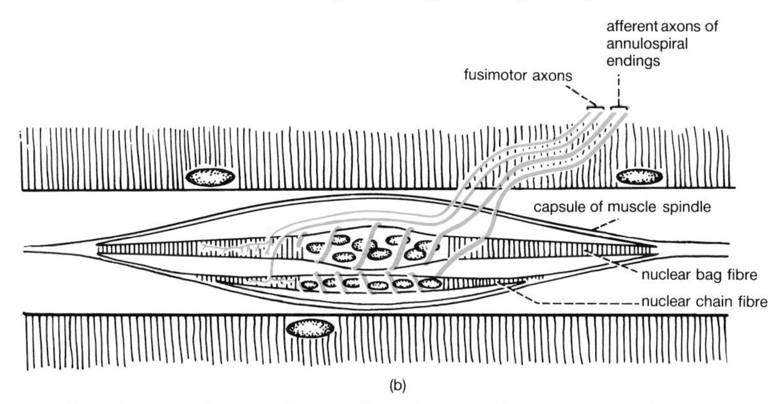

FIG. 10.1. (b) Diagram of a muscle spindle, showing the two types of intrafusal muscle fibre. The nuclear bag fibre is larger and has a cluster of nuclei. The nuclear chain fibre is shorter and slimmer, with its nuclei in a single row. Each intrafusal muscle fibre is supplied by both a fusimotor (gamma) motoneuron and an annulospiral receptor ending.

nuclei forming a single longitudinal row in the non-striated middle of the fibre (Fig. 10.1(b)). Both types of fibre attach either directly or indirectly (e.g. via the spindle capsule) to the connective tissue of the surrounding extrafusal fibres (and therefore indirectly to bones, Fig. 10.1(a)). The nuclear region of both types of fibre is more stretchable than the rest of the fibre.

10.3 The mode of operation of the muscle spindle

By stretching of the muscle as a whole

When the muscle is at rest, as in Fig. 10.2, the annulospiral receptor is 'set' just below its threshold and is therefore silent. So long as the length of the stretchable middle section of the intrafusal muscle fibre is not increased, the receptor remains silent. If the length is increased, the receptor fires.

These characteristics of the muscle spindle account for the **patellar reflex**. When the patellar ligament of the resting limb is suddenly tapped, the muscle spindles in the vastus and rectus femoris muscles are slightly stretched. This makes their annulospiral receptors fire as in Fig. 10.3. The firing of an annulospiral receptor reflexly excites its skeletomotor (alpha)

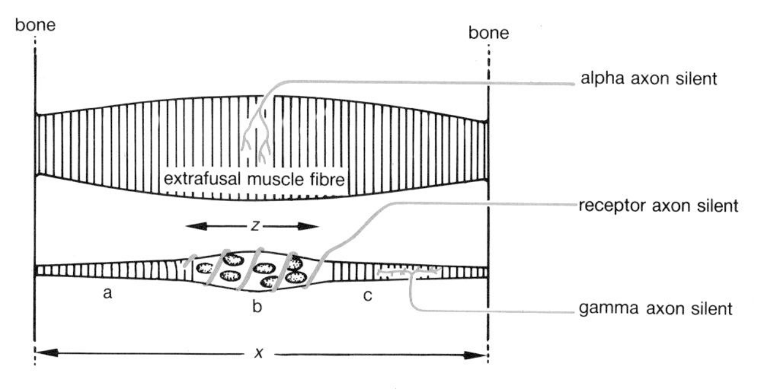

FIG. 10.2. Muscle spindle at rest. The annulospiral receptor ending, which is wound around b, is 'set' to be silent when b is of length z, or less than z.

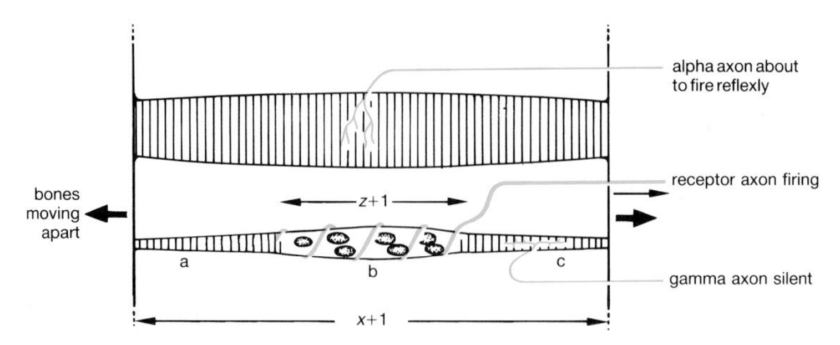

FIG. 10.3. Stimulation of the annulospiral receptor by moving the bones apart as in the patellar reflex. If the distance between the bones is increased to $x + 1$, this stretches b to $z + 1$, causing the receptor to fire. This reflexly fires the skeletomotor (alpha) neuron; the extrafusal muscle fibre then contracts returning the bones to their original positions, separated once again by the distance x.

neuron to fire. When the skeletomotor neuron fires, it induces contraction of the **extrafusal** muscle fibres in the vastus and rectus femoris muscles, and their contraction relieves the stretching of the spindle. This reflex response to stretch is known as the **myotatic reflex** or **stretch reflex**.

By stimulation of the fusimotor neuron

Contraction of a skeletal muscle such as rectus femoris (i.e. of its **extrafusal** fibres) can be initiated by its fusimotor (gamma) neurons. Firing of the fusimotor neurons makes the intrafusal muscle fibres contract, and this stretches their annulospiral receptors (Fig. 10.4). As before, stretching of the annulospiral receptor reflexly excites the skeletomotor neuron to fire. Hence the main muscle fibres (extrafusal muscle fibres) in turn contract. The fusimotor (gamma) neuron is itself excited or inhibited by the descending efferent pathways of the CNS. This can occur through the projections of either the pyramidal or the extrapyramidal systems (Fig. 14.1).

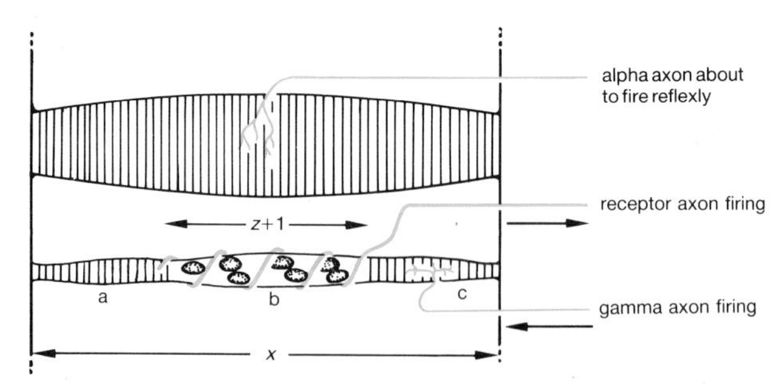

FIG. 10.4. Stimulation of the annulospiral receptor caused by contraction of the intrafusal muscle fibre. When the gamma neuron fires, the bellies (a and c) of the intrafusal muscle fibre contact. This stretches b to $z + 1$, exciting the annulospiral receptor to fire. The receptor reflexly activates the skeletomotor (alpha) neuron, causing the extrafusal muscle fibre to contract.

The contraction of the intrafusal muscle fibres is in itself exceedingly feeble. It causes no recordable change in the tension of the muscle as a whole. It is strong enough only to stretch the annulospiral receptor.

10.4 Role of muscle spindles in posture and movement

Muscle spindles can be 'set' to detect and correct any small changes arising in the angles of weight-bearing joints (Fig. 10.5(a), (b)). They can also be employed to induce shortening of the muscles of locomotion and thereby achieve locomotion (Fig. 10.5(c)); however, this entails initial stimulation of the fusimotor neurons by the motor pathways descending from the brain (e.g. the pyramidal or extrapyramidal pathways).

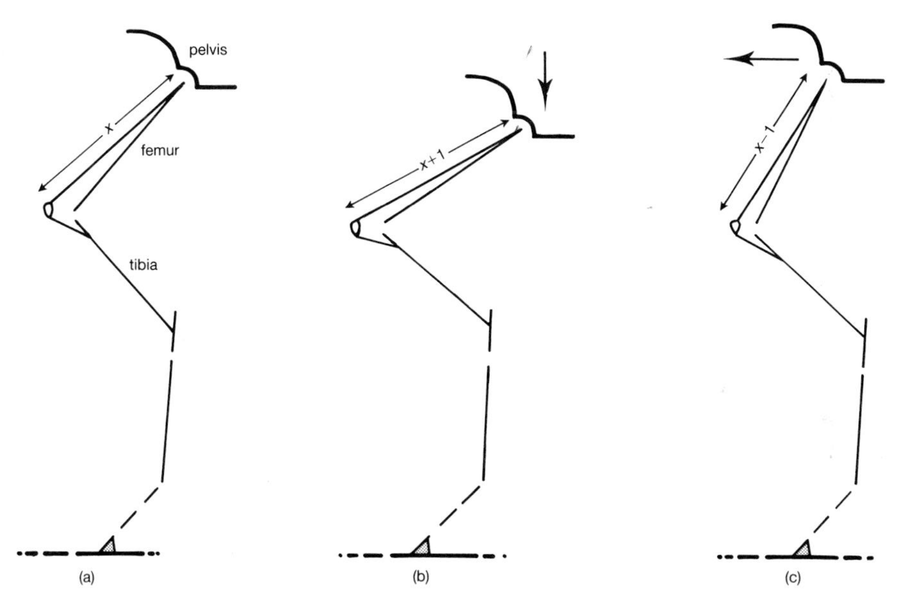

FIG. 10.5. Diagrams showing the role of the muscle spindle in maintaining posture and in locomotion.

(a) Vastus muscle in limb at rest. The spindles have been 'set' to maintain the length of the vastus muscle at *X*.

(b) Femorotibial joint flexed by the weight of the body. The flexing of the femorotibial joint stretches the vastus muscle. This excites the annulospiral receptors in the vastus muscle, and these in turn reflexly excite the skeletomotor neurons which control the ordinary extrafusal muscle fibres of vastus. This results in extension of the femorotibial joint, restoring it to its original position. In this way, the muscle spindles reflexly maintain posture against the force of gravity. Such a reflex is called a myotatic reflex.

(c) Body moved forward by contraction of vastus muscle. If the fusimotor neurons to the muscle spindles in the vastus muscle are fired by the higher motor centres, the intrafusal muscle fibres contract. This excites their annulospiral receptors, and these in turn reflexly excite skeletomotor (alpha) neurons. The vastus muscle as a whole then contracts, and the body is moved forward. In this way, the spindles induce the movements of locomotion.

10.5 Golgi tendon organs

The Golgi tendon organ (fusus neurotendineus) is a specialized receptor formation of a sensory axon. The receptor ending is much branched and beaded at its end, resembling a flower spray. The branching terminals are embedded in the muscle–tendon junction. Like the annulospiral receptor of a muscle spindle the Golgi tendon organ is a muscle proprioceptor. The Golgi tendon organ measures the **tension** produced in the muscle, in contrast to the annulospiral receptor which measures length. A Golgi tendon organ activates an inhibitory interneuron which **inhibits** the alpha motoneuron that is causing the tension in the muscle. Thus Golgi tendon organs may protect the muscle from overloading. They probably also promote the smooth working of the muscle.

The afferent fibres of muscle spindles and Golgi tendon organs have collaterals which control antagonistic muscles. Thus spindle afferents arising in the vastus muscles act (i) on alpha motoneurons innervating the vastus muscles, and (ii) on inhibitory interneurons which cause relaxation of the muscles which are the antagonists of the vastus muscles. The afferents from Golgi tendon organs in the vastus muscles act (1) on inhibitory neurons which cause relaxation of the vastus muscles, and (2) on excitatory interneurons which cause the antagonistic muscles to contract. Thus the spindle afferents have directly opposite actions to Golgi tendon organ afferents.

10.6 Muscle tone

In the awake animal many of the skeletal muscles are, at any given moment, in a state of contraction known as muscle tone. This applies particularly to the muscles which are opposing the effects of gravity. The degree and distribution of this tone determines the posture of the body. This postural muscle tone is maintained involuntarily (and with minimal fatigue) by the activity of the alpha neurons in the ventral horn. Neurons supplying adjacent motor units fire asynchronously, so that these adjacent units are alternatively active and resting and yet the muscle as a whole is able to exert a steady pull.

Muscle tone is basically of two types. It is due mainly to the activity of the gamma (fusimotor) neurons, which reflexly stimulate the skeletomotor neurons by activating their muscle spindles. It is also due partly to a direct stimulation of the skeletomotor neurons without the spindle reflex being involved (the vestibulospinal tract being an important source of this kind of muscle tone).

The clinician assesses muscle tone by the ease with which the joints may be passively flexed and extended. Unduly stiff resistance to moving the joints indicates too much tone or **hypertonus**, which may amount to **spasticity** or **rigidity**. When a **spastic** muscle is stretched there is at first no increase in resistance; this is the 'free interval'. Then the resistance increases, and finally it fades away again. In **rigidity** there is no free interval, but the other signs are the same as for spasticity. Reduced tone or **hypotonus**, or complete loss of tone (**atonus**, also known as **flaccidity**) can be recognized clinically by detecting reduced resistance to joint movement, and sometimes also by palpation of the affected muscle belly which has a soft, fluid texture. In veterinary neurology, a very marked loss of tone, amounting virtually to atonia, is relatively easy to recognize by palpating the muscle and shaking the limb to reveal the floppy characteristics of the joints; but hypertonia is difficult to distinguish from voluntary resistance to handling.

10.7 Motor unit

A motor unit is the group of skeletal muscle fibres innervated by all the axon terminals of a single alpha (skeletomotor) neuron. An anatomical muscle consists of several thousand motor units, each unit being controlled by its own alpha neuron. The muscle fibres of any one motor unit are not assembled close together within a muscle, but are separated and scattered among the muscle fibres of other motor units.

In one of the muscles of the eyeball, where great accuracy of movement is needed, there is believed to be a separate axon for each muscle fibre. At the other extreme, in muscles (such as the gluteal muscles) carrying out less precise actions, a thousand or more muscle fibres are innervated by only one alpha neuron.

All the muscle fibres of one motor unit contract **simultaneously** and **at the same force**, when their motor neuron fires. But the force of contraction of all these fibres of one motor unit can be varied from moment to moment. This **variation in the force** of contraction of the whole motor unit is achieved by varying the frequency with which the controlling neuron fires. High frequencies of firing cause the maximum force of contraction of the fibres of the unit. This may be many times greater than the force of contraction induced by lower frequencies.

10.8 Recruitment of motor units

It is very rare for all the motor units in a muscle to be employed at maximum capacity simultaneously. If this were to happen, the force of contraction would be excessive. Normally there is a phasic contraction, i.e. some motor units in a muscle contract while others relax. Some sort of central psychologial inhibition apparently prevents the simultaneous contraction of all the motor units of a muscle. The big athletic occasion reduces this block, so further 'recruitment' of motor units can occur. Thus the 'superhuman strength' of emergency is strength which is always there but not normally used.

10.9 Summary of ways of increasing the force of contraction of a muscle

(i) Increased frequency of firing of its ventral horn neurons.
(ii) Recruitment of its motor units.

10.10 Algebraic summation at the final common path

To fire or not to fire, that is the question. The **final common path** is the skeletomotor neuron. This neuron receives synapses (Fig. 14.1) from various pyramidal and extrapyramidal pathways (Chapters 11 and 12), and from monosynaptic spindle reflex arcs and the reflex arcs of Golgi tendon organs. Some of these pathways will be excitatory and others inhibitory. The summation of these many excitatory and inhibitory impulses decides the ultimate action of the final common path: this means firing or not firing (see Section 5.12).

10.11 Renshaw Cell

This cell is directly excited by the firing of an alpha neuron, by means of a collateral of that alpha neuron (Fig. 14.1). But when the Renshaw cell fires, it **inhibits** the alpha neuron. This is an example of **inhibitory (negative) feedback**.

Renshaw cells are mainly associated with those alpha neurons which innervate the extensor muscles of the limbs. Tetanus toxin binds specifically to Renshaw cells and blocks the release of their inhibitory neurotransmitter. Extensor rigidity of the limbs is therefore a principal clinical sign of tetanus.

10.12 Lower motor neuron

This is an alternative name for the alpha (skeletomotor) neuron, or final common path, and is commonly used in clinical neurology.

Destruction of the final common path (i.e. of the alpha neuron) paralyses all the muscle fibres which belong to its motor unit. Such destruction can result from a lesion in the ventral horn of the spinal cord, causing **lower motor neuron disorder**. Total destruction of the ventral horn in a segment of the spinal cord always causes a loss of tone (**hypotonia**) and some degree of **flaccidity** and **rapid wasting** of the muscles innervated by that segment, with at least a **flaccid paresis**. Spinal reflexes are reduced (**hyporeflexia**). The severity of these changes depends on whether the affected muscle or muscles are exclusively supplied by that particular segment of the spinal cord alone. In the limbs an individual muscle may receive motoneurons from the ventral horn of several segments of the spinal cord. For example the rectus femoris muscle and vastus muscle are innervated by the femoral nerve, and this receives motor fibres from at least two lumbar nerves (L4 and 5) and sometimes also L3 and 6. Therefore destruction of the ventral horn of more than one segment of the spinal cord might be required to produce **total flaccid paralysis, total** loss of reflexes (**areflexia**), and **total wasting** of the rectus femoris and vastus muscles together.

However, these clinical signs are not proof of injury to the cell body of the alpha neuron, i.e. a central lesion. Similar signs can also follow (i) injury to the **axon** of the alpha neuron or (ii) a disorder at the neuromuscular junction, i.e. peripheral lesions. The term lower motor neuron disorder covers all parts of the lower motor neuron, and therefore includes not only central injury to the cell body but also peripheral injury to its axon.

10.13 Integration of the two sides of the neuraxis

Integration of the left and right sides of the central nervous system is achieved by the presence of collaterals and interneurons which cross from one side of the neuraxis to the other. In the brain a notable example of such integrating fibres is formed by the many axons which cross from one side of the forebrain to the other in the corpus callosum (see Section 17.13). At the spinal level there are many interneurons which cross the midline and these form the basis of the primitive **walking reflexes**, which are mediated entirely by the spinal cord. A decapitated hen can run, and even a cat can be capable of walking after transection of the spinal cord in the upper part of the neck. The co-ordinated swinging of the arms by a man when walking represents a vestigial primitive locomotory reflex in which the two sides of the body are integrated.

The **white commissure** of the spinal cord (Fig. 12.3) is the site where the axons of these numerous integrating neurons cross over. The white commissure is the zone of white matter which lies in the midline between the ventral fissure of the spinal cord ventrally and the central zone of grey matter dorsally. The white commissure is also the place where many long nerve fibres decussate, such as the ventral spinocerebellar fibres (segment L6, in Fig. 9.1). There is no dorsal white commissure.

Derangement of the white commissure can result in disorganization of the left and right limbs during locomotion. In the dog, a failure of the white commissure to develop in the embryo results in a **'bunny hopping' gait** in which the left and right limbs work simultaneously rather than alternatingly, giving a typical bouncing movement. The forelimbs or hindlimbs are affected, depending on which part of the spinal cord is involved. **Dysraphism** is another cause of anormality of the white commissure in the dog. In this developmental anomaly of the spinal cord, grey matter may occupy the site of the white commissure, or the central canal may be abnormally dilated thus disturbing the white commissure.

Although the axons of many interneurons remain within their own segment of the spinal cord (and co-ordinate events taking place within that segment), there are may other interneurons which send their axons several segments up or down the spinal cord, and then re-enter the grey matter. In this way they form an extensive neuronal network which integrates the activities of neighbouring segments. These axons occupy the region of the white matter which adjoins the grey (Fig. 12.3), where they form a layer of numerous fibres known as the **propriospinal system** (or **fasciculus proprius**). Such fibres occur in all three of the funiculi of the spinal cord, and are therefore known as the dorsal fasciculus proprius, lateral fasciculus proprius, and ventral fasciculus proprius. The propriospinal fibres in the ventral part of the lateral column may be involved in the transmission of both pricking pain (see Section 7.6) and true pain (see Section 9.9).

11. Pyramidal system

The pyramidal system comprises (i) all corticospinal fibres that travel in the medullary pyramid, and (ii) all corticonuclear fibres projecting to the motor nuclei of those cranial nerves that innervate the striated muscles of the head.

11.1 Pyramidal pathways

Neuron 1

The cell location is in the primary motor area (Fig. 11.1, Cerebrum) next to the primary somatic sensory area (Figs. 8.2(a), 8.3).

In man two other smaller and separate motor areas are present. The **supplementary motor area** lies on the medial surface of the cerebral hemisphere, adjacent to the leg and foot area of the primary motor area, the body being horizontal with the head forwards. The **second motor area** lies at the ventral end of the central sulcus (Fig. 8.3), where it is continuous with a **second somatic sensory area**. A corresponding supplementary motor area has been reported in the cat. Yet other areas of the cortex also contribute to the pyramidal system.

The primary motor area of man is arranged somatotopically in a **motor 'homunculus'**. In the human homunculus the body is upside-down and the head is the right way up (Fig. 11.3). The area of the cortex allocated to each part of the body is proportional to the importance of the function of that area. The orientation and bodily representation of the primary somatic sensory area of man, with its sensory homunculus, is closely similar. In the cerebral cortex of the cat the primary motor area and primary somatic sensory area are each represented as an upside-down **'felunculus'** (Fig. 8.2(b)), the area of cortex allocated to each bodily component corresponding, as in man, to the functional importance of each region.

The axon of neuron 1 passes through the internal capsule (Fig. 11.1) into the cerebral crus, somatotopically arranged. The **internal capsule** is a major roadway for axons passing to and from the cerebral cortex, and is vulnerable to damage in 'stroke'. The **cerebral crus** forms the ventral region of the midbrain (Fig. 11.1). It consists of the massive bundles of axons belonging to the pyramidal system, and of the axons of the corticoponto-cerebellar feedback pathway (Fig. 11.2). The pyramidal system contains two groups of axons, namely, (i) corticonuclear axons controlling the striated muscles of the head, and (ii) corticospinal axons controlling the skeletal muscles of the body.

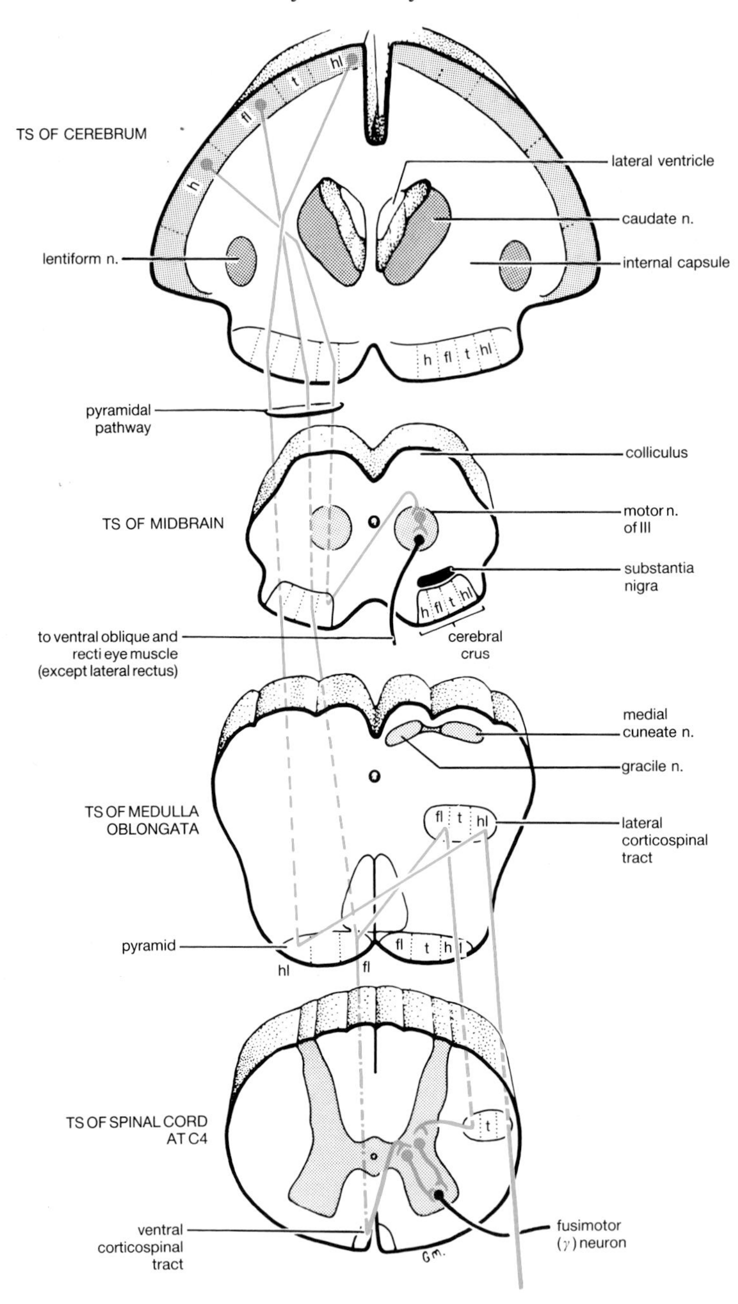

FIG. 11.1. The pyramidal system. All pyramidal pathways (green) contain three neurons. The first neuron is the longest, and crosses over; the second is an interneuron; the third is a fusimotor (gamma) neuron. Somatotopically the hindlimb is always lateral in the corticospinal path, as in the spinothalamic tract (Fig. 7.2) and the medial lemniscus after it has fallen on its face (Fig. 7.3). The dot–dash line, which continues the forelimb pathway into the spinal cord at C4 (at the bottom of the diagram) represents the ventral corticospinal tract. h, head; fl, forelimb; t, trunk; hl, hindlimb; n., nucleus.

(i) The **corticonuclear axons** which control striated muscles of the head peel off and decussate as they pass close to the motor nuclei of all those cranial nerves which innervate **striated** muscle—i.e. III, IV, V, VI, VII, IX, X–XI, and XII (e.g. Fig. 11.1, Midbrain). These nuclei include both the somatic motor and the special visceral motor nuclei. The corticonuclear fibres of the pyramidal system do not reach the pyramid, but it is convenient to regard them as part of the pyramidal system, since they travel in company with the corticospinal axons which do pass through the pyramid. They control voluntary movements of the eyes, jaws, facial musculature, tongue, pharynx and larynx.

(ii) The **corticospinal axons** continue in the **pyramid,** somatotopically arranged. All pyramidal axons decussate sooner or later. Many of them do so in the pyramid (Fig. 11.1, Medulla oblongata), forming the **lateral corticospinal tract**. Those that do not decussate in the pyramid continue as the **ventral corticospinal tract**, but these too decussate just before ending in the spinal cord (Fig. 11.1, Spinal cord). They control the voluntary movements of the limbs, trunk, and, in mammals fortunate enough to have one, the tail. (For the approximate positions of these tracts in the spinal cord see Fig. 12.3.)

Neuron 2

Neuron 2 is a short interneuron. In the spinal cord its cell location is typically in the base of the dorsal horn (Fig. 11.1, Spinal cord), As just stated, in the pathways which innervate the muscles of the head, the cell location of neuron 2 is in the motor nuclei of cranial nerves, for example in the motor nucleus of the oculomotor nerve (Fig. 11.1, Midbrain).

Sometimes this neuron is missing in the pyramidal pathway to the primate hand and foot, and then there are only two neurons in the chain. This reduction enables faster transmission and suggests a trend towards greater specialization.

Neuron 3

Head. The cell location is in the appropriate cranial nerve nuclei, i.e. the motor nuclei of the oculomotor nerve (Fig. 11.1, Midbrain), of the trochlear nerve, of the trigeminal nerve, of the abducent nerve, and of the facial nerve, the nucleus ambiguus, and the motor nucleus of the hypoglossal nerve.

Body. The cell location is in the ventral horn (Fig. 11.1, Spinal cord). Neuron 3 is usually the gamma neuron, but sometimes it can be the alpha neuron.

11.2 Feedback of the pyramidal system

In general, the higher motor centres project to the cerebellum and in return receive 'feedback' projections from the cerebellum. By this means the higher motor centres inform the cerebellum of their **intended** actions, and the cerebellum is then able to regulate these actions as they take place. These regulatory functions of the cerebellum are discussed in Section 15.9.

Corticopontocerebellar path

This pathway is the first half of the feedback circuit; it runs from the cerebral cortex to the cerebellum, and comprises two neurons:

Neuron 1

The primary motor area of the cerebral cortex is the cell location of the first neuron of both the pyramidal and the corticopontocerebellar path (Fig. 11.2, Cerebrum). The axons of neuron 1 of the corticopontocerebellar path pass caudally in the lateral, and medial regions of the **cerebral crus**; they are somatotopically arranged, the axons of the hindlimb and trunk being lateral and those of the forelimb and head being medial (Fig. 11.2, upper two diagrams), like the pyramidal axons lying in the midregion of the cerebral crus (Fig. 11.1).

Neuron 2

The cell location of the seond neuron in the corticopontocerebellar path is in the **pontine nuclei** (Fig. 11.2, Pons). The axon of this neuron nearly always decussates. The decussating fibres form the **pons**, and then continue into the **middle cerebellar peduncle**. The pons is therefore a transverse bundle, which continues directly into the middle cerebellar peduncle (Fig. 7.4).

Return pathway from cerebellum to cerebral cortex

The return component of pyramidal feedback passes from the cerebellum to the cerebral cortex. It consists of three more neurons.

Neuron 3

This is in the cerebellar cortex (Fig. 11.2, lowermost diagram).

Neuron 4

This neuron is in a cerebellar nucleus, for example in the dentate nucleus (Fig. 11.2, lowermost diagram). Its axon decussates in the cerebellum and escapes in the rostral cerebellar peduncle.

Neuron 5

The fifth and final neuron in the pyramidal feedback pathway is in the thalamus (Fig. 11.2, middle diagram), projecting from there to the cerebral cortex; it lies in the ventral group of thalamic nuclei.

Strictly, it lies in the ventrolateral thalamic nucleus (see Section 17.17).

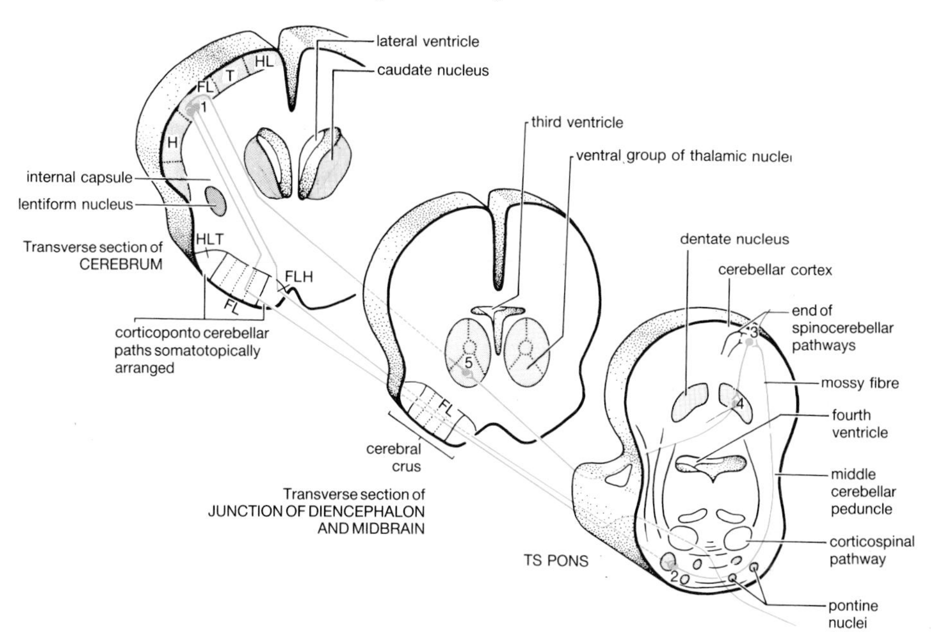

FIG. 11.2. Feedback pathways of the pyramidal system. The outgoing (corticofugal) component of the feedback circuit comprises two neurons. The cell body of the first neuron (No. 1) is in the primary motor area. The second neuron has its cell location in the pontine nuclei (No. 2); its axon decussates to the opposite side of the cerebellar cortex. The return pathway consists of three neurons (Nos. 3, 4, and 5), of which neuron 4 also decussates. The final neuron (No. 5) has its cell body in a ventral thalamic nucleus, and projects to the primary motor area. Thus the **right cerebellar** cortex regulates the **left cerebral** cortex. H, head; FL, forelimb; T, trunk; HL, hindlimb.

From these pathways it follows that the right side of the cerebellum directly regulates the left cerebral cortex. Since the left cerebral cortex initiates the movements of the right side of the body, it follows that the **right** side of the cerebellum indirectly regulates the movements of the **right** side of the body (see Section 15.11). Damage to one side of the cerebellum therefore affects the movements of the **ipsilateral** side of the body (see Section 15.13).

11.3 Species variations in the primary motor area of the cerebral cortex

In mammals in general, the primary motor area is in the rostral region of the cerebral hemisphere. In subprimate mammals the primary motor area and the primary somatic sensory area are relatively extensive in comparison with the rest of the cerebral cortex (Fig. 8.2(a)); in man they occupy a

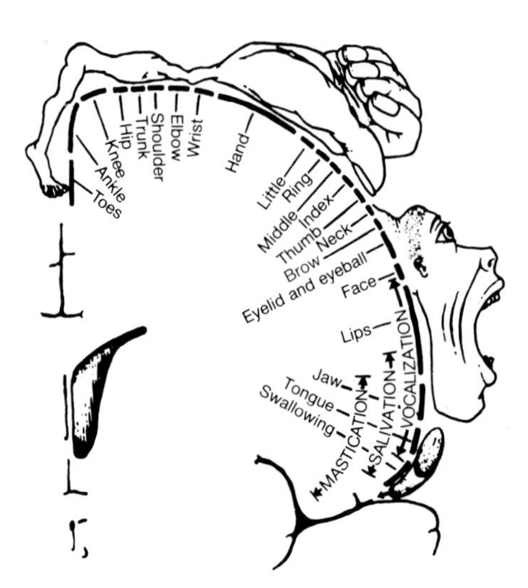

FIG. 11.3. **Penfield's motor homunculus.** This delightful manikin represents the several parts of the head and body in varying sizes according to the extent of their motor representation in the primary motor area of man, and hence indicates their functional importance. The relatively great areas allocated to the tongue and the prehensile thumb reflect the significance in hominid evolution of speech, and of the capacity to hold tools and weapons. (From the *Cerebral cortex of man* (1950) by Wilden Penfield and Theodore Rasmussen, by courtesy of the MacMillan Publishing Company and the curator of the Penfield Papers.)

relatively small area (Fig. 8.3), because the association areas are much more extensively developed (see Section 17.6).

11.4 Species variations in the pyramidal system

Only mammals possess the pyramidal system. It is absent in birds, reptiles, amphibia, and fish (see Section 18.11). Among the mammalian orders there is great variability in the pyramidal spinal pathways, and this is probably due to the relatively recent phylogenetic origin of the system.

Phylogenetically ancient pathways such as the vestibulospinal tract are relatively constant in their anatomy.

Primates and carnivores

The pyramidal pathway is best developed in primates and carnivores, its general arrangement in these two groups being essentially the same, except that carnivores tend to have an even less substantial ventral tract that that of man. The decussation at the pyramid is 100 per cent in dogs, so that there is no ventral corticospinal tract at all; in the cat there is a small ventral tract, but it reaches only the upper segments of the neck.

In man about 85 per cent of the axons of the pyramidal system decussate in the pyramid; the 15 per cent that go straight on in the ventral corticospinal tract end in the first thoracic segments. Thus in both primates and carnivores the ventral tract is of relatively little importance. The corticospinal tracts form about 10 per cent of the total white matter in the spinal cord of the dog, about 20 per cent in monkeys, and about 30 per cent in man.

Ungulates

In the other domestic mammals, that is the horse, ox, sheep, and pig (i.e. the hoofed animals or ungulates), the entire pyramidal system is small and ends anatomically in the cervical region. In these species the fibres apparently decussate in the pyramid, but then mainly descend as a ventral corticospinal tract alongside the ventral fissure. Injury to the ventral aspect of the cervical spinal cord can involve the ventral corticospinal tract and this may then contribute to the signs arising from the lesion.

There is a small lateral corticospinal tract in ungulates. Usually there is also a very small **dorsal corticospinal tract** in the dorsal funiculus (Fig. 12.3), which only goes as far as about C5. All of these pyramidal fibres eventually cross over. In the sheep and goat, there is physiological evidence that the pyramidal system is continued functionally down the cord by non-specific neurons. In most **rodents** the majority of the pyramidal fibres travel in the dorsal funiculus.

Pyramidal feedback pathways

These are essentially the same in man and the domestic mammals.

11.5 The function of the pyramidal system

It has long been supposed that the pyramidal system is responsible for skilled voluntary movements, these being superimposed on involuntary postural and locomotory activities mediated by the extrapyramidal system. It now seems likely, however, that other tracts must also be involved in skilled voluntary movements, notably the cortico–rubrospinal pathway.

11.6 Effects of lesions in the pyramidal system

Lesions of this type may be included among **upper motor neuron lesions** (see Section 13.6).

As would be expected, **primates** show severe motor disability after destruction of the primary motor area of the cortex (the motor cortex). Such destruction may be caused in man by haemorrhage, thrombosis, or an embolism, as in a **'stroke'**. **Flaccid paralysis** of the contralateral side of the body immediately follows a stroke, with total loss of muscle tone (**atonus**) and abolition of all reflexes on the affected side. After a period lasting a few hours or days, reflexes return and become exaggerated (**hyperreflexia**);

muscle tone also returns, with spasticity (severe **hypertonus**). Strokes more often arise from lesions in the **internal capsule** than from lesions in the motor cortex, and in the former case involve many motor and sensory pathways besides the pyramidal system itself. Stroke is evidently one of the most common neurological syndromes encountered by medical practitioners.

Lesions of the **supplementary motor area** result in contralateral hyperreflexia and hypertonia with spasticity.

In the **domestic mammals** ischaemic lesions of the motor cortex do occur but are uncommon, and therefore genuine 'strokes' are seldom seen in veterinary practice. However, it is known that in the dog there is relatively little disturbance soon after experimental destruction of the motor cortex. The animal can walk, choose one food from another and eat it, and respond to stimulation by growling and barking. However, there is a greater or lesser degree of paresis, and there is also **hypertonus** with **spasticity** or **rigidity** and **hyperreflexia**, these effects being mainly contralateral. The term **paresis** means partial paralysis; it is generally regarded as a moderate motor deficiency, resulting in a locomotor deficit.

The commonest cause of **stroke** in man is ischaemia in the internal capsule, caused by haemorrhage, thrombosis, or embolism in the territory supplied by the middle cerebral artery. The most obvious clinical signs are those arising from lesions in corticofugal fibres, especially those of the pyramidal system. However, many other types of corticofugal fibres may be involved, including corticorubral fibres to the red nucleus and corticoreticular fibres to the reticular formation. Also involvement of corticopontine fibres among the corticofugal fibres may manifest itself by cerebellar disturbance. Moreover, the presence of numerous thalamocortical (corticopetal) fibres means that sensory deficits quite often occur in strokes, although they not infrequently remain unnoticed.

Since the motor cortex is such an immensely complex network of neurons, receiving and giving many projections, attempts have been made, by means of **experimental lesions of the pyramid**, to study the effects of eliminating only the pyramidal pathway itself. Hence the medullary pyramid has been experimentally transected in monkeys. This results in longlasting hypotonia and no hyperreflexia (thus differing from naturally occurring lesions in the motor cortex or internal capsule in man); there are initial deficits in limb movements and a permanent loss of skilled movements of the hand and fingers. Not much information has been gained from experimental transection of the pyramid in the cat, which has yielded only small and mainly transient deficits in movement, and impairment of some postural responses such as the tactile placing reaction. It has been pointed out that care is needed in interpreting the results of apparently conclusive experiments such as transection of the medullary pyramid. In the acute phase of the experiment, haemorrhage, oedema, or vascular spasm may affect adjacent pathways or centres. In the chronic stages, recovery processes may conceal defects created by the experiment.

11.7 Validity of the distinction between pyramidal and extrapyramidal systems

As knowledge advances it becomes apparent that the distinction between the pyramidal and extrapyramidal systems, which was based on man, has largely outlived its strict validity both for mammals in general and for man in particular. No doubt it will eventually be necessary to abandon it altogether and accept the two components as parts of one integrated motor system. However, the distinction is generally retained for descriptive convenience, in both neuroanatomy and clinical neurology.

12. Extrapyramidal system

This system is open to several different definitions. Here it means all of the descending somatic motor pathways, **except** the corticospinal fibres that pass through the medullary pyramid and except also for the corticonuclear fibres which project to the motor nuclei of the cranial nerves. In contrast to the pyramidal system, the extrapyramidal system is phylogenetically primitive and represented in all but the lowest vertebrates. It consists essentially of (i) a series of motor 'command centres', which are either facilitatory or inhibitory, (ii) spinal pathways, and (iii) feedback circuits.

Motor centres

The extrapyramidal system can be regarded schematically as containing **nine** main motor 'command centres', situated at three levels of the brain, namely in the forebrain, midbrain, and hindbrain (Fig. 12.1). They are:

Forebrain: 1. cerebral cortex; 2. basal nuclei.
Midbrain: 3. midbrain descending reticular formation; 4. red nucleus; 5. tectum.
Hindbrain: 6. pontine motor reticular centres; 7. lateral medullary motor

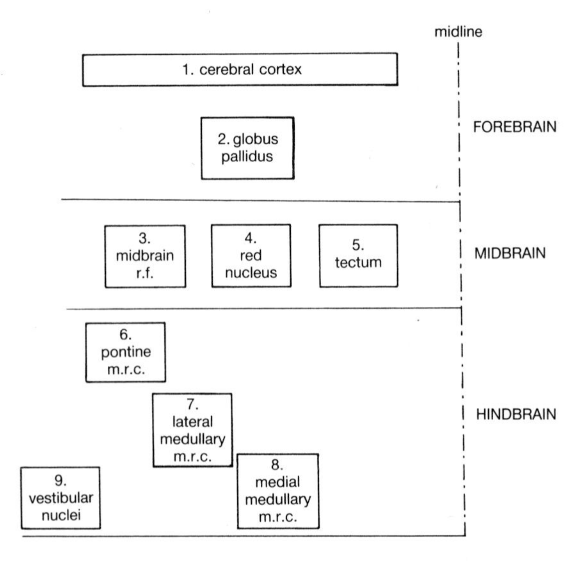

FIG. 12.1. Diagram of the nine motor command centres of the extrapyramidal system on the left side. r.f. reticular formation; m.r.c., motor reticular centre. The globus pallidus is the focal point of the basal nuclei (ganglia), and in this diagram is used to represent the basal nuclei.

reticular centres; 8. medial medullary motor reticular centres; 9. vestibular nuclei.

This is a very substantial simplification of the extrapyramidal centres, but it illustrates the principles.

12.1 The cerebral cortex

Extrapyramidal pathways arise from almost the whole of the cortex. These cortical areas control the **descending reticular formation**, mainly but not exclusively by inhibition. They also drive the **red nucleus**, presumably being mainly facilitatory.

12.2 Basal nuclei and corpus striatum

There is no agreement about the definition of either of these terms, but in this book they are regarded as synonymous. They include several telencephalic grey areas, of which two are of special importance, namely the **caudate nucleus** and **lentiform nucleus** (Figs. 11.1, 21.16). The anatomy of these and other components of the basal nuclei is discussed in Section 21.29, **Grey Matter**. The basal nuclei have long been known as the basal 'ganglia'.

The **globus pallidus**, which is part of the lentiform nucleus, is a major focal point of the basal nuclei; it receives converging fibres from all the other basal nuclei, and is the only component of the basal nuclei which sends fibre projections outside the basal nuclei themselves. In Figs. 12.1 and 12.2 it represents the basal nuclei as a whole.

These structures, notably the globus pallidus, may be regarded as the top of the **descending reticular formation**. They exert a mainly facilitatory influence on the descending reticular formation (Figs. 12.1, 12.2), which is discussed immediately below under the headings midbrain reticular formation, pontine motor reticular centres, lateral medullary motor reticular centres, medial medullary motor reticular centres, and reticulo-spinal tracts.

For a long time it has been widely believed that these descending projections of the reticular formation towards the spinal cord are the main outlet of the basal nuclei, but recent work has shown that most of the efferent axons of the basal nuclei project in fact to the thalamus. It is now becoming apparent that the basal nuclei are mainly engaged in collaborating with the cerebral cortex via the thalamus (through relays in the ventral group of thalamic nuclei, see Fig. 13.1 and Section 17.17); thus their chief effects on motor functions are somewhat indirect, being exerted initially on the thalamus and thence on the cerebral cortex. In short, **most of the activity of the basal nuclei is applied to the cerebral cortex and not to the**

lower centres. It is still far from clear how these mechanisms work, but they seem to bring about the complicated automatic actions which an animal performs every day of its life whilst changing its posture, walking or running, feeding, and defending itself.

In the living human animal the activity of the basal nuclei seems to express itself as initiating and orchestrating the vast array of subtle motor activities which actually achieve our complicated patterns of movement on level ground, up- or downhill, over or through obstacles, and at varying speeds during everyday life. It appears as though the basal nuclei are programmed to carry out these functions automatically, and can perform them accurately without requiring conscious directions from the cerebral cortex. Presumably any such 'programming' develops as a child learns first to crawl, then to walk, and finally to move with precision over varying terrains and at varying speeds. Possibly a similar programming occurs during the development of the immature carnivore, which has gradually to learn the movement patterns of hunting and defence. On the other hand the young of Equidae and other fleet-footed herbivores run with their mother within few hours of birth and rely on this to escape predators. Presumably their basal nuclei are fully programmed genetically.

12.3 Midbrain reticular formation

Much of the central region of the midbrain is formed by this component (Fig. 21.13), which is also known as the mesencephalic reticular formation.

12.4 Red nucleus

The red nucleus lies under the colliculi, buried in the midbrain reticular formation (Fig. 21.13).

12.5 Mesencephalic tectum

The rostral and caudal **colliculi** constitute the mesencephalic tectum. The rostral colliculus is joined to the lateral geniculate body by its arm or brachium; a similar brachium joins the caudal colliculus to the medial geniculate body (Fig. 7.4). (The geniculate bodies belong to the diencephalon, and not to the tectum.)

FIG. 12.2. **Diagram of the descending projections of the extrapyramidal system.** The diagram shows facilitatory (green) and inhibitory projections (black). Some of these projections pass between the motor centres (1 to 9) of the extrapyramidal system. Others project from the brainstem motor centres down the spinal cord, upon interneurons in the ventral horn. Each final interneuron receives converging projections from above, as shown in principle in the diagram. The diagram seems to suggest that, in the spinal cord, the inhibitory pathways are heavily outnumbered by the facilitatory, but probably the reverse is the case; the

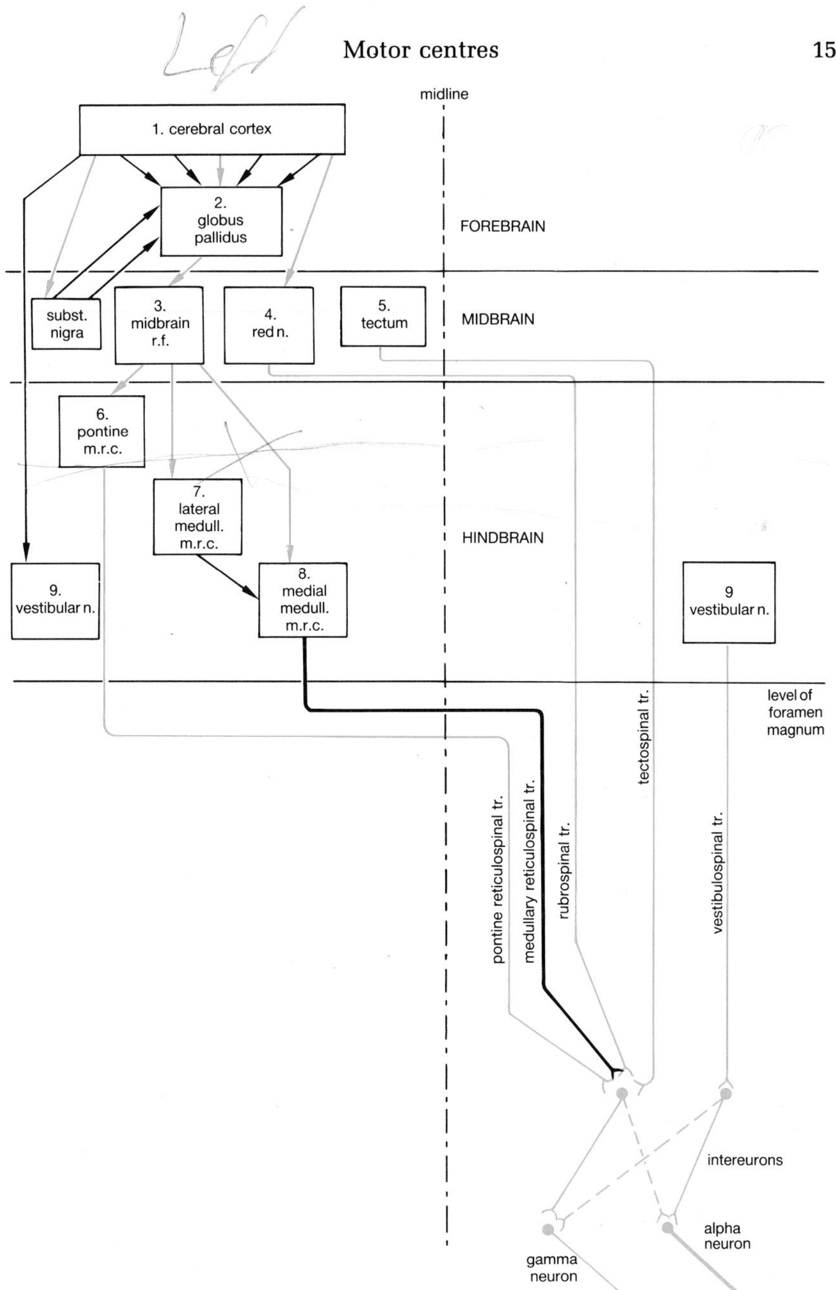

large medullary reticulospinal tract from the medial medullary motor reticular centres is extensively inhibitory. The interneurons (at the bottom of the diagram) project on to either a gamma or an alpha neuron, the majority being indicated by continuous lines, and the minority by broken lines. medull., medullary; m.r.c., motor reticular centre; n., nucleus; r.f. reticular formation; subst. nigra, substantia nigra; tr., tract. The globus pallidus represents the basal nuclei (ganglia). Its main projections (not shown) are to the thalamus (see Fig. 13.1).

12.6 Pontine motor reticular centres

These centres exert a mainly facilitatory drive on fusimotor (gamma) neurons of the spinal cord, via the **pontine reticulospinal tract** (Fig. 12.2).

12.7 Lateral medullary motor reticular centres

These have no direct pathway to the reticulospinal tracts, but work indirectly through the medial medullary motor reticular centres. They inhibit the medial medullary motor reticular centres (Fig. 12.2), and are therefore **facilitatory** by means of 'disinhibition' (see Section 5.13).

12.8 Medial medullary motor reticular centres

The medial medullary motor reticular centres project directly down the spinal cord via the medullary reticulospinal tract, exerting a massive **inhibitory** influence mainly on gamma (fusimotor) neurons throughout the length of the spinal cord (Fig. 12.2).

12.9 Vestibular nuclei

The vestibular nuclei are not always considered to be part of the extra-pyramidal system, but it is simpler to include them. They are strongly **facilitatory**, via the vestibulospinal pathways, mainly upon extensor alpha (skeletomotor) neurons of the spinal cord (Fig. 12.2, on the right side of the diagram).

Spinal pathways

There are five extrapyramidal spinal pathways, namely the pontine and medullary reticulospinal tracts, the rubrospinal tract, the vestibulospinal tract, and the tectospinal tract. They all have essentially three neurons, neuron 1 being in one of the motor command centres, neuron 2 being a short intercalated neuron in the ventral horn, and neuron 3 being mainly the gamma neuron but sometimes also the alpha neuron in the ventral horn (Fig. 12.2).

12.10 Pontine and medullary reticulospinal tracts

The pontine reticulospinal tract arises from the pontine motor reticular centres (Fig. 12.2). The medullary reticulospinal tract arises from the medial medullary motor reticular centres, and is strongly inhibitory on gamma neurons in the ventral horn of the spinal cord (Fig. 12.2).

The reticulospinal tracts have projections from one side of the midline to the other and therefore can be regarded as a midline system. (Fig. 12.2 shows an initial decussation of the reticulospinal tracts, arising from the pontine and medial medullary reticular motor centres; the subsequent projections from side to side have been omitted.)

12.11 Rubrospinal tract

The rubrospinal tract arises from the red nucleus. It decussates in the midbrain (Fig. 12.2). It is quite well developed in the cat and in man, in both of which it controls semiskilled movements. It is even better developed in the lower domestic animals, being important in postural control and locomotion. It is somatotopic, the hind limbs being **lateral** (Fig. 12.3), as in the pyramidal tract and in all ascending tracts except the cuneate and gracile fascicles.

12.12 Vestibulospinal tract

This pathway arises from neurons in the vestibular nuclei (Fig. 12.2). It does **not** decussate. It reaches the end of the spinal cord, and projects mainly to **extensor skeletomotor** neurons throughout the length of the cord. This pathway is strongly facilitatory, being normally inhibited by the cerebrum and cerebellum. In the absence of the cerebrum and cerebellum, it produces strong **decerebrate rigidity**. For further details of the pathways of balance, including the effects of disease, see Section 8.3.

There are two vestibulospinal pathways. The vestibulospinal tract (which has just been described) is much the larger and extends throughout the whole length of the spinal cord. It arises from the **lateral vestibular nucleus**. Fibres from the **medial vestibular nucleus** also descend the spinal cord, but in the vestibulospinal part of the **medial longitudinal fasciculus**, which lies in the most dorsal part of the ventral funiculus (Fig. 12.3); these fibres end in the cranial part of the thoracic spinal cord, and control cervical motoneurons thus regulating the position of the head and hence participating in maintaining equilibrium. Some authors call the vestibulospinal tract the lateral vestibulospinal tract, and the smaller pathway the medial vestibulospinal tract.

12.13 Tectospinal tract

This tract completely decussates in the midbrain (Fig. 12.2). It has its first neuron in the rostral and caudal **colliculi**. These neurons receive projections from the second neuron in the visual and auditory paths, respectively (see Figs. 8.4, and 8.5). They then project to interneurons, and then mainly to fusimotor neurons, in the **cervical part** of the spinal cord. Hence the tectospinal tract controls reflex movements of the head and neck in response to visual or auditory stimuli.

12.14 The position in the spinal cord of the tracts of the extrapyramidal system

The main area lies in the ventral funiculus (Fig. 12.3), and contains the vestibulospinal tract and medial longitudinal fasciculus, the pontine reticulospinal tract, and the tectospinal tract. The lateral funiculus contains the rubrospinal and medullary reticulospinal tracts (Fig. 12.3).

As already stated (see Section 7.11), the ascending and descending tracts of the spinal cord are not discrete, but are more or less intermingled with each other.

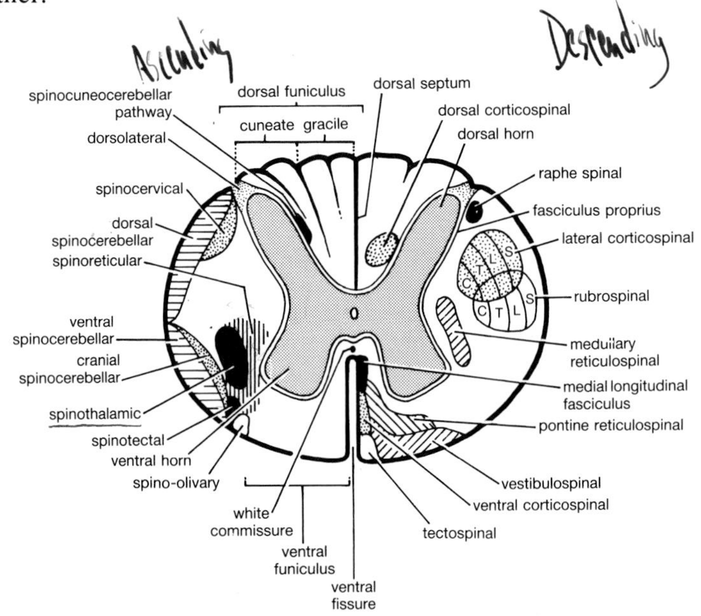

FIG. 12.3. Diagrammatic transverse section of the cervical spinal cord of a hypothetical domestic mammal to show the spinal tracts. In reality the various tracts are not distinctly separated, but mingled together. Left side of the diagram, ascending tracts; right side, descending tracts. C, cervical; T, thoracic; L, lumbar; S, sacral.

12.15 Summary of the tracts of the extrapyramidal system

(i) There are essentially three neurons in each pathway. These resemble the three neurons of the corticospinal tracts, the second neuron being an interneuron.

(ii) The rubrospinal and tectospinal tracts decussate: the vestibulospinal tract does **not** decussate: The medullary and pontine reticulospinal tracts are best regarded as a **midline** system in which decussation is not meaningful.

(iii) The rubrospinal, tectospinal, and reticulospinal tracts project mainly to gamma (fusimotor) neurons: the vestibulospinal tract projects mainly to alpha (skeletomotor) neurons.

(iv) In general, the gamma (fusimotor) neuron receives more extrapyramidal projections than the alpha (skeletomotor) neuron. The gamma neuron receives more pyramidal projections also. For these reasons, the gamma neuron is the principal link in the motor pathways. Nevertheless, the skeletomotor neuron remains the **final common path**.

(v) Inhibitory pathways are very numerous in the extrapyramidal system, and tend to dominate the facilitatory (excitatory) pathways. The main inhibitory component of the extrapyramidal system in the spinal cord is the medullary reticulospinal tract; through this the extrapyramidal system exerts a substantial damping influence on the activity of the gamma neurons, and thence, indirectly, on the alpha neurons of the spinal cord. On the other hand, the important vestibulospinal tract is strongly facilitatory, and presumably the rubrospinal and the less important tectospinal tracts are also essentially facilitatory.

13. Extrapyramidal feedback: extrapyramidal disease and upper motor neuron disorders

Feedback of the extrapyramidal system

13.1 Neuronal centres of the feedback circuits

Three major centres are involved in extrapyramidal feedback, namely the olivary nucleus, the cerebellum, and the thalamus.

Olivary nucleus

This is an intermediate station on the pathway from the higher extra-pyramidal command centres to the cerebellum (Fig. 13.1). (It corresponds to the pontine nuclei in the pyramidal feedback circuit, Fig. 11.2.) The site of the olivary nucleus in the medulla oblongata is shown schematically in Fig. 7.3 (see also Fig. 21.7).

Cerebellum

Two components of the cerebellum, i.e. the cerebellar cortex and the cerebellar nuclei (such as the **dentate nucleus**), are involved in extra-pyramidal feedback circuits. Note the similarity to the pyramidal feedback circuit (Fig. 11.2).

Thalamus

This is the intermediate station between the cerebellum and the cerebral cortex, in those feedback circuits which return to the forebrain. The **ventral group** of thalamic nuclei controls all thalamic feedback circuits.

Strictly, it is the **ventrolateral** thalamic nucleus, as in the pyramidal feedback circuits (Fig. 11.2) (see Section 17.17).

13.2 Feedback circuits

All of the nine motor 'command centres' of the extrapyramidal system have feedback circuits through the cerebellum. As for the feedback circuits of the pyramidal system (see Section 11.2), the function of these extrapyramidal feedback pathways is to inform the cerebellum of the intended motor actions of the extrapyramidal motor centres and enable the cerebellum to regulate these actions as they progress (see Section 15.9).

One of the major motor centres of the extrapyramidal system, the basal nuclei (basal ganglia), has a shorter feedback circuit directly through the thalamus to the cerebral cortex. This feedback circuit enables the basal nuclei to carry out their main role, namely collaboration with the cerebral cortex via the thalamus (see Section 12.2).

Feedback between extrapyramidal command centres and cerebellum

The cerebellar feedback circuits of the nine extrapyramidal motor command centres (Nos. 1 to 9) are as follows:

Centres 1 and 2: the cerebral cortex and **globus pallidus**. The feedback pathways to these two command centres are similar, projecting (in sequence) to the olivary nucleus, to the cerebellar cortex, to a cerebellar nucleus, to the thalamus, and finally back to the cerebral cortex (Fig. 13.1). These pathways decussate (either before or after the olivary nucleus) on their way to the cerebellar cortex, and then decussate again on the way back. The circuit for the globus pallidus is finally completed by projections from the cerebral cortex back to the globus pallidus (Fig. 13.1).

These pathways enter the cerebellum through the caudal cerebellar peduncle, and exit through the rostral peduncle (see Sections 15.3, 15.5). The relay in the thalamus takes place in the ventral group of thalamic nuclei (in the ventrolateral nucleus, see Section 17.17).

Centres 3 and 4: the midbrain reticular formation and **red nucleus**. Again these feedback circuits project first to the olivary nucleus, then to the cerebellar cortex, onwards to a cerebellar nucleus, and finally back to the midbrain reticular formation and red nucleus (Fig. 13.1). As before, they also decussate twice, i.e. on the way in and the way out of the cerebellum.

Entrance to the cerebellum is gained through the caudal peduncle, and the outlet is via the rostral cerebellar peduncle, as for centres 1 and 2.

Centres 5 and 9: the tectum and **vestibular nuclei**. These are the extrapyramidal motor centres that are associated with information from the special senses. Their feedback pathways miss out the olivary nucleus, projecting via the caudal cerebellar peduncle directly to the cerebellar cortex. They receive return pathways from a cerebellar nucleus (Fig. 13.1). The projections of the vestibular nuclei to and from the cerebellum are

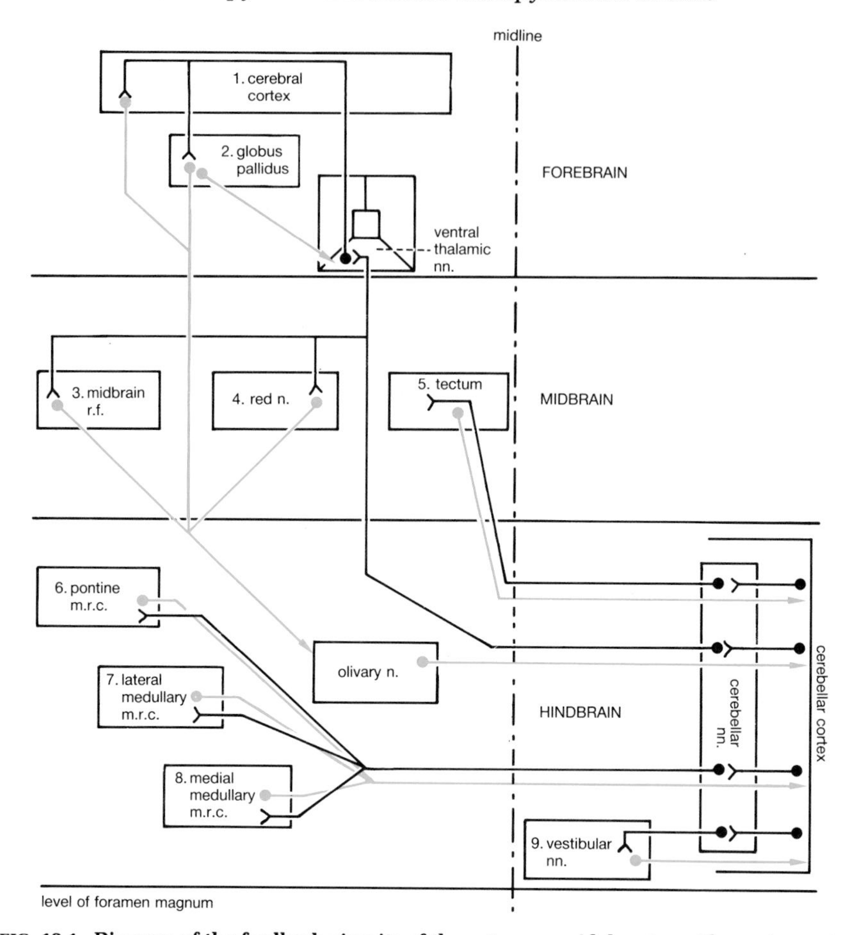

FIG. 13.1. Diagram of the feedback circuits of the extrapyramidal system. The motor centres are numbered 1 to 9. The green and black projections are the feedback circuits: the green projections lead **to** the cerebellum, and the black lines return **from** the cerebellum. The globus pallidus is the focal point of, and in this diagram represents, the basal nuclei. The globus pallidus has a feedback circuit through the thalamus and cerebral cortex, which enables the basal nuclei to collaborate with the cerebral cortex. n., nucleus; nn., nuclei; m.r.c., motor reticular centre; r.f., reticular formation.

ipsilateral, and this is unique among the feedback pathways of the extrapyramidal centres; i.e. the left vestibular nuclei project to, and receive return projections from, the left side of the cerebellum. Correlate this with the fact that the vestibulospinal tract does **not** decussate (see Section 8.3).

The tectal pathways to and from the cerebellum pass through the rostral cerebellar peduncle; the afferent and efferent vestibular pathways use the caudal cerebellar peduncle (see Sections 15.3, 15.5).

Centres 6, 7, & 8: the pontine motor reticular centres the **lateral**

medullary motor reticular centres, and the **medial medullary motor reticular centres**. The command centres in the reticular formation of the hindbrain project to the contralateral cerebellar cortex and receive return pathways via a cerebellar nucleus (Fig. 13.1).

All of these pathways from the hindbrain command centres of the reticular formation to the cerebellum travel in the caudal cerebellar peduncle (see Section 15.3).

Feedback between basal nuclei and cerebral cortex

This feedback circuit runs from the globus pallidus, to the ventral group of thalamic nuclei, to the cerebral cortex, and finally back to the globus pallidus (Fig. 13.1).

Of the ventral group of thalamic nuclei, the ventrolateral thalamic nucleus is mainly involved in this feedback circuit.

Extrapyramidal disease

13.3 Balance between inhibitory and facilitatory centres

As indicated at the beginning of Chapter 12, some of the motor centres of the extrapyramidal system are facilitatory and others are inhibitory. The normal functioning of the system depends on a perfect balance between these two antagonistic components.

Facilitatory components

These are shown in Fig. 12.2 (green), the most important being: (i) the **cerebral cortex** itself (relatively restricted parts of it); (ii) large areas of the **basal nuclei**, especially the globus pallidus; (iii) the **red nucleus, tectum**, and the **reticular formation** of the midbrain; (iv) the **pontine motor reticular centres**; (v) the **lateral medullary motor reticular centres** of the medulla, which are strictly inhibitory but produce facilitation by disinhibition of the medial medullary motor reticular centres (see Section 12.7); and (vi) the **vestibular nuclei**.

Inhibitory components

The main inhibitory components of the extrapyramidal system are shown in Fig. 12.2 (black projections). (i) On the whole, the influence of the **cerebral cortex** on the lower extrapyramidal centres appears to be inhibitory rather than facilitatory. (ii) The **substantia nigra** is a dense area of grey matter in the cerebral peduncle of the midbrain (Figs. 11.1, 21.13). It receives projections from the cerebral cortex, and in turn projects inhibitory pathways to the basal nuclei (Fig. 12.2). Thus it normally damps down the activity of the basal nuclei. This is an important function, controlling the strong facilitatory drive of the basal nuclei, via the globus pallidus, upon the lower motor centres of the brainstem, and also controlling the influence of the basal nuclei on the cerebral cortex (see Section 12.2). (iii) The **medial medullary motor reticular centres** exert a massive inhibitory drive upon the descending reticular formation of the spinal cord, by means of the medullary reticulospinal tract (see Section 12.8).

13.4 Clinical signs of lesions in extrapyramidal motor centres in man

General principles

Lesions of the extrapyramidal motor centres, notably in the basal nuclei, tend mainly to knock out inhibitory components: too much facilitation results, causing increased muscle tone. Much of this **hypertonus** seems to arise from the increased firing of gamma neurons, once the inhibitory control from above has been lifted. There are postural and locomotory deficits, which typically include involuntary movements, the latter being known as **hyperkinesia**. Usually spinal reflexes are exaggerated (**hyperreflexia**). When the lesions are unilateral, the signs tend to be contralateral, though they may sometimes be bilateral.

If the lesions happen to affect mainly excitatory components of the extrapyramidal motor centres then muscle tone will be reduced, and this does happen in some clinical cases.

Lesions in the basal nuclei

In man the single most characteristic sign of lesions of the basal nuclei is the presence of **hyperkinesia**. This consists of involuntary movements ranging from continuous tremor to sudden jerking movements, and may even extend to uncontrollable flailing movements of the arm and leg on the opposite side to the lesion (**hemiballismus**).

Tremor is a particular feature of lesions of the **substantia nigra** (see below, Parkinson's disease). The much more violent hyperkinesia of hemiballismus is usually associated with a vascular lesion of the contralateral **subthalamic nucleus** (see Section 21.23, **Subthalamus**).

Hypertonus, hyperreflexia, and hyperkinesia seem to be among the relatively simple expressions of lesions in the human basal nuclei. Much more elaborate manifestations may reveal themselves. For example during walking there may be gradual acceleration, so that what starts as a somewhat tottery gait turns into uncontrollable running and ends in crashing over. Sometimes a patient appears to be brought to a total standstill when confronted by a doorway, or when half-way up the stairs.

Parkinson's disease

This is a particularly well-known example of extrapyramidal disease. It occurs in man but has no close parallel in domestic animals. The balance between facilitatory and inhibitory areas is disturbed, the facilitatory components dominating. Increased fusimotor neuron activity, and hence increased muscle tone, sometimes amounting to rigidity, is consequently the main sign. There is also a fine tremor of the hands, abolished by sleep. Usually the lesion is mainly in the **substantia nigra**.

The mechanisms by which the substantia nigra controls motor functions are not known. However, it has been established that the cells of the substantia nigra are dopaminergic, and that the signs of Parkinson's disease can be relieved by L-dopa. Although nigrostriatal projections appear to be the largest **dopaminergic pathway** in the central nervous system, there are others extending from the substantia nigra to the midbrain reticular formation, and these may account for some of the undesirable side-effects of L-dopa.

13.5 Clinical signs of lesions in the basal nuclei in domestic animals

It seems likely that the clinical signs arising from lesions in the basal nuclei in domestic animals would, in principle, resemble those of man, i.e. hypertonus, hyperreflexia, locomotory and postural deficits, and hyperkinesia. Experimental lesions do sometimes produce hyperkinesia. However, there is not much firm information about the clinical effects of naturally occuring lesions in the domestic species.

Lesions in the basal nuclei are evidently quite common in the dog, but it seems very difficult if not impossible at present to relate them to specific neurological signs. Attempts to produce specific signs in experimental animals have also been rather unsuccessful. However, small unilateral lesions in the **caudate nucleus** in cats have been followed by continuous exaggerated movements of the limbs in the form of alternating flexion and extension of the paws of the forelimb (resembling the 'knitting' movements of affection); these movements disappeared with sleep and during locomotion. Although the lesions were unilateral, the hyperkinesia affected both forelimbs, possibly because of the midline pathways of the reticular formation which descend from the basal nuclei.

The clinical signs of **spastic paresis in the Friesian bull** may be partly due to the lesions which are quite consistently scattered throughout the higher centres of the extrapyramidal system, including the basal nuclei and red nucleus (as well as being found in various other parts of the CNS). The characteristic clinical signs include rigidity, or hyperkinesia, the hindlimb being thrown backwards when movement is attempted. The immediate cause is hypertonus of the gastrocnemius and superficial digital flexor muscles, and in some cases of the quadriceps femoris muscles.

The hyperkinesia which characterizes **stringhalt** and **shivering** in horses is suggestive of basal nuclei lesions. Ingestion of plants of the genus Centaurea by horses results in bilateral, sharply circumscribed, necrosis of the substantia nigra or globus pallidus, or both, with facial rigidity and rhythmic tongue and jaw movements but not much involvement of the limbs; the hypertonia of the facial muscles resembles that of Parkinson's disease in man.

13.6 Upper motor neuron disorders

The term **upper motor neuron** is widely used in both human and veterinary clinical neurology. It is sometimes confined to neurons of the pyramidal tract, but is often extended to include all the pyramidal and extrapyramidal pathways which control voluntary activity. Thus there are a great many possible 'upper motor neurons', and the term is so ill-defined as to be a source of considerable confusion; indeed it has been suggested that it 'might with advantage be avoided'. Nevertheless, the term is strongly entrenched in the clinical literature in the concept of 'upper motor neuron disorder'.

In veterinary neurology, **lesions of the spinal cord**, which typically involve many of the descending motor pathways, are considered to give classical signs of upper motor neuron disorder. These are: **paralysis**, or in less severe cases **paresis**, on the side of the lesion, gait and posture being obviously or drastically affected; **hypertonus** of the affected limb(s), often severe enough to be termed **spasticity**; **hyperreflexia** of the affected limbs(s); some **muscle wasting** in long-term cases (due to atrophy of disuse, such as occurs in a limb immobilized in a plaster cast). In contrast, the classical signs of **lower motor neuron disease** are: paralysis; hypotonus, often amounting to flaccidity; hyporeflexia or areflexia; and severe and relatively rapid muscle wasting (see Section 10.12).

Hypertonus is commonly encountered in upper motor neuron disorders. Indeed, in long-term, naturally occurring, disorders of the higher motor centres or their tracts, in both man and domestic mammals, any changes in tone are nearly always in the direction of hypertonus. There are a few exceptions, an example being the hypotonus of **spinal shock** (see immediately below); however, this phenomenon is essentially transient even in man and is generally too short-lasting in domestic animals to be observed clinically.

The mechanism of hypertonus in upper motor neuron disorders is not entirely understood. However, an essential general feature appears to be a **release** of fusimotor neurons from inhibition. Thus the corticospinal, rubrospinal, and reticulospinal tracts evidently include fibres which normally have a tonic inhibitory effect on fusimotor neurons. When this descending inhibition is removed, fusimotor neurons become active and reflexly induce a continuous partial contraction of the extrafusal muscle fibres in, for example, the muscles of the limbs. This is perceived by the clinician as an increased resistance to the manual movement of the joints, i.e. as increased tone.

In severe lesions of the spinal cord, widespread destruction of the white matter of the lateral and ventral funiculi on one side of the cervical spinal cord causes total paralysis of the ipsilateral fore- and hindlimbs (**hemiplegia**). If both sides of the spinal cord are damaged all four limbs may be totally paralysed (**tetraplegia**). Comparable lesions caudal to the second thoracic segment of the spinal cord affect the hindlimbs only, and can give paralysis of one hindlimb (**monoplegia**), or both hindlimbs (**paraplegia**). Less severe lesions give rise to paresis rather than paralysis (i.e. **hemiparesis, tetraparesis, monoparesis,** or **paraparesis**).

Spinal shock occurs in mammalian species immediately after complete transection of the spinal cord. Caudal to the lesion there is flaccid paralysis, total disappearance of muscle tone (atonus), and complete loss of reflexes (areflexia), and there is also atony of the bladder and rectum with retention of urine and faeces. In man this phase gradually converts during the following weeks into a stage of reorganization, in which the spinal cord below the lesion slowly resumes its reflex functions. In lower animals this reorganization occurs far more rapidly, so that reflexes return within an hour in the dog and within minutes in the chicken and frog. Evidently the lower the phylogenetic status of the animal the greater the reflex independence of its spinal cord. In veterinary practice, the signs of spinal shock have usually disappeared before the case is examined clinically.

The **Schiff–Sherrington phenomenon** is known to occur only in the dog. Sudden complete transection or compression of the spinal cord in the thoracolumbar region (such as may occur when a dog is run over) is followed immediately by the complete flaccid paralysis and areflexia, caudal to the lesion, which typify spinal shock. In addition, however, there is severe and relatively longlasting **hypertonus of the forelimbs**, with extensor rigidity. This disappears usually within two weeks of the injury. Clinically, this involvement of the forelimbs may misleadingly direct attention to the cervical region as the site of the lesion, rather than the thoracolumbar spinal cord. The Schiff–Sherrington phenomenon is believed to be caused by interruption (at the site of the thoracolumbar transection) of fibres which normally arise in the ventral horn of the lumbar spinal cord, ascend in the **fasciculus proprius**, and inhibit extensor skeletomotor neurons in the cervical enlargement of the spinal cord.

14. Summary of somatic motor systems

There are two great somatic motor systems in the neuraxis, the pyramidal system and the extrapyramidal system.

14.1 Pyramidal system

The pyramidal system is a relatively recent evolutionary development. It is poorly developed in lower mammals, and absent in all vertebrates below mammals. Because of its phylogenetic youth, its anatomy varies considerably among the mammalian orders and may still be in an active state of evolution.

It is called the pyramidal system because it runs through the well-defined pyramid on the ventral surface of the medulla oblongata; actually the 'pyramidal' pathways to the striated muscle of the head (corticonuclear pathways) leave the system before it reaches the pyramid. The **components** of the pyramidal system are corticonuclear pathways, corticospinal pathways, and feedback circuits. (i) The **corticonuclear pathways** arise from the primary motor area of the cerebral cortex and project to the motor nuclei of the cranial nerves that innervate the striated muscles of the head. There are three neurons in the chain, of which the first decussates. (ii) The **corticospinal pathways** also arise from the primary motor area and project to the striated muscle in the body. These pathways take the form of two main **corticospinal tracts**, each with three neurons. In man and carnivores the well-developed lateral corticospinal tract is the principal pathway, and runs the whole length of the spinal cord; the ventral corticospinal tract is relatively insignificant and virtually confined to the neck. In 'ungulates' the ventral corticospinal tract is the main pathway and the lateral corticospinal tract is small; also a very small dorsal corticospinal tract lies in the dorsal funiculus. However, in the ungulate species the entire pyramidal system is poorly developed and ends anatomically in the cervical spinal cord. In all of these corticospinal pathways the first neuron decussates sooner or later, usually in the pyramid. (iii) The **feedback circuits** arise from the primary motor area of the cerebral cortex, and proceed through the pontine nuclei to the cerebellum, thence to the thalamus, and finally back to the primary motor area. The first half of such a circuit, i.e. from the cerebral cortex to the cerebellar cortex, is known as the corticopontocerebellar path. The circuit decussates on the way to the cerebellum and on the way back.

14.2 Extrapyramidal system

The extrapyramidal system is phylogenetically primitive, and well-developed in all but the lowest vertebrates. It includes all those other descending, somatic motor, spinal pathways the fibres of which do not pass through the pyramid of the medulla oblongata (though this is not a definition which is universally used). At the top of the brain it arises from extensive, perhaps even all, areas of the cerebral cortex (other than the primary motor area). These areas exchange projections with a sequence of motor centres in the lower levels of the brain, which culminate in spinal pathways.

The main **components** of the extrapyramidal system are its motor centres, spinal pathways, and feedback circuits.

(i) There are nine **higher motor centres** (the cerebral cortex, basal nuclei, midbrain reticular formation, red nucleus, tectum, pontine motor reticular centres, lateral medullary motor reticular centres, medial medullary motor reticular centres, and vestibular nuclei). Some of these centres are facilitatory but others are inhibitory, these opposing influences normally being accurately balanced.

The basal nuclei (notably via the globus pallidus) exert a mainly facilitatory influence on the descending reticular formation. The basal nuclei also send even more numerous projections to the ventral thalamic nuclei and thence to the motor cortex, thus strongly regulating the motor functions of the cortex. The regulation of the motor cortex via the thalamus appears to be the main role of the basal nuclei.

(ii) Five **descending spinal tracts** are present (rubrospinal, tectospinal, pontine reticulospinal, medullary reticulospinal, and vestibulospinal tracts). Of these, the first two decussate, the third and fourth are essentially midline, and the fifth is entirely ipsilateral. These spinal pathways usually involve a chain of three neurons. The cell body of the first neuron is in one of the motor centres of the brain. The second neuron is an interneuron in the spinal cord. The third neuron is a motoneuron in the ventral horn, usually a gamma neuron but sometimes an alpha neuron.

(iii) The **feedback circuits** of the highest extrapyramidal motor centres, e.g. the cerebral cortex, resemble the pyramidal feedback circuit, except that they project to the cerebellum through the olivary nucleus in place of the pontine nuclei. The extrapyramidal centres at the intermediate level, e.g. the red nucleus, still relay in the olivary nucleus on the way to the cerebellum, but they omit the thalamus on the way back. The extrapyramidal centres involved with the special senses, and the hindbrain centres, omit both the olivary nucleus and the thalamus; these centres therefore project direct to the cerebellum and receive a direct projection in return. All of these feedback circuits decussate on the way to the cerebellum and again on the way back, except for the vestibulocerebellar and cerebellovestibular pathways which remain entirely ipsilateral.

14.3 Distinction between pyramidal and extrapyramidal systems

The pyramidal and extrapyramidal systems are all one integrated system, but for descriptive purposes the distinction is still convenient, anatomically, functionally, and clinically.

14.4 Functions of the pyramidal and extrapyramidal systems; effects of injury of the motor command centres

The **extrapyramidal system** is responsible for the more or less automatic muscular activity which maintains both posture and also the semi-volitional deep-rooted somatic activities of daily life, such as locomotion, feeding and defence. The **pyramidal system** superimposes voluntary, detailed, muscular movements upon the semi-automatic postural and locomotory background of the extrapyramidal system.

A considerable capacity for voluntary movement survives **destruction of the primary motor area** of the cortex in many mammals (see section 11.6). For instance, in the dog there is relatively little disturbance soon after destruction of the primary motor area, the main abnormalities being some degree of paresis accompanied by hypertonus and hyperreflexia in the contralateral limbs. On the other hand, primates show severe motor disability after destruction of the primary motor area, with flaccid paralysis (atonus) and areflexia on the contralateral side, this being followed some hours or days later by hyperreflexia and spasticity (hypertonus).

In man, **lesions in the extrapyramidal motor centres** and particularly in the basal nuclei generally produce hypertonus, hyperreflexia, and postural and locomotory deficits. The single most characteristic sign of basal nuclei lesions in man is involuntary movement (hyperkinesia). Unilateral lesions tend to give rise to contralateral signs, but bilateral disturbances may result. The hypertonus may be due to the activity of gamma neurons which have been released from the inhibitory control of the higher centres. Little is known about the effects of naturally occurring lesions in the basal nuclei in domestic mammals, but they might be expected to produce clinical signs which, at least in principle, resemble those observed in man; experimental lesions sometimes produce hyperkinesia.

14.5 Upper motor neuron

The term 'upper motor neuron' is widely used in clinical neurology. It can be confined to neurons of the pyramidal tract, but is usually extended to include all the pyramidal and extrapyramidal pathways which control voluntary activity. Upper motor neuron disorders include both pyramidal and extrapyramidal lesions. Interruption of the upper motor neuron by **spinal lesions** is characterized by paralysis or paresis, with hypertonus (spastic paralysis, spastic paresis), hyperreflexia, and gradual muscle wasting.

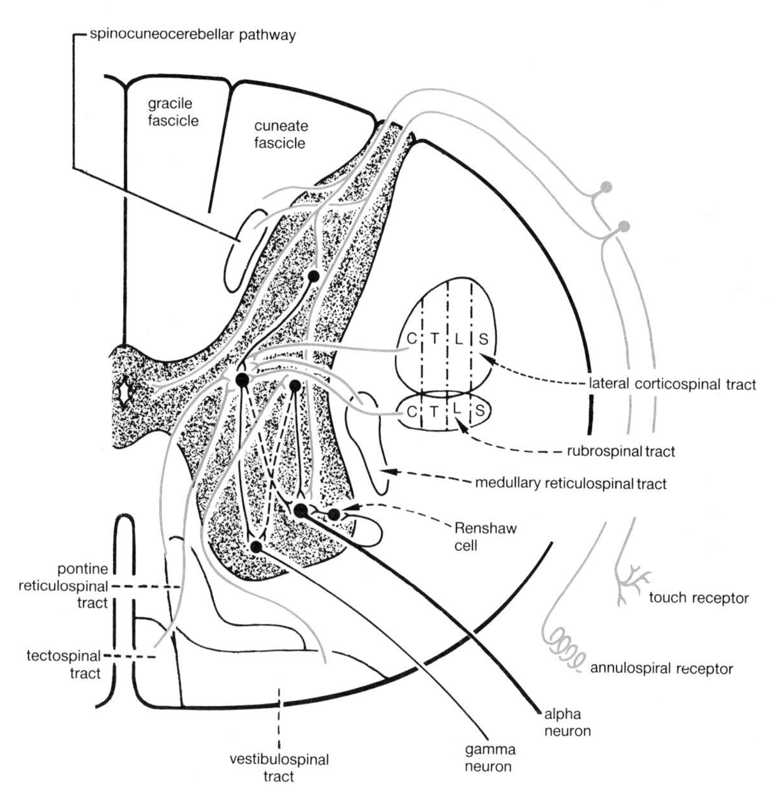

FIG. 14.1. Diagram summarizing the main projections on motor neurons in the ventral horn. The diagram is based on a transverse section of the spinal cord at about the eighth cervical segment. Pyramidal and extrapyramidal neurons and primary afferent neurons (shown in green) project on interneurons (black). Most of these interneurons (continuous lines) project on the gamma neuron; a minority of interneurons (broken lines) project on the alpha neuron. Interneurons in broken lines are the less common terminal pathways. Typically there are several, perhaps seven or more, interneurons to each motor neuron, not one to one as shown here. Many of the interneurons are excitatory, but others are inhibitory (e.g. the Renshaw cell). The relatively great thickness of the alpha neuron reflects its greater diameter and hence conduction velocity when compared with the gamma neuron. The spinal projections of the extrapyramidal system are shown in more detail in Fig. 12.2. C, cervical; T, thoracic; L, lumbar; S, sacral.

14.6 Lower motor neuron

The lower motor neuron is the alpha neuron of the ventral horn. Destruction of the lower motor neuron causes paralysis or paresis with hypotonus (flaccid paralysis, flaccid paresis), reduced reflexes (hyporeflexia) or loss of reflexes (areflexia), and severe and rapid muscle wasting.

14.7 Summary of projections on the final common path

The final common path is the **skeletomotor neuron (alpha neuron)**. It receives the following projections (Fig. 14.1).

(i) Monosynaptic reflex arcs (no interneuron being present) from the annulospinal receptors of muscle spindles. These are excitatory to the final common path.

(ii) The Renshaw cell, which is inhibitory to the final common path.

(iii) A majority of the projections of the vestibulospinal tract, via interneurons. These projections are highly facilitatory to extensor skeletomotor neurons.

(iv) A minority of the pyramidal, tectospinal, rubrospinal, and reticulospinal projections, via interneurons in each instance. Most of these are facilitatory, but the relatively numerous medullary reticulospinal fibres (from the medial medullary reticular motor centres) are extensively inhibitory.

(v) A minority of the interneurons of the spinal reflex arcs of touch, pressure, temperature, and pain. Some of these will be excitatory and some inhibitory to the final common path.

(vi) The interneurons substantially outnumber the alpha motor neurons. Some of them are excitatory. Others are inhibitory, the Renshaw cell being an example of an inhibitory interneuron. The Golgi tendon organ works through an inhibitory interneuron, which inhibits the alpha neuron.

The **fusimotor neuron (gamma neuron)**, however, receives more projections than the final common path. The projections on to the fusimotor neuron are (Fig. 14.1): a majority of the pyramidal, and tectospinal, rubrospinal, and reticulospinal projections, via interneurons; and a majority of the interneurons of the spinal reflex arcs of touch, pressure, temperature, and pain.

15. Cerebellum

The cerebellum **cannot initiate** movement, but it **co-ordinates** all somatic motor activity. For this to be possible it must receive full information about **intended** movement and about movement which is **actually in progress**, and must then project its own corrections to all the somatic motor centres of the brain. These activities are made possible by (i) sensory projections to the cerebellum from the muscles of the body, and (ii) feedback circuits which project to and fro between the cerebellum and the motor command centres of the pyramidal and extrapyramidal systems. These projections of all kinds are organized into afferent pathways to the cerebellum, and efferent pathways from the cerebellum.

Afferent pathways to the cerebellum

15.1 Ascending from the spinal cord

The four **spinocerebellar pathways** (Fig. 15.1) transmit proprioceptive information from muscle spindles and Golgi tendon organs (Fig. 9.1, and see Section 9.1). All of these pathways project to the cerebellar cortex, ending essentially on the same side as that on which they began. Their final projections on the cerebellar cortex are somatotopically arranged (see Section 15.7 below).

These are a major component of the so-called **'mossy fibres'** (see below, Functional histology of cerebellum).

15.2 Feedback input into the cerebellar cortex

From the pyramidal system

The corticopontocerebellar feedback pathway projects from the primary motor area, via the **pontine nuclei**, to the contralateral cerebellar cortex (Figs. 11.2, 15.1, and see Section 11.2).

These terminal projections from the pontine nuclei are also 'mossy fibres'.

From the extrapyramidal system

The principles of the neuronal pathways of extrapyramidal feedback to the

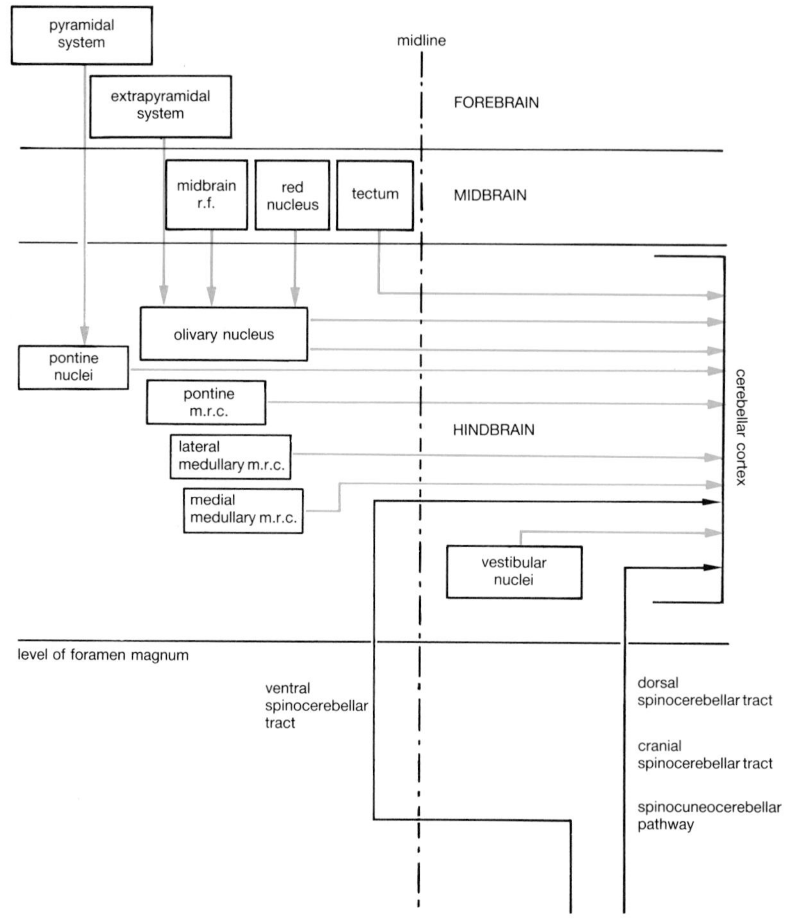

FIG. 15.1. Diagram of the main afferent pathways to the cerebellum. The projections from centres in the brain are shown in green, and the spinal projections in black. m.r.c., motor reticular centre; r.f., reticular formation.

cerebellum are outlined in Fig. 13.1 (see Sections 13.1, 13.2). They resolve themselves into two types of pathway.

(i) The first type consists of **indirect**, contralateral, pathways via the olivary nucleus (Fig. 13.1).

These are the **'climbing fibres'**.

(ii) The second variety comprises **direct** pathways from the vestibular nuclei (balance) and tectum (vision and sound) (Fig. 13.1). The pathway from the tectum is contralateral, but the feedback from the vestibular nuclei is unique in being **ipsilateral**, e.g. the left vestibular nuclei project to the left side of the cerebellum. The projections between the vestibular nuclei and cerebellum are particularly important in maintaining posture, in view of the strong extensor activity of the vestibulospinal tract (see Section 12.12).

The projections from the vestibular nuclei and tectum form yet another component of the 'mossy fibres'.

The projections from the tectum and vestibular nuclei can be regarded not only as part of the extrapyramidal feedback into the cerebellum, but also specifically as **projections from the special senses** of vision and balance to the cerebellum. The senses of vision and balance provide the cerebellum with information about the position of the head and thus assist the cerebellum in maintaining posture.

Efferent pathways from the cerebellum

The efferent pathways are the cerebellar output of the feedback circuits (Fig. 15.2). All such output from the cerebellum is via two neurons:

Neuron 1 has its cell station in the cerebellar cortex. This is the **Purkinje cell**. It projects on neuron 2.

Neuron 2 lies in a cerebellar nucleus, for example the lateral (dentate) nucleus, beneath the cerebellar cortex.

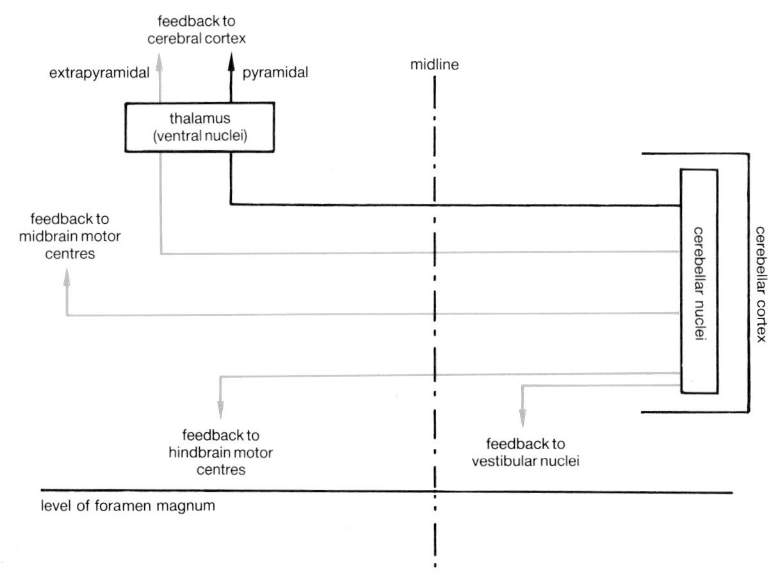

FIG. 15.2. Diagram summarizing the efferent pathways from the cerebellar nuclei. The output consists exclusively of feedback projections returning to the motor centres. The pathway returning to the pyramidal motor centre, i.e. to the primary motor area, is shown in black. The pathways to the extrapyramidal motor centres are shown in green.

In the dog three cerebellar nuclei are embedded within the white matter of the cerebellum. The **lateral nucleus** and **interpositus nucleus** are known to project to the thalamus and red nucleus, via the rostral cerebellar peduncle. The **fastigial nucleus** projects to the vestibular nuclei and reticular motor centres through the caudal cerebellar peduncle.

Neuron 2 may project to any of the following components: (i) It may project to the **thalamus**, where it synapses with a neuron that then projects to the cerebral cortex and globus pallidus. These are the return pathways of pyramidal and extrapyramidal feedback (Fig. 15.2); the **ventral group** of thalamic nuclei is involved (*the* ***ventrolateral nucleus***). The projection from the thalamus to the globus pallidus is not direct but indirect via the cerebral cortex (Fig. 13.1). (ii) Alternatively, neuron 2 may project to the **midbrain motor centres** of the **extrapyramidal system** (Fig. 13.1). These are the midbrain reticular formation (centre 3 in Fig. 13.1), the red nucleus (centre 4), and the tectum (colliculi) (centre 5). (iii) Finally neuron 2 may project to the **hindbrain motor centres** of the **extrapyramidal system** (Fig. 13.1). These are the pontine motor reticular centres (centre 6), the lateral medullary motor reticular centres (centre 7), the medial medullary motor reticular centres (centre 8), and the vestibular nuclei (centre 9).

With one exception, all output from the cerebellum to the pyramidal and extrapyramidal motor centres **decussates**. The exception is the cerebellar feedback output to the vestibular nuclei; like all vestibular–cerebellar connections, this feedback output is **ipsilateral**.

The feedback projections to the cerebral cortex and basal nuclei (centres 1 and 2 of the extrapyramidal system) cross over as the **decussation of the rostral cerebellar peduncles**. The projections to extrapyramidal motor centres 3 to 8 of the midbrain and hindbrain have rather complex decussations, sometimes within the cerebellum itself, but they all do decussate.

Note the **complete absence** of **descending** spinal paths from the cerebellum. The cerebellum is **unable** to **initiate movement**.

Summary of decussation of the feedback circuits of the cerebellum

Except for the vestibular feedback circuits, all feedback circuits, both pyramidal and extrapyramidal, decussate both on the way into the cerebellum and on the way out of the cerebellum; thus they return to the side from which they started. The vestibular feedback circuits remain always on the same side.

Summary of pathways in the cerebellar peduncles

Most of the fibres in the caudal peduncle are afferent to the cerebellum, whereas most of those in the rostral peduncle are efferent from the cerebellum. The middle peduncle is allocated exclusively to afferent fibres, i.e. the incoming fibres of pyramidal feedback.

15.3 Caudal cerebellar peduncle

This carries a large group of **afferent** fibres: (i) olivocerebellar, (ii) vestibulo-cerebellar, (iii) medullary reticulocerebellar fibres, and (iv) pontine reticulocerebellar fibres; (v) the dorsal spinocerebellar fibres, (vi) some cranial spinocerebellar fibres, and (vii) the fibres of the cuneocerebellar tract. It also contains some **efferent** fibres, i.e. (i) cerebellovestibular fibres, and (ii) cerebelloreticular fibres.

15.4 Middle cerebellar peduncle

The middle peduncle is entirely **afferent**. Its main component is formed by the axons of neuron 2 in the corticopontocerebellar pathway.

15.5 Rostral cerebellar peduncle

The principal **afferent** fibres are those of the ventral spinocerebellar tract, and some of the cranial spinocerebellar fibres; tectocerebellar fibres also enter by this peduncle. Most of the fibres in this peduncle, however, are **efferent**, forming efferent feedback pathways from the cerebellar nuclei, i.e. to the contralateral thalamus, red nucleus, and tectum.

Representation in the cerebellum

Functionally (and phylogenetically) the cerebellum can be divided into three regions, according to its afferent input.

15.6 Vestibular areas

The **flocculonodular lobe** is known as the **'vestibulocerebellum'** (or archi-cerebellum), since it receives the projections from the vestibular nuclei. It lies at the caudal end of the cerebellum (Fig. 15.3).

15.7 Proprioceptive areas

The **vermis** of the cerebellum (except for its midregion and the nodulus) receives the projections from the spinocerebellar pathways (Fig. 15.3), and therefore constitutes the **'spinocerebellum'** (or paleocerebellum).

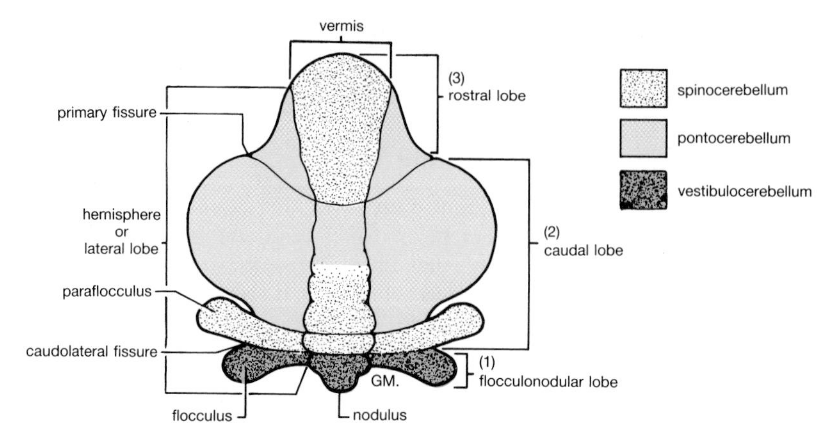

FIG. 15.3. Diagram of the feline cerebellum, rolled out flat and viewed from its dorsal aspect. Developmentally, the cerebellum can be divided into (1) the flocculonodular lobe, (2) the caudal lobe, and (3) the rostral lobe. Lobes (1) and (2) are separated by the caudolateral fissure, which is the first fissure to develop; lobes (2) and (3) are separated by the primary fissure, the second to develop. Topographically, the adult cerebellum is divided into (a) the vermis, which runs the whole length of the cerebellum in the midline, including the nodulus, and (b) the left and right cerebellar hemispheres (or lateral lobes); the rostral and caudal extremities are tucked ventrally under the middle part, and are therefore concealed when the intact cerebellum is seen from the dorsal aspect. Functionally, (and phylogenetically) the cerebellum can also be divided into: (i) the vestibulocerebellum (or archicerebellum), which is the flocculonodular lobe and is the oldest region phylogenetically; (ii) the spinocerebellum (or paleocerebellum), comprising the rostral and caudal regions of the vermis (but not the nodulus), and the paraflocculus; and (iii) the pontocerebellum (or neocerebellum), which is formed by the midpart of the vermis plus the rest of the two hemispheres, and is the most recent region phylogenetically. The prefixes vestibulo-, spino-, and ponto-, indicate the sources of afferent projections into these three functional regions. In the *Nomina Anatomica Veterinaria* the caudolateral fissure is the uvulonodular fissure.

15.8 Feedback areas

Feedback pathways from the pyramidal and extrapyramidal systems project into the **'pontocerebellum'** (or neocerebellum), which consists of the majority of the two hemispheres and the midpart of the vermis (Fig. 15.3).

This is an oversimplification. The **olivary nuclei** have widespread projections into many parts of the cerebellum.

Functions of the cerebellum

15.9 Co-ordination and regulation of movement

This is achieved by the proprioceptive projections from the muscles and the feedback circuits, which together enable the cerebellum to **compare response with command** in both the pyramidal system and the extra-pyramidal system. It can then adjust command appropriately. All this is done in two steps: (i) By exerting **pre-control** over the movement which is about to take place. The motor 'command centres' (e.g. the cerebral cortex, red nucleus, etc.) initiate the movement and simultaneously inform the cerebellum via the pontine nuclei or olivary nuclei about what they are intending to do. By means of its feedback projections to the motor centres, the cerebellum can instantly modify the intended movement. (ii) By **regulating movements**, once they are actually in progress. As soon as the movement begins, the events taking place in the muscles are measured and reported to the cerebellum by the spinocerebellar pathways; by means of the feedback circuits, the cerebellum then regulates the motor 'command centres' to make the movement perfect.

15.10 Control of posture

This is achieved as for the co-ordination of movement, but with the additional aid of information reaching the cerebellum from the special senses, particularly balance and vision.

15.11 Ipsilateral function of the cerebellum

The activities of the cerebellum are always essentially ipsilateral. Thus the right side of the cerebellum controls movements on the right side of the body. This is because the right side of the cerebellum regulates the left motor centres of the pyramidal and extrapyramidal systems, apart from the vestibular nuclei; except for the vestibular nuclei, the left motor centres of the pyramidal system and extrapyramidal system initiate movement on the right side of the body. Also the right side of the cerebellum controls the right vestibular nuclei, which further initiate movement on the right side of the body. Furthermore, the right side of the cerebellum is being continually informed by muscle proprioceptors about muscular activity on the right side of the body.

In summary, this essentially **ipsilateral** function of the cerebellum occurs because: (i) both the input and the output of the feedback circuits decussate, except for the vestibular feedback; (ii) the vestibular feedback circuits and vestibular tract remain always on the same side of the body; (iii) all of the spinocerebellar pathways end on the same side of the cerebellum as the side of the body from which they arose.

15.12 Summary of cerebellar function

The cerebellum is informed of virtually all the somatic motor activity of the nervous system, and is thereby able to co-ordinate all somatic motor activities. In engineering parlance, it is the **control box** in the feedback circuitry of the somatic motor nervous system. The cerebellum is not a part of the pyramidal or the extrapyramidal systems, but is superimposed on both of these two systems.

The cerebellum has been regarded as more or less completely cut off from visceral influences, but visceral pathways have recently been discovered. Their overall function is uncertain.

Disease of the cerebellum

15.13 The three cerebellar syndromes

In both man and lower mammals there are three basic cerebellar syndromes, namely the vestibulocerebellar, spinocerebellar and pontocerebellar syndromes; these are founded on the three functional regions summarized in Sections 15.6–15.8 above. The clinical signs of these three syndromes are based on the afferent input to each region, as follows: (i) Destructive lesions of the **'vestibulocerebellum'** (for example, caused by tumours or hydrocephalus) cause **disturbance of equilibrium**, with swaying when standing, falling to either side despite a wide-based stance, unsteadiness when walking, and positional nystagmus, but **no change in muscle tone** and **no tremor**. (ii) Destructive lesions of the **'spinocerebellum'** cause **hypertonus** and **exaggeration of the postural reflexes**. These reflexes aim at preserving the normal posture of the body, and are required for changing basic postures (e.g. going from lying, to sitting, to standing). Lesions of the spinocerebellum are therefore disclosed by postural tests such as hopping, wheelbarrow, hemiwalking, tactile placing, and extensor postural thrust. The postural disturbances evidently arise through the **hypertonus**, which causes rigidity in extension (and hence impairment of the synergistic actions required by the four limbs when any movement to change or maintain posture is attempted). (iii) Destructive lesions in the **'pontocerebellum'** interfere with feedback pathways between the cerebellum and the higher motor centres, causing **asynergia**; this means that the components of a movement are no longer synchronous and harmonious, but isolated and out of proportion. Such lesions therefore reveal themselves during movement, e.g. as an erratic length and height of stride (**dysmetria**), jabbing movements of the head (overshooting) during feeding, **tremor** when attempting a precise movement of the head or a paw, and **hypotonia**.

A **unilateral lesion** produces these disturbances on the **same side** of the body, because the cerebellum controls the somatic musculature **ipsilaterally**.

15.14 Cerebellar disease in domestic mammals and man

Even in **man**, cerebellar lesions are not often so clearly defined that only one of the three basic syndromes is evident. In veterinary practice, an animal is seldom presented at such an early stage that a lesion is still restricted to only one of the three functional regions of the cerebellum. Usually the lesion will be relatively widespread, so that the clinical signs will be a combination of more than one of the three basic syndromes. Thus in the **domestic mammals**, typical signs are loss of balance with wide-based staggering gait ('vestibulocerebellum'); exaggerated and erratic reflex responses to postural tests such as hopping ('spinocerebellum'); a high-stepping action (hypermetria), jabbing movements of the head when feeding, and tremor ('pontocerebellum'). As can be seen from the conflicting changes in tone (hypertonia in lesions of the 'spinocerebellum', and hypotonia in lesions of the 'pontocerebellum'), in a composite lesion there may be either hyper- or hypotonia, but **muscle tone is usually increased** in the domestic animals. See also p. 109.

In human neurology the term **ataxia** is applied to incoordination arising from either a cerebellar disturbance, or an afferent kinaesthetic deficit as from a lesion of the dorsal column: it is not applied to the incoordination from a motor deficit. In domestic animals it is difficult to distinguish between incoordination from a cerebellar or an afferent deficit on the one hand, and weakness of voluntary movements from a motor deficit (paresis) on the other hand; consequently, the term ataxia is sometimes used in veterinary neurology to mean inadequacies of gait, of either cerebellar, kinaesthetic, or motor origin.

Alcohol particularly affects cerebellar neurons, and readily-induces 'drunken gait'. **'Drunken gait'** is also a common sign of injury to the cerebellum in man, and may lead to arrest for drunkenness. The signs are swaying and falling over. With a unilateral lesion, falling is to the affected side. A less severe lesion can produce a tremor, but this tremor **only occurs during movement** and becomes progressively worse in precise movements as they advance to completion. Consequently it is called **'intention tremor'**. Compare **Parkinson's disease**, in which the main sign is **continuous** muscular rigidity and the tremor goes on **all the time**, except in sleep. The **nystagmus** which may occur in cerebellar disease may also be a form of intention tremor; the subject 'intends' to direct his eyes continuously towards a fixed object, and fails. In contrast to the domestic mammals, cerebellar lesions in man are usually accompanied by a **decrease** in muscle tone.

Functional histology of the cerebellum

Histological sections of the cerebellar cortex, examined from the depth towards the surface, show three layers: (i) the **granular cell layer**, consisting of small nerve cell bodies (the granular cells); (ii) the **Purkinje cell layer** (piriform cell layer) formed by a single stratum of large flask-shaped cells (the Purkinje cells, or piriform cells); and (iii) the **molecular layer** composed mainly of the unmyelinated fibres of the Purkinje and granular cells, together with the cell bodies of a few interneurons. In the molecular layer

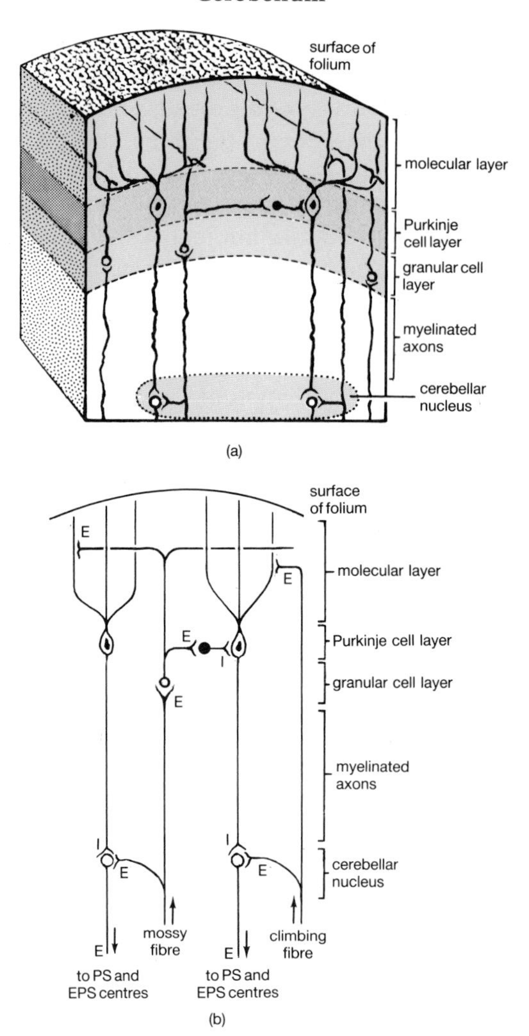

FIG. 15.4. (a) Diagram of the cerebellar cortex. The cortex consists of the granular cell layer, the Purkinje cell layer, and the molecular layer. Deep to the granular cell layer are incoming and outgoing myelinated axons. The axons of the granular cells bifurcate in the molecular layer, passing parallel with the long axis of the folium and at right angles to the dendritic fields of the Purkinje cells. Apart from the two granular cells at the right and left hand sides of the diagram, the neurons are shown as in Fig. 15.4 (b). The cerebellar folia run transversely in relation to the long axis of the brainstem.

(b) Diagram summarizing the main pathways in the cerebellar cortex. The climbing fibres are excitatory (E) to the dendrites of the Purkinje cells. The mossy fibres are excitatory to the granular cells. The granular cells are excitatory to the Purkinje cells and also to interneurons (black) which are inhibitory (I) to the Purkinje cells. The Purkinje cells are inhibitory to the neurons of the cerebellar nuclei. The Purkinje cells will fire or be silent depending on the balance of the E and I projections which they receive. Likewise the neurons of the cerebellar nuclei will be silent or will fire, depending on the balance of the projections which they receive. If they fire they will excite the motor centres of the pyramidal and extrapyramidal systems.

the axons of the granular cells bifurcate and run parallel with the long axis of the cerebellar folia, intersecting at right angles the dendritic fields of the Purkinje cells (Fig. 15.4(a)). Beneath the granular cell layer there are massive bundles of incoming and outgoing myelinated axons.

Of the incoming fibres to the cerebellum (Fig. 15.4(b)) those from the olivary nuclei are called **climbing fibres** (neurofibrae ascendentes); those from the pontine nuclei, vestibular nuclei, and spinocerebellar tracts are known as **mossy fibres** (neurofibrae muscoideae) because they look like moss. As shown in Figs. 11.2, 13.1, and 15.1, all of these afferent fibres project to the cerebellar cortex; in the cortex they are all **excitatory**.

The climbing fibres project directly upon the dendrites of Purkinje cells. They are directly **excitatory** to the Purkinje cells. The mossy fibres project upon the granular cells, which in turn project upon the dendrites of Purkinje cells. This pathway is again excitatory, causing **excitation** of the Purkinje cells, though indirectly. The granular cells project not only upon the dendrites of Purkinje cells but also upon inhibitory interneurons (black cell bodies in Fig. 15.4(b)). These interneurons (which include the **basket cells**), are **inhibitory** to the Purkinje cells.

Thus the Purkinje cells may be excited or inhibited (Fig. 15.4(b)) depending on the balance between the excitatory influences of (a) and (b) on the one hand, and the inhibitory influences of (c) on the other hand. The Purkinje cells themselves are entirely **inhibitory**, and project to the cerebellar nuclei below the cortex. The neurons in the cerebellar nuclei are all **excitatory** to the motor command centres of the pyramidal and extrapyramidal systems (Fig. 15.2).

One other detail is needed to explain how these combinations of cells work. The incoming climbing fibres and mossy fibres project not only to the cerebellar cortex but also directly **to the deep cerebellar nuclei**; these projections are **not** shown in Figs. 11.2 and 13.1 but they are shown in Fig. 15.4(a), (b). The action of these incoming **direct** projections is to **excite** the nerve cells in the cerebellar nuclei, in opposition to the **inhibition** caused by the Purkinje cells. These opposing influences upon the cells of the cerebellar nuclei may cancel out, in which case the neurons of the cerebellar nuclei are silent. On the other hand, if the **direct** projections on to the cerebellar nuclei are dominant, the neurons of the cerebellar nuclei will fire. Thus the output from the cerebellar nuclei to the command centres of the pyramidal and extrapyramidal systems can be either nil or excitatory, but never inhibitory.

Summary of functional cerebellar histology

(i) All afferent fibres (climbing and mossy fibres) to the cerebellar cortex are excitatory.

(ii) The only outgoing fibres from the cerebellar cortex are the axons of the Purkinje cells. These are inhibitory to the cerebellar nuclei.

(iii) The Purkinje cells are directly excited by the afferent (climbing and mossy) fibres to the cerebellar cortex, but are also indirectly inhibited by interneurons within the cortex. They respond according to the balance between these excitatory and inhibitory influences.

(iv) The neurons of the cerebellar nuclei are all excitatory, to the motor command centres of the pyramidal and extrapyramidal systems.

(v) The neurons of the cerebellar nuclei receive inhibitory projections from Purkinje cells, and excitatory projections directly from the afferent fibres entering the cerebellum.

(vi) The cerebellar nuclei can be excitatory or silent, but never inhibitory, to the command centres of the pyramidal and extrapyramidal systems.

16. Autonomic components of the central nervous system

Each of the three great divisions of the brain (the forebrain, midbrain, and hindbrain) contains components with substantial autonomic functions. In the largest subdivision of the forebrain (the telencephalon or cerebral hemisphere), these components include the neocortex itself and the hippocampus. In the other great subdivision of the forebrain, the diencephalon, the hypothalamus plays a major role in autonomic functions. In the midbrain and hindbrain, the reticular formation contains many nuclei and 'centres' which are extensively engaged in autonomic activities. In the spinal cord, tracts in the white matter and regions of the grey are associated with the thoracolumbar (sympathetic) and sacral (parasympathetic) autonomic pathways.

16.1 Cortical components

In man, the premotor area of the frontal lobe of the neocortex (Fig. 17.2(a)) exchanges two-way projections with the hypothalamus, including an indirect pathway via the thalamus (Figs. 16.3, 17.3). Thus the frontal–hypothalamic connections form a two-way path. By these projections the cerebral cortex restrains the emotions of rage and aggression which emanate from the hypothalamus. Other cortical projections excite autonomic centres lower in the brainstem. For example, thinking about a frightening situation can accelerate the heart, raise the arterial pressure, change the pattern of breathing, and dilate the pupils.

16.2 Hippocampus

This is a large component of the **limbic part** of the **rhinencephalon** (see Section 8.6), consisting essentially of primitive motor cortex and occurring in vertebrates generally. Yet its functions are uncertain. It appears to be closely associated with the motor manifestations of emotion and only remotely with olfaction. Its rich two-way projections to the hypothalamus (Figs. 16.3 and 17.3) suggest an involvement in autonomic functions. These projections evidently enable it to inhibit the **rage** reactions of the hypothalamus, thus reinforcing the inhibitory action of the frontal lobe of the neocortex (see Section 17.11). Damage to these hippocampal–hypothalamic projections by the rabies virus may account for furious symptoms in the rabid dog; the hippocampus is routinely tested for virus in suspected cases of rabies. The hippocampus is also involved in **learning** and **memory**, through its connections (Fig. 17.3) with the temporal lobe of the cerebral cortex and hypothalamus; learning is linked to autonomic functions, in that learning improves when the emotion of **interest** is aroused (see Section 17.9). Destruction of the hippocampus in man destroys new learning and memory, but old memory is retained; in cats it has been shown to produce rage. There are also grounds for regarding the hippocampus as the anatomical site of **instinct** (see Section 17.2).

A major pathway for two-way projections between the hippocampus and hypothalamus is formed by the **fimbria** and **fornix** (see Section 21.29, **Grey matter**). The hippocampus is especially liable to loss of neurons in two diseases of major importance, namely the quite common form of dementia in people known as Alzheimer's disease, and scrapie in sheep which is characterized by excessive skin irritation. One of the typical signs of dementia in man is a severe loss of short-term memory and capacity for new learning, and this would be consistent with the hippocampal role in memory. Since memory and learning are not strong points in sheep, such effects of hippocampal lesions are less evident in that species.

16.3 Hypothalamus

The hypothalamus is the principal diencephalic component of the autonomic nervous system. It is linked to the adenohypophysis by a portal circulation, and is directly attached to the neurohypophysis by axons; thus the hypothalamus and hypophysis are closely associated with each other. Caudally the hypothalamus continues directly into the midbrain reticular formation, and indirectly into the hindbrain reticular formation which contains the parasympathetic nuclei of cranial nerves as well as an array of autonomic 'centres'. By means of these various connections, direct and indirect, the hypothalamus integrates the mechanisms of **homeostasis**.

The hypothalamus is also linked directly and indirectly with several other major autonomic centres in the forebrain, notably the amygdaloid body, septal nuclei, and habenular nuclei (see Sections 16.4 and 16.5). These connections extend still further the dominating influence of the hypothalamus over homeostasis.

Components of the hypothalamus

There are three main components, namely the medial part of the hypothalamus, the lateral part, and the mamillary body (Figs. 16.1, 16.2). The most important of these is the **medial part**, since it is here that the majority

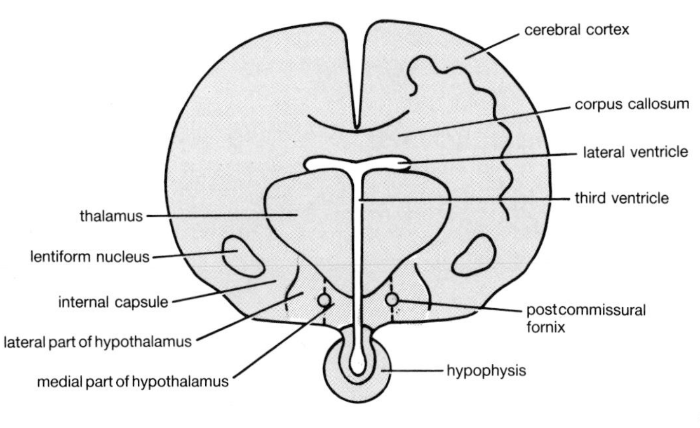

FIG. 16.1. Diagrammatic transverse section through the forebrain in the region of the hypothalamus. The hypothalamus (green) is divided (by the postcommissural fornix) into a medial and a lateral zone. The medial zone contains most of the hypothalamic nuclei, including the supraoptic and paraventricular nuclei.

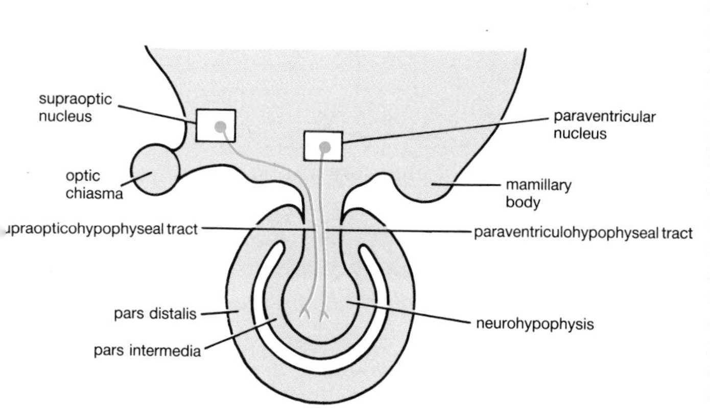

FIG. 16.2. Diagrammatic sagittal section through the hypothalamus. Neurons in the supraoptic and paraventricular nuclei form the antidiuretic hormone and oxytocin. These hormones migrate down the axons into the neurohypophysis, and are stored there in the axon terminals. The pars distalis and pars intermedia are components of the adenohypophysis.

of the main hypothalamic nuclei with established autonomic functions reside.

Autonomic nuclei of the hypothalamus

There are many hypothalamic nuclei. Two of the most important nuclei lie essentially in the **medial part** of the hypothalamus, i.e. in the part nearest to the hypophysis. These are: (i) the **supraoptic nucleus** which lies immediately dorsal to the optic chiasma, hence its name (Fig. 16.2); and (ii) the **paraventricular nucleus** which lies alongside the third ventricle, again a position which gives it its name (Fig. 16.2).

Twenty-two hypothalamic nuclei are named by the NAV.

The axons of these two nuclei form the supraopticohypophyseal and paraventriculohypophyseal tracts. Their axons deliver antidiuretic hormone and oxytocin to the neurohypophysis (posterior lobe) (see ii below).

Neurons containing antidiuretic hormone are relatively more numerous in the supraoptic nucleus, whereas neurons containing oxytocin are relatively more abundant in the paraventricular nucleus.

The autonomic functions of the hypothalamus

As just stated, the hypothalamus exercises a general **integrating** control over homeostasis. Some of its special functions are as follows:

(i) **Bladder contraction and gut motility** are known to be controlled by hypothalamic nuclei.

(ii) Both the supraoptic and the paraventricular nuclei are involved in forming the **antidiuretic hormone** (ADH) and **oxytocin**, in response to changes in the osmotic pressure of the blood which supplies these nuclei. The hormones migrate down the axons to the neurohypophysis, where they are stored in the axonal terminals. These are the only two hormones which are liberated by the neurohypophysis. They may be regarded as transmitter substances of the neurons which produce them.

(iii) **Temperature regulation** is achieved mainly by vasomotor, and respiratory control, and also by sudomotor control in mammals such as man and the horse. Parts of the hypothalamus are directly responsive to variations in the temperature of the blood which supplies them (see Section 1.6). Projections through the reticular formation (Fig. 16.3) enable them to regulate the activities of the cardiovascular and respiratory centres (see section 16.6), thus controlling cutaneous vasodilation and panting.

(iv) Centres in the hypothalamus respond to **hunger** and **thirst**. These have been demonstrated experimentally by installing a stimulator in the appropriate area of the hypothalamus; application of a stimulus in the conscious animal induces immediate eating or drinking, which stop as soon as the stimulus is switched off.

(v) The **adenohypophysis** (anterior lobe of the hypophysis) is controlled by substances known as releasing factors, which are secreted by nerve terminals of the hypothalamus. These substances enter the capillaries of a venous portal blood flow and travel to cells of the adenohypophysis, making these cells secrete.

(vi) The powerful **emotions** of **rage** and **aggression** seem to originate in the hypothalamus. In cats in which all the upper parts of the brain have been removed, leaving only the hypothalamus connected to the lower levels of the CNS, trifling stimuli which do not normally elicit rage induce violent reactions including biting, scratching, snarling, and facial grimaces (all of which, of course, are **somatic** responses). The rage reactions of the hypothalamus appear to be normally inhibited by the hippocampus (see Section 16.2), and in man by the frontal lobe of the cerebral cortex also (see Section 17.11).

(vii) **Sleep** is regulated by the hypothalamus. In rats destruction of rostral regions of the hypothalamus produces **insomnia**. Similar changes have been

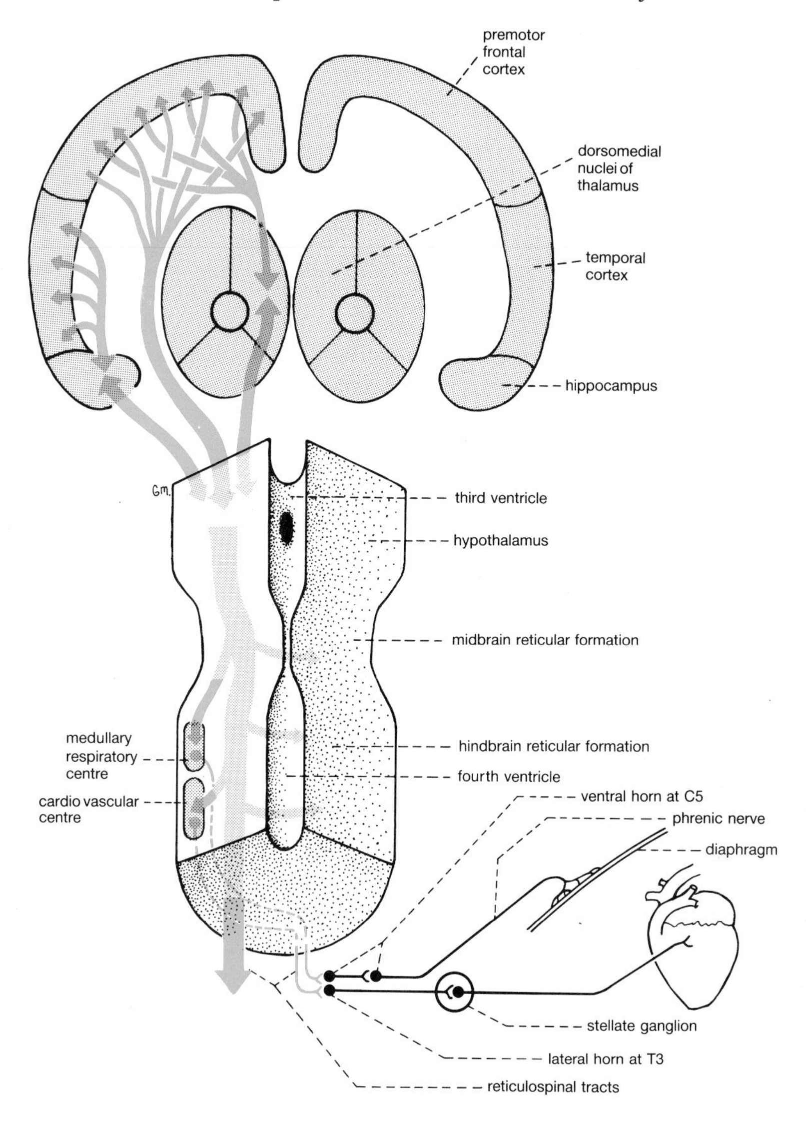

FIG. 16.3. **Diagram showing the relationships of cerebral, hypothalamic, brainstem, and spinal cord components of the autonomic nervous system.** Projections between these components are shown in green. The emotional urges of the hypothalamus are suppressed by two-way connections with the premotor frontal cortex and hippocampus. Two-way connections between the hippocampus and temporal cortex are involved in memory and learning. Caudally the hypothalamus projects into the reticular formation of the midbrain and hindbrain. There it influences functions such as respiration and circulation through the respiratory and cardiovascular centres. These and other medullary autonomic centres project downwards through the reticulospinal tracts. At the foot of the diagram these descending spinal pathways are shown projecting to a neuron of the phrenic nerve, and to a preganglionic neuron of a cardiac accelerator pathway. The anatomical structures in the diagram are not drawn to scale.

observed following clinical lesions in the same region in man. These observations suggest that this area of the hypothalamus is involved in sleep, probably by inhibiting the ascending reticular formation (see Section 9.9). The caudal hypothalamus seems to contain a **waking centre**.

In the sleeping mammal two forms of sleep succeed each other cyclically, slow-wave (deep) sleep and paradoxical (rapid eye movement) sleep. **Slow-wave sleep** is characterized by high-amplitude, low-frequency, synchronized EEG waves. **Paradoxical sleep** shows low-amplitude, high-frequency, desynchronized EEG waves and rapid movements of the eyes; **dreams** occur during paradoxical sleep. Paradoxical sleep is a highly activated state with many features resembling those of alert wakefulness, notably the EEG waves and the eye movements. The mechanisms which switch on the episodes of paradoxical sleep reside somewhere in the pons (possibly in the rostral and lateral part of the floor of the fourth ventricle, the **locus ceruleus**). The inhibitory action of the medial medullary reticular nuclei is ultimately responsible for the muscle atonus which typifies sleep. Arousal from paradoxical sleep is usually rapid and without confusion. The onset and maintenance of **slow-wave sleep** is regulated by much more extensive components of the brain. These include groups of nerve cells in the midline of the medulla oblongata (the **raphe nuclei**, the cells of which contain **serotonin**), although it seems to be the hypothalamic area that has the most basic influence on the sleeping–waking rhythm. Arousal from slow-wave sleep is slow and confused. **Hibernation** in mammals may be physiologically homologous to slow-wave sleep. Body temperature is lowered in slow-wave sleep as a result of resetting the hypothalamic neural controls at a lower level of sensitivity, thus conserving energy. Hibernation begins from slow-wave sleep, and is a much more profound way of saving energy since the body temperature falls well below the level of normal sleep. The hypothalamus is the basis of the neural control of hibernation. If the hypothalamus is damaged, the hibernator cannot arouse itself when its body temperature falls below the normal minimal level (which is only a few degrees above 0°C), nor can it spontaneously terminate its period of hibernation.

(viii) **Sexual functions.** These are extensively controlled by the hypothalamus but the detailed mechanisms and pathways are not fully understood.

16.4 Amygdaloid body and septal nuclei

By means of their afferent and efferent connections with the hypothalamus these two components of the telencephalon are involved in many autonomic activities. Those of the amygdaloid body are especially wide-ranging, and include respiratory, circulatory, iridal, and piloerector manifestations during emotional reactions. The septal nuclei receive olfactory projections but appear to be concerned with sexual and aggressive behaviour. Both the amygdaloid body and the septal nuclei belong to the limbic system (see Section 8.6).

16.5 Habenular nuclei

The habenular nuclei belong to the diencephalon, and are also a part of the limbic system (see Section 8.6). Their afferent fibres come from the septal nuclei via the **habenular stria**, and are partly olfactory and partly limbic; since the septal nuclei are connected with the hypothalamus, the hypothalamus is indirectly linked with the habenular nuclei. The efferent fibres of the habenular nuclei connect indirectly (via the **fasciculus retroflexus, interpeduncular nucleus**, and **fasciculus longitudinalis dorsalis**) with the pontine and medullary reticular formation, and hence influence autonomic reticulospinal pathways and the parasympathetic (cranial nerve) motor nuclei in the medulla oblongata. By these pathways the habenular nuclei are involved in the autonomic and visceral responses to olfactory and emotional stimuli (see Section 21.23 for anatomical details). A similar autonomic link between the forebrain and hindbrain is formed by the **medial forebrain bundle** (see Section 8.5); by this pathway, olfactory stimuli induce salivation and gastric secretion.

16.6 Hindbrain autonomic areas

The autonomic areas of the hindbrain include cardiovascular, respiratory, and alimentary 'centres', but these are really no more than ill-defined neuronal pools in the medullary reticular formation, and are certainly not discrete nuclei. Often the pools overlap each other, one being superimposed on another. They consist of primitive neurons of the reticular formation, each neuron having many collaterals and numerous connections with other neurons. Downward projections enter the spinal cord in the reticulospinal tracts.

Cardiovascular centres

There is evidence from stimulation experiments that there are two regions in the reticular formation of the medulla oblongata which control the heart and blood vessels. One is known as the **vasomotor depressor centre**, and the other as the **vasomotor pressor centre**. Collectively they may be called the **medullary cardiovascular centre** (Fig. 16.3).

Neurons with a depressor effect on arterial pressure and heart rate have been found in the central (medial) group of nuclei in the reticular formation of the medulla oblongata (see Section 21.11, **Reticular formation**); pressor neurons with excitatory cardiovascular effects have been encountered in the lateral group of nuclei in the medullary reticular formation. More recent investigations have shown that the original concept of a depressor centre and a pressor centre requires extensive revision.

Respiratory centres

In the medulla oblongata, two regions have been discovered in the reticular formation which influence breathing. One induces inspiration and has become known as the **inspiratory centre**. The other induces expiration, and has been named the **expiratory centre**. Together they are called the **medullary respiratory centre** (Fig. 16.3).

The inspiratory neurons lie mainly in the central (medial) group of nuclei of the reticular formation of the medulla oblongata (see Section 21.11, **Reticular formation**); the expiratory neurons tend to be situated in the lateral group of nuclei in the reticular formation of the medulla. The details of their location and function are still controversial.

In the pons there is a **pneumotaxic centre**, which inhibits the inspiratory centre by negative feedback. Also in the pons, but more caudally situated, is an **apneustic centre**, which applies a steady excitatory drive on the inspiratory centre.

Alimentary centres

In sheep, there is a ruminoreticular centre and an oesophageal centre in the medulla oblongata.

16.7 Autonomic motor pathways in the spinal cord

Some of the neurons in the 'cardiovascular centre' project to the vagal motor nucleus; others project into the spinal cord in the **reticulospinal tracts** and synapse with preganglionic sympathetic neurons in the lateral horn in segments T1 to L2, thus controlling the sympathetic cardiac and vasomotor outflow (Fig. 16.3, the lower right side of the diagram).

The lateral horn extends caudally to L3 in man, and to L5 in the dog.

The neurons in the 'medullary respiratory centre' again project into the spinal cord through the **reticulospinal tracts**. In segments C5, C6, and C7 of the spinal cord their axons make synapses with interneurons, which in turn activate ventral horn motoneurons forming the phrenic nerve (Fig. 16.3). Similar projections are made to motoneurons in segments T1 to 13 controlling the intercostal muscles. The reticulospinal tracts also carry fibres regulating the pelvic viscera. These axons synapse in the first three sacral segments of the spinal cord with preganglionic parasympathetic neurons; their axons pass through the **pelvic nerve** and project to post-ganglionic neurons in the pelvic viscera, where they regulate glands and smooth muscle including the muscles of the bladder and rectum. The **reticulospinal tracts** are therefore shared by **visceral motor** paths as well as by extrapyramidal, i.e. **somatic motor**, pathways (see Section 12.10).

The reticulospinal tracts have many projections from one side of the body to the other, and for this reason autonomic motor effects may be bilateral rather than unilateral in their distribution to the periphery (see Section 12.10).

16.8 Ascending (afferent) visceral pathways in the spinal cord and brainstem

The cell body of the primary afferent neuron is, of course, always located in a dorsal root ganglion, or in the equivalent ganglion of a cranial nerve (e.g. vagal ganglion).

Visceral pain pathways

In man, some of these pathways ascend in the spinothalamic tract and eventually in the medial lemniscus (see Section 7.6); these are arranged somatotopically, alongside the somatic pain afferents. Others ascend in the spinoreticular tract (see Section 9.9).

Visceral afferent pathways not concerned with pain

These pathways typically enter the CNS via the pelvic nerve, or via the vagus. Those entering via the pelvic nerve (arising for example from stretch receptor endings in the wall of the bladder) ascend the cord in the spino-reticular tract. Those entering the CNS via the vagus (e.g. arising from baroreceptors in the aortic wall or from Hering–Breuer stretch receptors in bronchial walls) project immediately into the nucleus of the solitary tract; their next relay projects into the reticular formation of the brainstem, in the regions of the medullary cardiovascular centres and medullary respiratory centres.

16.9 Effects of lesions in autonomic pathways

Knowledge of the structure and function of the autonomic nervous system is still so incomplete that not many diseases, or syndromes of clinical signs, can be firmly attributed to lesions within the autonomic nervous system, even in man. Nevertheless, disturbance of autonomic pathways within the central nervous system undoubtedly does contribute to the signs and symptoms in many neurological cases.

Lesions in the hypothalamus

In man the hypothalamus has been damaged by fractures of the base of the skull, tumours, and particularly by infections of the brain, but vascular accidents are rare possibly because of the exceptionally profuse vascularity of the hypothalamus (Section 1.10). The observed clinical consequences can be anticipated from the functions of this region of the brainstem and include alimentary disorders, diabetes insipidus (due to loss of antidiuretic hormone, and hence to a failure of water resorption by the distal convoluted tubule), defects in thermoregulation (either hyperthermia or hypothermia), emotional changes, and abnormal sleep (usually excessive sleepiness).

Lesions in the pontine, medullary, and spinal reticular formation

Lesions in the pontine and medullary reticular formation could cause **cardiovascular** and **respiratory disorders**.

Vasomotor, sudomotor, and piloerector disturbances can result from lesions in the autonomic pathways which descend from the medulla oblongata in the reticulospinal pathways. In man and probably in other mammals the most conspicuous result of chronic interruption of these spinal autonomic pathways is **vasodilation** (revealed by the skin becoming warmer). The spinal pathways of vasomotor control tend to be bilateral, as would be expected from the essentially midline organization of the descending reticular pathways (see Section 16.7 above); consequently vasomotor changes may be bilateral even when there is a unilateral lesion in the spinal cord.

Lesions in the spinal cord within the segments of the thoracolumbar outflow may not only interrupt the descending reticulospinal pathways but may also destroy the preganglionic neurons in the lateral horn. In general the clinical signs of lesions involving vasomotor (and sudomotor and pilo-erector) changes tend to diminish with the passage of time.

The thoracolumbar outflow extends from T1 to L2 in the cat, T1 to L3 in man, and T1 to L5 in the dog.

Severe defects in the control of the bladder and rectum may arise from lesions in the sacral segments of the spinal cord which form the pelvic nerve. The muscle of the wall of the bladder will be paralysed and the bladder then becomes distended, with dribbling **incontinence**. The rectum and descending colon are similarly paralysed. No automatic emptying of the bladder can occur, since the reflex arcs, which are normally activated by stretch receptors, are interrupted in the spinal cord itself. Lesions in the spinal cord at more cranial levels may interfere with voluntary control of the bladder and rectum by damaging the reticulospinal tracts. In man automatic emptying of the bladder then presently supervenes. In the dog this may not happen if the lesion paralyses the hindlimbs, as for example in a severe disc protrusion at the thoracolumbar junction; it has been suggested that reflex micturition may then fail because the animal is unable to adopt the right posture.

Horner's syndrome is one group of autonomic clinical signs which is well recognized in man and has also been clearly identified in the domestic species. The main features are constriction of the pupil (**miosis**), and vaso-dilation in the region of the face. The vasodilation can be detected by the warmth of the ears and congestion of the mucous membranes. There is also a slight **ptosis** (drooping of the upper eyelid) due to paralysis of smooth muscle in the upper lid. Furthermore, there is a protrusion of the **nictitating membrane**, caused by paralysis of the smooth muscle which retracts the membrane. In man there is a loss of sweating by the face; in the horse, the opposite effect occurs, i.e. profuse sweating over the face, probably as a result of cutaneous vasodilation since horse's sweat glands secrete in response to increased blood flow. The lesion which causes these signs may have interrupted the sympathetic pathway somewhere between the hindbrain and the target organs in the head. Injury to the descending reticular formation in the caudal part of the brainstem, or a lesion in the reticulospinal tracts of the cervical spinal cord, or destruction of the preganglionic cell bodies in the first few thoracic segments, are all possible **central** causes. **Peripheral** causes include injury to the cervical sympathetic trunk or cranial cervical ganglion, arising for example in the horse after a faulty intravenous injection into the external jugular vein so that an irritative injection mass spreads alongside the vein. Fungal infection of the guttural pouch (diverticulum of the auditory tube) is also a possible cause in the horse. In the cat, infection of the middle ear can produce all the signs of Horner's syndrome except peripheral vasodilation (and excluding also sweating changes, which are not applicable in this species). This is because the sympathetic pathways to the orbit pass through the middle ear cavity. A peripheral lesion necessarily produces its clinical signs on the same side of the head as the lesion.

16.10 Summary of descending autonomic pathways

Essentially the autonomic pathways project downwards by means of the **reticular formation** of the brainstem and the **reticulospinal tracts** of the spinal cord.

(i) The autonomic areas of the cerebral cortex and hippocampus exchange two-way projections with the hypothalamus.

(ii) The hypothalamus is closely associated with the hypophysis and directly continuous with the midbrain reticular formation. The hypothalamus integrates the various mechanisms of homeostasis.

(iii) The midbrain reticular formation projects into the hindbrain reticular formation, thus forming a link between the hypothalamic autonomic centres and the medullary autonomic centres.

(iv) The medullary autonomic centres project, via the reticulospinal tracts, to the lateral horn in the segments forming the sympathetic outflow, to the ventral horn in the segments innervating the diaphragm and intercostal muscles, and to the parasympathetic neurons in the sacral segments of the spinal cord.

17. Cerebral cortex and thalamus

Cerebral cortex

The cerebral 'cortex' is divided into three components: the neocortex, the archicortex, and the paleocortex. The archicortex and paleocortex form respectively the hippoçampus and the piriform lobe, both of which belong to the rhinencephalon; the hippocampus is a major component of the limbic part of the rhinencephalon, and the piriform lobe is an important constituent of the olfactory part of the rhinencephalon (see Section 8.6). This chapter is concerned primarily with the **neocortex**, and particularly with the increasing importance in the highest vertebrates of one particular component of the neocortex, namely the **association cortex**.

17.1 Projection areas and association areas

The primary somatic sensory area, visual and auditory areas, and primary motor areas are called **projection areas**. The cerebral cortex can then be subdivided into three components: projection areas, rhinencephalic (olfactory and limbic) areas, and association areas. The association areas become progressively more and more important in those mammals which are phylogenetically advanced. The association areas (association cortex) and projection areas together constitute the **neocortex**.

Association (which is, of course, the function of the association cortex) consists of receiving items of information, distinguishing between their relative importance, comparing them with previous experience, selecting a suitable response, and predicting its consequences. In its highest forms, association makes possible complex problem-solving and abstract creative thinking. The commonest neurons in the association areas are not neurons forming long fibre tracts, but cells with short axons and very widespread dendrites. These are virtually interneurons connected into a vast meshwork of nerve cells without a beginning or end. The networks which these neurons form are not confined to one layer, but link layer with layer, gyrus with gyrus, and area with area.

17.2 Instinct

Instinct is a pattern of behaviour which is inflexible and automatic rather than flexible and reasoned, being activated by a particular stimulus and genetically determined. According to this definition, instinct excludes conditioned reflexes and behaviour which has been learned. The anatomical site of instinct is difficult to identify, but the co-ordinated behavioural patterns of instinct suggest that it originates in regions involved in memory: the most likely site appears to be the **hippocampus** (the archicortex).

17.3 Cerebral cortex in primitive mammals

In the rabbit and rat the cerebral cortex consists solely of projection areas and rhinencephalic cortex (Fig. 17.1). These are sufficient to fit the animal adequately to life in its own particular environment; the responses to its environment will depend essentially on instinct, suggesting domination by the hippocampus.

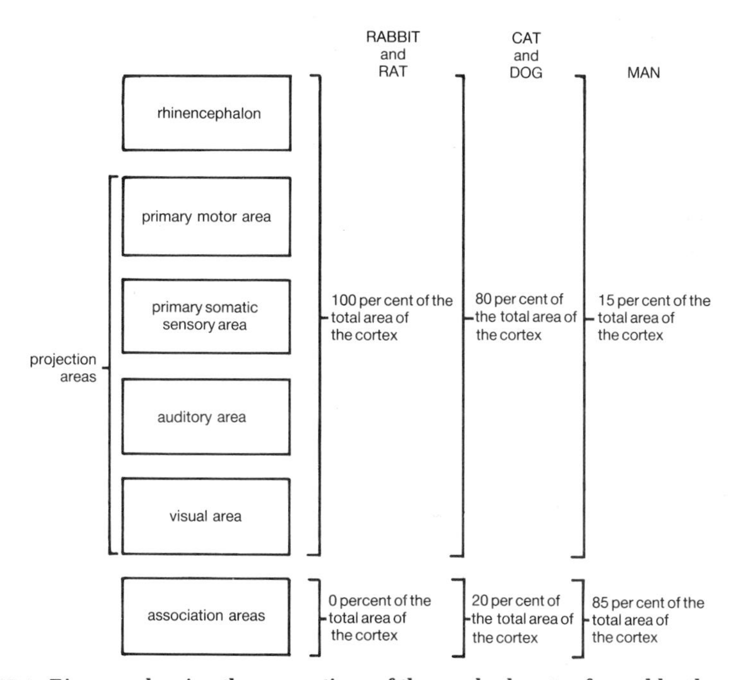

FIG. 17.1. Diagram showing the proportions of the cerebral cortex formed by the projection areas and association areas. Three types of mammal are represented: man; relatively advanced lower mammals (cat and dog); and relatively primitive lower mammals (rabbit and rat).

17.4 Cerebral cortex in cat and dog

The projection areas and rhinencephalic cortex make up about four-fifths of the total cortex (Figs. 8.2(a), 17.1). The remaining areas are association areas. Since the premotor frontal area is still only very slightly developed, the main motor and sensory projection areas are at the rostral end of the cortex.

The association areas in these species are sufficiently developed to enable them to analyse the information they receive, compare it with past experience, and elaborate an appropriate course of action. The varying manner in which an individual cat and dog follows these procedures endows it with a distinctive **personality** of its own. Although the functional capacities of the human association areas are much greater than those of the cat and dog, the difference lies only in the more advanced degree of sophistication in man rather than in any fundamental changes in the way the cortex works.

17.5 Conditioned reflexes

The basic role of conditioned reflexes in the mechanisms of learning in the highest mammals was established by experiments on the dog. The ringing of a bell caused the animal to prick its ears. When food was given at the same time as the bell was rung, the dog salivated. Subsequently, the ringing of the bell induced salivation even though food was not presented. Thus there are innumerable potential neuronal circuits in the association areas, which can be activated and then firmly established by appropriate usage (or training). In this way an animal can **learn** by experience, and consequently vary its responses in the most appropriate manner.

17.6 Cerebral cortex in man

The projection areas and rhinencephalic cortex amount to less than one-sixth of the total cortex (Figs. 8.3, 17.1). The association areas have become relatively much larger, forming four vast cortical areas, the frontal, temporal (interpretative), and parietal and occipital (cognitive) association areas (Fig. 17.2(a)). Of these, a well-developed premotor frontal association area appears for the first time in primates. The four association areas are related to the frontal, temporal, parietal, and occipital **lobes** (Figs. 17.2(a), 17.3). The grooves which separate these four areas, or lobes, are among the first to appear on the surface of the embryonic brain. Each cortical lobe has one of the projection areas in its territory, as well as one of the association areas (Fig. 17.2(a)).

The association areas endow the human brain with its unique capacity for steering the activity of the animal into constructive channels. The size of the

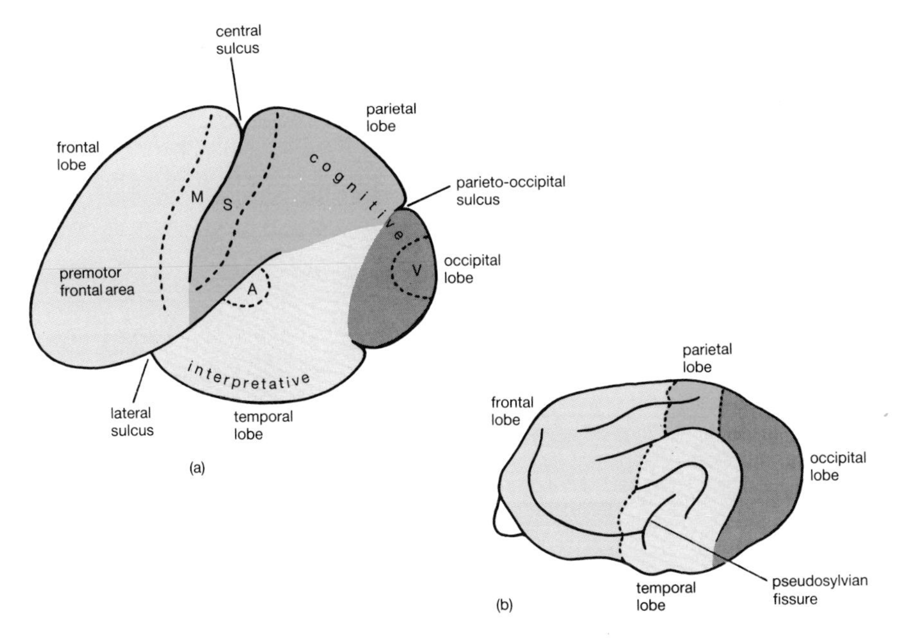

FIG. 17.2. (a) Diagram of the four lobes of the cerebral cortex of man. Projection areas: M, primary motor area; S, primary somatic sensory area; V, visual area; A, auditory area. The term premotor frontal area means the whole of the frontal lobe except the motor area. The association areas are labelled cognitive, interpretive, and premotor frontal areas.

(b) Diagram of the four lobes of the cerebral cortex of the dog. (Based on McGrath, J. (1960). *Neurological examination of the dog*, by courtesy of the author and Lea and Febiger, Philadelphia.)

association areas in general, and of the all-important frontal area in particular, varies much less than the apparent intellectual capacities of individuals or races of mankind.

17.7 Cognitive association area in man

The **parietal** and **occipital lobes** together form the cognitive association area (Figs. 17.2(a), 17.3). Specific afferent stimuli are projected by the thalamus to the primary sensory area. In man this enables **conscious recognition** of localized cutaneous stimulation, as when a coin is held between finger and thumb. Diffuse projections from the primary somatic sensory area to the cognitive association areas (Fig. 17.3) enable the contours and surface of the coin to be recognized as those of a penny. The lateral group of thalamic nuclei also project diffusely to the cognitive areas (Fig. 17.3), reinforcing the recognition of the coin.

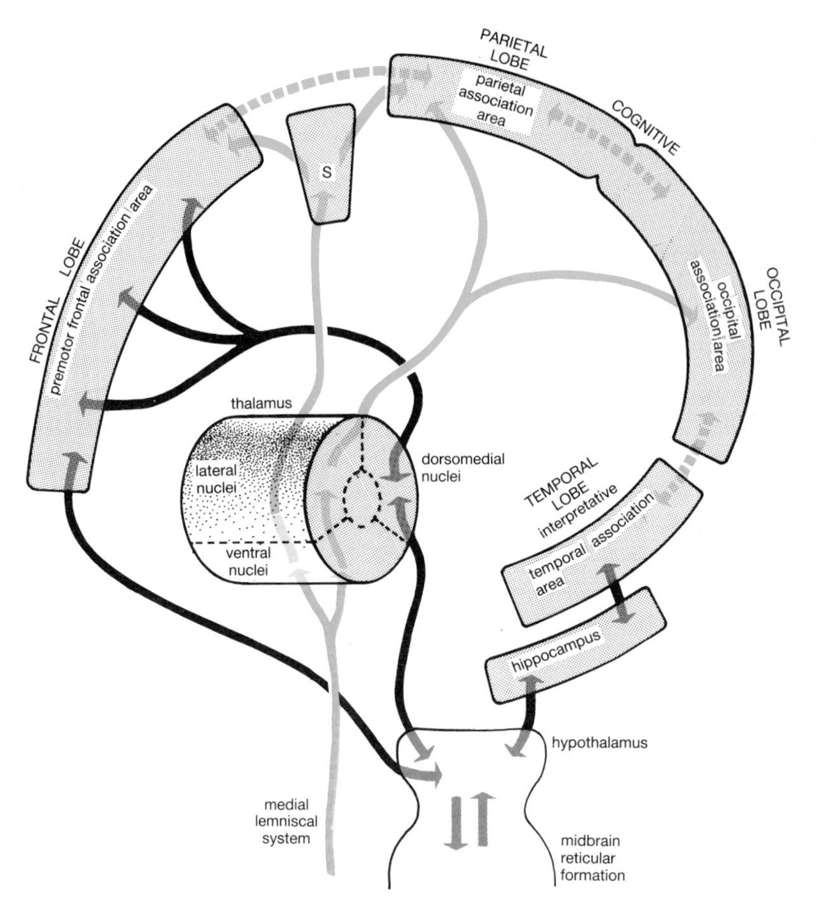

FIG. 17.3. Highly schematic diagram of the association areas of the human cerebral cortex, and some of their main connections. Shown in black are two-way projections between the frontal association area and the hypothalamus. Also shown in black are two-way projections between the hippocampus and hypothalamus. Two-way projections between the hypothalamus and midbrain reticular formation at the bottom of the diagram are again represented in black. Ascending projections of the lemniscal system through the thalamus to the cortex are given in green. Also shown in green, but as broken lines, are the association fibre pathways which interconnect the association areas. S, primary somatic sensory area. The black projections are shown in Fig. 16.3.

17.8 Cognitive association area in carnivores

In the cat and dog, parietal and occipital association areas are present, but it is difficult to evaluate their functions and the effects of lesions within them. In the carnivore brain the frontal, parietal, occipital, and temporal lobes (Fig. 17.2(b)) are not delineated by sulci but correspond roughly to the overlying skull bones (see Section 21.27).

17.9 Interpretative association area in man

The **temporal lobe** forms the interpretative association area (Fig. 17.2(a)). Diffuse connections between the cognitive and interpretative areas (Fig. 17.3) enable a coin to be assessed in terms of what it can buy. This involves having learned by previous experience, and remembered, the value of this particular coin. The temporal lobe is therefore concerned with **learning** and **memory**. It stores memories of past events and releases this information to enable present events to be interpreted. Ablation of the temporal lobe results in loss of **long-term memory**.

The temporal lobe is also concerned with **dreams** and **hallucinations**. During sleep the experiences stored within it may express themselves in the disorder typical of dreams. Damage to the temporal lobe can dislocate the integration of past and present experience, and so produce hallucinations.

The temporal lobe connects with the hippocampus, and the hippocampus has rich projections to the hypothalamus (Figs. 16.3, 17.3). This may explain the relationship between memory and the emotion of **interest** (see Section 16.2); when interest is aroused, memory is good—and the reverse is also true. Destruction of the hippocampus destroys new learning and memory, but old memory persists.

17.10 Interpretative association area in carnivores

The cat and dog have a temporal association area, presumably with functions similar to that of man, i.e. learning and memory which contribute extensively to the individual behaviour and personality of each particular animal. Lesions in the temporal lobe in cats and dogs may cause bilaterally diminished hearing, expressions of panic, undue docility or viciousness, and epilepsy.

17.11 Frontal association area in man

This forms about one-third of the surface area of the whole human cortex (Fig. 17.2(a)). The frontal association area is believed to have special importance for essentially human attributes. It possesses a complex tangle of fibre connections, largely reciprocal, with many cortical and subcortical structures. They include direct and indirect two-way connections with the hypothalamus (Figs. 16.3, 17.3) and with other parts of the limbic system (amygdaloid body, septal nuclei, and cingulate gyrus), and these projections participate in emotional and behavioural functions. There are also vast numbers of diffuse reciprocal connections with the temporo–parieto–occipital association cortex by association fibres (Fig. 17.3), and these are evidently involved in intellectual functions.

The brain injuries of the First World War threw the first light on the functions of the frontal lobe. The first experiments were made in 1935 on the frontal lobe of apes, and shortly after this operations began on the frontal lobe of man.

Inhibition of the emotional urges of the hypothalamus is achieved in man by the two-way connections (Figs. 16.3, 17.3) between the frontal association area and the hypothalamus, including the indirect reciprocal pathways of the dorsomedial nucleus of the thalamus. (The strong emotional urges of the hypothalamus are also inhibited by the **hippocampus**, section 16.2.) This inhibition of the spontaneous and often violent emotional responses of **rage** etc. is beneficial because it enables man to set aside continual minor distractions and concentrate on long term objectives. However, over-activity of the frontal lobe may be a factor in severe chronic depression and anxiety. Interruption of these frontal–hypothalamic–thalamic–frontal projections (by cutting beneath the frontal cortex) has been used to treat this condition. Although this may give relief, it may also create other problems such as social irresponsibility and loss of concentration.

17.12 Frontal association area in carnivores

In the dog (and probably also the cat) the frontal association area is present though not well developed. Nevertheless it evidently contributes quite substantially to the general alertness, intelligence, and temperament of the individual animal. Lesions in the frontal association area of the cat and dog cause changes in behaviour and personality, leading to stupidity and inactivity, or viciousness and hyperactivity. Epilepsy may be expected.

17.13 Corpus callosum

The corpus callosum is the main connection between the left and right cerebral cortex, comprising millions of axons. Nevertheless, cutting of the corpus callosum in man and experimental animals produces surprisingly small disturbances which can be detected only by careful testing. However, it is clear that learning in one hemisphere is usually inaccessible to the other hemisphere if the corpus callosum is cut. The corpus callosum allows the two hemispheres to share learning and memory. Thus a cat can be trained to distinguish the surface textures of two pedals. In a normal cat, whatever is learned with one paw is transferred to the other paw. But an animal in which the corpus callosum is cut must learn every task separately for each paw.

Other functions are:

(i) Correlation of images in the left and right cortical visual areas. However, this applies only to animals in which the eyes are directed forward and the optic fibres do not entirely decussate at the chiasma.

(ii) Learning to co-ordinate skilled motor activities of the left and right limbs simultaneously.

17.14 Effects of extensive damage to the cerebral hemisphere in domestic mammals

The clinical effects of relatively localized lesions in the pyramidal pathways were reviewed in Section 11.6. The following survey considers the clinical consequences of large, space-occupying, lesions in the cerebral hemisphere. Extensive damage to one or both of the cerebral hemispheres occurs in the sheep as a result of intracranial cysts of the parasite Coenurus cerebralis, and in the dog as a result of hydrocephalus and intracranial tumours. Similar lesions are less common in the horse, ox, pig, and cat, but tumours or abscesses may be encountered occasionally in any of the domestic species.

Cases with massive unilateral lesions often show **locomotor** and/or **postural disturbances**. The most conspicuous of these are **circling** and **aversion of the head**, but wheelbarrow or hemiwalking tests may reveal a locomotory deficit. A unilateral locomotor deficit is consistent with a contralateral cerebral lesion. However, in some cases there may be circling but no locomotor deficit; in others there may be a locomotor deficit but no circling. The direction of circling is not a reliable indicator of the side of the lesion.

Of sheep with a Coenurus cyst in one cerebral hemisphere, only about one-third circle. When **circling** does happen, it is slightly more often towards the side of the lesion. Furthermore, only about one-third of affected sheep show **aversion of the head** (see Section 8.3 for definitions of head aversion and head tilt). In these cases, the ear goes down on the side opposite to the lesion slightly more commonly than it goes down on the same side as the lesion, but almost as often the ear goes down inconsistently first on one side and then on the other. For these reasons, the direction of circling and head aversion are not reliable indicators of the side of the lesion, at least in sheep. The wheelbarrow and hemiwalking tests for locomotory, postural, and proprioceptive deficits are much better indicators of the side of a unilateral cerebral lesion in sheep; since the majority of fibres in these systems decussate, a unilateral deficit (as revealed by a wheelbarrow or hemiwalking test) would indicate a contralateral lesion. On the other hand not all cerebral cysts in sheep give locomotory or postural deficits. If there is a unilateral cerebral cyst with a contralateral locomotory-postural deficit, the deficit will become much more obvious if the animal is blindfolded, since this prevents the animal from using visual information to compensate for proprioceptive deficits. **Dogs** with cerebral tumours may circle, and it is generally stated that typically they do so towards the side of the lesion.

The mechanism of circling in unilateral cerebral lesions is not clear, but it is almost certainly due to involvement of deep components of the cerebral hemisphere, and not to injury of the cerebral cortex itself. Experimental stimulation of the caudate nucleus on one side in unanaesthetized cats may cause circling to the opposite side; this suggests that complex disturbances within the caudate nucleus, or the basal nuclei in general, contribute to the circling which characterizes major damage to a cerebral hemisphere. However, it is possible that mechanisms of circling may lie even deeper within the cerebral hemisphere, since circling has been observed with lesions in the rostral region of the thalamus. Circling has also been reported after more caudal lesions of the brainstem, for example lesions of the vestibular nuclei or of the middle cerebellar peduncle; the latter has been explained by interference with the corticopontocerebellar pathway. Yet other possible mechanisms of circling exist. For example in rats, experimental stimulation has indicated that circling can be induced from a longitudinal bundle of fibres passing in the midline between the caudal midbrain and the tegmentum of the pons; tectospinal fibres and/or the medial longitudinal fasciculus may be involved.

Consistent with the probability that a large cerebral lesion will destroy some of the association areas of the cortex is the fact that the animal may undergo a **change of personality**. A bad-tempered dog may become more tractable, and a good-tempered animal may become vicious. Furthermore, there may be epileptic fits (see below).

There may also be **visual deficits**. With unilateral lesions the animal may appear to be blind on the side opposite to the lesion, with a loss of the contralateral visual placing response. However, if the lesion is unilateral, a careful examination of a dog or cat should reveal **hemianopia**, i.e. a reduced nasal field of vision on the side of the lesion and loss of the temporal field of vision on the side opposite to the lesion, as would occur in a lesion of the optic tract. In the dog the eye on the side opposite to the lesion is much the more severely affected, since in this species about three-quarters of the optic fibres decussate (see Section 8.1); it is this that would make a dog seem blind on one side only (the side opposite to the lesion). (To work out the rationale of these visual signs examine Fig. 8.1).

A reduced nasal field of vision on one side and loss of the temporal field on the other side is known as **homonymous hemianopia**. If light coming from the left side fails to induce a response in (a) the nasal half of the left retina, and (b) the temporal half of the right retina, the condition is known as **left** homonymous hemianopia.

Cutaneous sensation may be reduced on the contralateral side of a unilateral cerebral lesion.

When a very large area, or even the whole, of both of the cerebral hemispheres is destroyed an animal can suck, chew, and swallow, but is unable to prehend food. Consequently, solids and liquids have to be introduced into the mouth. Once there, they induce the reflex arcs of chewing and swallowing. The effects in the dog of destruction of the primary motor area alone are much less severe (see Section 11.6).

17.15 Epileptic fits

Epileptic fits are also known as convulsions, seizures, epilepsy, or simply as 'fits'. They tend to be abrupt in onset, short-lasting, and recurrent, and to be followed by a period of depressed nervous activity. The basic neural abnormality is a wave of spontaneous excitation which originates in a small group of neurons of the motor cortex and then progressively spreads more or less extensively over the motor cortex. The excitation may remain quite focal, causing no more than rhythmic muscular movements within part of a single limb. Confinement of the signs to one side of the body may indicate a lesion in the contralateral cortex. In other instances where the excitation remains focal the only signs may be altered behaviour such as sleepiness, aggressiveness, or panic, and sometimes apparent blindness. If the excitation of the cortex becomes generalized (as is usual in the dog), many lower motor neurons on both sides of the brain and spinal cord may eventually become involved; in such instances an inital phase of minor behavioural abnormality is followed by rigidity, falling over, and bilateral running actions, with chewing movements, salivation, and sometimes also urination and defecation. The whole episode often lasts only a few minutes. The animal generally loses consciousness.

Epileptic fits may arise in metabolic or toxic diseases, or as a result of intracranial changes which disturb the actual structure of the cerebral cortex. The latter group of disorders appears to be the most common cause; encephalitis of canine distemper is a notable example, and tumours, hydrocephalus, and trauma are also important possibilities which should be assessed by careful neurological examination in the intervals between seizures. In the large farm animals metabolic disorders such as hypomagnesaemia ('grass staggers'), and toxic conditions such as lead poisoning in calves and salt poisoning in pigs are important possibilities.

Epileptic fits also occur in animals and man without any apparent pathological cause. This form of epilepsy (idiopathic epilepsy) has no known cause, but since it is relatively common in certain canine and bovine breeds it is belived to be influenced by genetic factors.

17.16 Histology of cerebral cortex

The great majority of the nerve cells in the cerebral cortex are either the relatively small **stellate nerve cells** (also known as **granular cells**) or **pyramidal nerve cells**; these two types of neuron are involved with input and output respectively. They are arranged essentially in six parallel layers in the same plane as the outer surface of the cortex (Fig. 17.4). The outermost layer is the **molecular layer**, layer I. It consists of the dendrites of pyramidal cells, the fibres of stellate cells, and the afferent endings of projection fibres and association fibres. Layers II and IV are the **external** and **internal granular layers**. These two layers consist mainly of stellate cells, the axons of which are confined to the cortex, ramifying mainly vertically (both centripetally and centrifugally) and also horizontally (layer II also contains many small pyramidal cells, of which the axons may terminate in the cortex). Layers III and V are the **external** and **internal pyramidal layers** and these contain mainly pyramidal cells. The apices of **pyramidal cells** point towards the surface, each giving rise to a thick apical dendrite which extends towards or into the molecular layer; the lateral angles of the pyramidal cells form shorter horizontal dendrites. The axons of pyramidal cells are much thinner than the main dendrites. Some of the

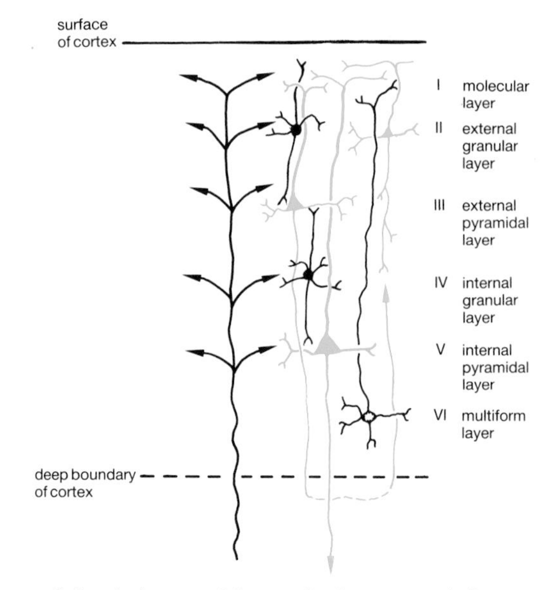

FIG. 17.4. Diagram of the six layers of the cerebral cortex and the neurons which characterize them. The green neurons are pyramidal nerve cells, medium-sized in layer III and large-sized in layer V. The axons of the medium-sized pyramidal cells end in the cortex, or enter the white matter and then return to the cortex (as in the diagram). The large pyramidal cells form corticofugal axons, including corticospinal (pyramidal) pathways. The black neurons in layers II and IV are stellate (granular) cells. Their axons ramify mainly vertically (both upwards and downwards) and remain in the cortex. The small (unshaded) neurons in layer VI form axons which go mainly to the more superficial layers. On the left is an incoming thalamo-cortical fibre projecting into all six layers.

axons of the medium-sized pyramidal cells which typify layer III may end in the deeper layers of the cortex; many others enter the white matter but return to the cortex, where they ramify as association or commissural fibres. The large pyramidal cells which characterize layer V form corticofugal fibres. These project, for example, to the basal nuclei, to brainstem nuclei, and to the spinal cord, the latter projections forming corticospinal pathways; others are association fibres, which leave the cortex and go to more or less distant cortical regions. Layer VI is named the **multiform layer**. It consists mainly of small nerve cells with axons which are distributed chiefly in the more superficial layers of the same or the contralateral side of the cortex. In the somatic, auditory, and visual **sensory cortices** each incoming thalamic fibre is associated with a cylinder of cortical cells extending through all six layers.

Thalamus

A thalamus for collecting afferent information first appears in amphibia. This can be associated with the differentiation of the grey matter into paleo- and archicortex, which also occurs first in amphibia. As the cerebral cortex advances (Chapter 18), especially with the expansion of the neocortex in mammals, so the thalamus advances also.

Four groups of thalamic nuclei are of particular functional interest, ventral, lateral, central, and dorsomedial.

17.17 Ventral group of thalamic nuclei

This group has four functional components:

Component 1: This forms the cell station of the third and final neuron of the medial lemniscal system (see Section 7.3, and Fig. 7.3).

The cell station is in a subdivision of the ventral group named the ventrocaudal thalamic nucleus.

Component 2: The second component is made up of the **lateral** and **medial geniculate nuclei**, forming respectively the cell stations of the third neuron of vision and the third neuron of hearing (see Sections 8.1, 8.2, and Fig. 8.1).

Strictly, the geniculate nuclei constitute the metathalamus.

Component 3: The third component of the ventral group of thalamic nuclei is entirely concerned with feedback. It forms the cell station of the final neuron in the cerebellar–pyramidal feedback pathway (see Section 11.2), and the final neuron in the cerebellar–extrapyramidal feedback pathways to the highest extrapyramidal command centres (see Section 13.2). It also forms the feedback link between the basal nuclei (globus pallidus) and the extrapyramidal motor areas of the cerebral cortex (see Section 13.2). Furthermore it enables the basal nuclei to carry out their main function, namely collaborating with the cerebral cortex (see Section 13.2).

The name of the third component is the ventrolateral thalamic nucleus.

Component 4: The fourth component of the ventral group of thalamic nuclei projects only to the lateral group of thalamic nuclei, thus forming a link between the medial lemniscal system and the cognitive cortex (Fig. 17.3).

17.18 The lateral group

These nuclei receive afferents from the ventral group only (component 4, above). They project to the parietal and occipital (cognitive) areas of the cerebral cortex (Fig. 17.3). Thus, as just stated, they form one of the links between the sensory information arriving in the medial lemniscal system on the one hand, and the cognitive cortex on the other.

17.19 Central (or intralaminar) group

This group of nuclei projects from the ascending reticular formation, diffusely over the **whole** cerebral cortex (Fig. 9.3). This group of nuclei is the final step in the ascending reticular formation.

Strictly, the central group of thalamic nuclei is called the intralaminar group. This group also projects to the basal nuclei, and thus forms the main pathway for alerting the motor systems at the highest **non**-cortical level.

17.20 Dorsomedial group

The dorsomedial group of thalamic nuclei exchanges two-way projections with the hypothalamus, and with the frontal lobe of the cortex (Fig. 17.3). These reciprocal projections contribute to the two-way pathways between the frontal association area and the hypothalamus.

17.21 Summary of incoming afferent paths to the thalamus

(i) With the exception of olfaction, all the modalities of the special senses, and the modalities of touch, pressure, joint proprioception, pinprick pain, heat, and cold (the medial lemniscal system) eventually relay in the ventral group of nuclei in the thalamus (the lateral and medial geniculate nuclei being part of the ventral group of thalamic nuclei). Muscle proprioception projects extensively to the cerebellum, but it is now established that muscle proprioceptor pathways also reach the cerebral cortex via the medial lemniscus and ventral group of thalamic nuclei, and thus contribute to the sense of limb movement and position (kinaesthesia) (see Section 9.5).

(ii) The ascending reticular pathways relay in the central group of thalamic nuclei.

(iii) The pyramidal and extrapyramidal feedback circuits relay in the ventral group of thalamic nuclei. Most of them come from the cerebellum; others arise from the basal nuclei, notably from the globus pallidus.

17.22 Summary of the projections from the thalamus to the cerebral cortex

These are partly specific and partly diffuse.

Specific projections

(i) The classical modalities which arrive in the ventral group of thalamic nuclei, via the medial lemniscal system, are projected point-to-point to the primary somatic sensory projection area.

(ii) The projections of vision and hearing, arriving (via the optic and auditory radiations) from the lateral and medial geniculate nuclei, project to the visual and auditory areas.

(iii) The pyramidal feedback circuits project to the primary motor area.

Diffuse projections

(i) The central group of thalamic nuclei, which receives the ascending reticular formation, projects diffusely over the whole cerebral cortex.

(ii) The ventral group of thalamic nuclei projects extrapyramidal feedback diffusely over the extensive extrapyramidal motor areas of the cerebral cortex.

(iii) The lateral group of thalamic nuclei, forming a link between the ventral group and the cognitive association area, projects diffusely to the parietal and occipital association areas.

(iv) The dorsomedial group projects diffusely to the frontal lobe, forming a link between hypothalamus and frontal lobe.

Together, the point-to-point and diffuse projections from the thalamus to the cortex are like the spokes of an umbrella, the cerebral cortex being the cover of the umbrella and the thalamus the hub.

17.23 Summary of functions of the thalamus

(i) To alert the whole cerebral cortex diffusely, and then to project the specific point-to-point stimuli for conscious discrimination.

(ii) To distribute information to the cognitive (and frontal, when present) association areas.

(iii) To participate in regulation of the motor activities of the cerebral cortex by completing the feedback circuits from the cerebellum and basal nuclei, and by providing general connections from the basal nuclei to the cerebral cortex.

17.24 Clinical effects of lesions of the thalamus in domestic mammals

Massive lesions of the thalamus are likely to include the central (intralaminar) group of thalamic nuclei, and would then lead to impaired consciousness or, in severe cases, **coma**; however, coma can arise from damage to the ascending reticular formation anywhere in the higher levels of the brainstem (see Section 9.9, **Effects of destruction**).

Focal lesions of the thalamus could in theory produce more restricted deficits. For example, injury to the lateral geniculate nucleus would cause hemianopia; damage to the ventrocaudal thalamic nucleus would interrupt the medial lemniscal system and therefore lead to loss of somatic sensation on the opposite side of the body and also give a contralateral kinaesthesia deficit; a lesion in the ventrolateral thalamic nucleus should interfere with cerebellar feedback pathways to the motor centres of the forebrain, resulting in dysmetria and other signs of asynergia (see Section 15.13). However, it would be difficult clinically to localize such deficits to the thalamus, and focal lesions of the thalamus are believed to be rare anyway.

17.25 Clinical effects of lesions of the thalamus in man

Tumours of the diencephalon and mesencephalon are often accompanied by loss of consciousness, presumably through disruption of the ascending reticular formation. **Destructive lesions** of the ventrocaudal nucleus caused by vascular accidents or tumours give rise to striking clinical signs. A severe lesion is followed by anaesthesia on the opposite side of the body, as is predictable from the functions of the medial lemniscal system. But if the lesion is less severe there is usually a contralateral hyperaesthesia, so that a mild cutaneous stimulus such as contact with a piece of ice produces intolerable pain. If the nucleus is affected by an **irritative lesion**, for example due to a deteriorating blood supply, contralateral burning pain of very severe intensity may be started by merely moving the limbs or may occur spontaneously without any stimulus at all. Surgical destruction (coagulation) of the ventrocaudal nucleus changes the lesion from an irritative to a destructive lesion and results in hemianaesthesia. Thalamic lesions may also be accompanied by marked emotional instability with bouts of compulsive laughing and crying.

Growth of the human brain

The human brain goes on growing long after birth. The most rapid growth occurs during the first two years after birth. By the sixth year, the increase in the surface area of the whole cortex has been sixfold. The frontal lobe continues to grow long after the other areas of the brain have finished, finally completing its maximum enlargement by 10 years. At this stage the brain has reached its full size.

Sheer increase in surface area is not the whole story, however. The axons of the cerebral cortex do not all acquire their myelin sheaths at the same period of development, and it is generally believed that myelinated fibres cannot conduct nerve impulses until they possess a myelin sheath. Myelination can only proceed after the multiplication of glial cells, most of which occurs after birth in man. The first axons in the cerebral cortex to become myelinated are the afferent fibres passing to the primary somatic sensory projection area, these being completed by the eighth month of fetal life. The next to acquire myelin are the afferent fibres to the visual and auditory projection areas. Myelination of the pyramidal pathway is not completed until the third month after birth, so that any movements made before that age are mostly reflex from the spinal cord and brainstem. Subsequently the axons of the association areas become progressively myelinated. Until they are, a child cannot perform integrated voluntary acts or benefit fully from experience. Myelination is largely concluded by the end of the third year However, since the process is not completed in the association areas until the eighteenth year, full mental development may not be possible until then.

Experiments on rats indicate that the phase of glial proliferation is vulnerable to malnutrition, which permanently stunts the growth of the brain. This hazard almost certainly applies also to children growing up under famine conditions.

18. Comparative neuroanatomy

Evolution of the vertebrate forebrain

18.1 Primitive vertebrates

In very primitive vertebrates such as cyclostomes (for example, the lamprey) the whole cerebral hemisphere is an olfactory organ. The grey matter lines the **inner** surface of the ventricle of the brain, as it does in the spinal cord, and is little differentiated (Fig. 18.1). In the forebrain of primitive vertebrates the hypothalamus is also a major region.

In somewhat more advanced vertebrates, it seems likely that dorsal and ventral laminae became differentiated (Fig. 18.2), resembling the dorsal and ventral laminae of the neural tube of the mammalian embryo (Fig. 6.3). As evolution proceeded, the dorsal laminae probably developed into the paleo- and archicortex: the ventral laminae evidently formed the basal nuclei (Fig. 18.2).

18.2 Contemporary amphibian

The grey matter still lines the ventricle, but has divided into three zones

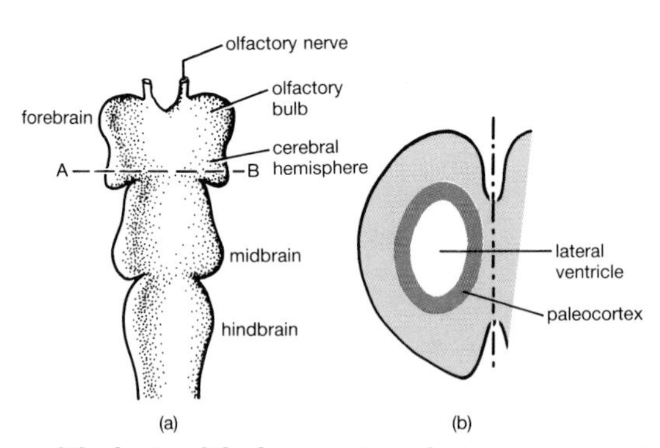

FIG. 18.1. Diagrams of the brain of the lamprey (a cyclostome), a very primitive vertebrate. The whole of the forebrain, including the cerebral hemisphere, is purely olfactory in function. (a) dorsal view. (b) Transverse section through cerebral hemisphere at A–B, showing that the grey matter is **internal**, lining the lateral ventricle. Only the paleocortex, olfactory in function, has been differentiated.

(Fig. 18.3): (i) the basal nuclei, ventrally; (ii) the paleopallium or paleocortex dorsolaterally, which remains largely olfactory; and (iii) the archipallium or archicortex dorsomedially, which retains some olfactory functions but is differentiating itself more towards emotion, and will eventually form the hippocampus.

The paleocortex and archicortex arise from the grey matter on either side of the dorsal half of the neural tube, and their position therefore corresponds to the dorsal laminae of the neural tube of the early mammalian embryo. The basal nuclei correspond in position to the ventral laminae of the embryonic neural tube. The afferent thalamus appears for the first time in this group.

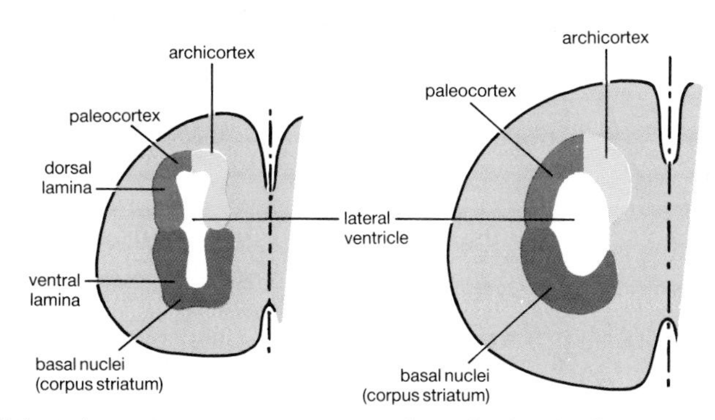

FIG. 18.2. (left) Schematic transverse section through the forebrain of a hypothetical primitive vertebrate. The dorsal and ventral laminae form the grey matter as in a mammalian embryo (Fig. 6.3).

FIG. 18.3. (right) Schematic transverse section through the forebrain of a contemporary amphibian. The grey matter is differentiated into three components, but is still adjacent to the lateral ventricle.

18.3 Contemporary advanced reptile

Three developments have occurred. (i) The **paleocortex** (still olfactory), and the **archicortex** (both olfactory and emotional), have detached themselves from the ventricle and moved centrifugally **towards the surface** (Fig. 18.4). (ii) A primordium of the **neocortex**, homologous to the mammalian neocortex, may have appeared for the first time. It may take the form of a small new type of multilayered cortex, separating the two older cortices; alternatively, and more probably, it is represented by neurons which have migrated (inward-pointing arrow in Fig. 18.4) into the dorsolateral region (external striatum) of the basal nuclei. (iii) The **basal nuclei** remain at the ventral aspect of the ventricle; as in birds (see Section 18.5), they consist of a ventromedial internal striatum (paleostriatum), and a dorsolateral external striatum which may well contain the neurons homologous to those of the mammalian neocortex.

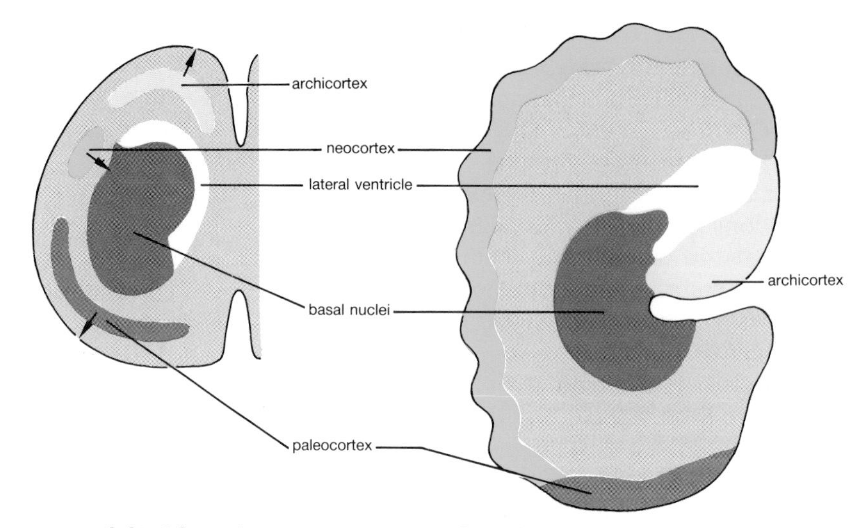

FIG. 18.4. (left) Schematic transverse section through the forebrain of a contemporary reptile. The paleocortex and archicortex are moving centrifugally towards the surface (outward-pointing arrows). The neocortex may have appeared for the first time, but is more likely to be represented by neurons which are migrating (inward-pointing arrow) into the dorsolateral region of the basal nuclei (corpus striatum).

FIG. 18.5. (right) Schematic transverse section through the forebrain of a contemporary mammal. The archicortex is rolled under the lateral ventricle by the expanding neocortex, forming the hippocampus. The paleocortex is forced into a ventral position by the neocortex, forming the piriform lobe, which is still olfactory in function. The basal nuclei (corpus striatum) remain adjacent to the lateral ventricle.

18.4 Mammal

Five major evolutionary changes have taken place (Fig. 18.5). (i) The **neocortex** has undergone a huge expansion, being multilayered and, in advanced types like carnivores and primates, folded to increase its surface area still further. (ii) The **paleocortex** is squashed into a purely ventral position, and is still olfactory in function (**piriform lobe**). (iii) The **archicortex** is rolled medially and beneath the lateral ventricle, and is now called the **hippocampus** (involved in the motor manifestations of emotion, and only remotely olfactory). (iv) The **paleo-, archi-,** and **neocortex**, are all entirely on the surface of the brain, i.e. truly cortical. (v) The **basal nuclei** remain at the ventral aspect of the ventricle, i.e. in the primitive deep position in which the grey matter surrounds the central canal.

18.5 Bird

The homologies of the avian forebrain are difficult to identify, but are beginning to become clear. (i) Although present, and on the surface (i.e. truly cortical), the **paleocortex** and **archicortex** are largely suppressed by the huge centrally situated enlargement of what has been regarded until recently as the **basal nuclei** (or corpus striatum) (Fig. 18.6). (ii) It has been suggested that a small area of **neocortex** may be present, but it now appears that birds do not have any neocortex as such. (iii) It is becoming apparent that the avian **'basal nuclei'** (corpus striatum) are not fully homologous to the basal nuclei of mammals. On the contrary, the massive dorsolateral component of the avian 'basal nuclei', namely the **external striatum**, has much in common with the mammalian neocortex; both of them receive visual and auditory projections, and both of them project motor pathways to the hindbrain and spinal cord. Consequently, it has been suggested that neurons homologous to those of the mammalian neocortex are present in the avian external striatum. This means that the vast expansion of the neocortex is not a totally new development in mammals after all; it also implies the existence of a neuronal population which occupies either the avian (and probably the reptilian) external striatum, or the mammalian neocortex. At present a missing link in this interpretation is the apparent absence of anatomical evidence for a substantial somatic sensory projection to the avian external striatum, comparable to the medial lemniscal projection to the somatic sensory areas of the mammalian neocortex, although there is physiological evidence for such a pathway in birds. The ventromedial region of the 'basal nuclei' (corpus striatum), known as the **internal striatum** (paleostriatum) in both birds and reptiles, is probably the sole true homologue of the mammalian basal nuclei.

The **external striatum**, in both reptiles and birds, consists of four components. Of these the hyperstriatum is the most dorsal and is divided into dorsal, ventral, and accessory regions. The ectostriatum, and more caudally the archistriatum, are lateral in position. The neostriatum forms the massive central region of the forebrain. The **internal striatum** consists of a medial region, the paleostriatum primitivum, and a larger and somewhat more dorsolateral region, the paleostriatum augmentatum.

18.6 Major homologies in mammals and birds

The following table summarizes the supposed homologies of some of the principal landmarks in the mammalian and avian brains.

Mammal	**Bird**
Lateral geniculate nucleus	Nucleus opticus principalis thalami
Medial geniculate nucleus	Nucleus ovoidalis
Rostral colliculus	Optic lobe
Caudal colliculus	Nucleus mesencephalicus lateralis, pars dorsalis
Corpus striatum	Internal striatum
Caudoputamen	Paleostriatum augmentatum
Globus pallidus	Paleostriatum primitivum
Amygdaloid body	Ventrocaudal part of hyperstriatum

The term **neostriatum** has been used in two different ways in vertebrate neuroanatomy. In the bird it applies to a major part of the **external** striatum, and is not homologous to any one structure in the mammalian forebrain. In the mammal the term neostriatum is applied to the caudoputamen, which is homologous to a part of the avian **internal** striatum. Thus the neostriatum of the bird is not homologous to the neostriatum of the mammal.

Evolution of the capacity to differentiate sensory modalities

18.7 Lower vertebrates, including amphibians

Lower vertebrates possess only a simple network of neurons in the brain and spinal cord. There are no long collaterals within the neuraxis for projecting specific impulses. Amphibians have a good array of peripheral receptors, but these discharge directly into the neuronal network of the neuraxis, so that specific pathways are entirely lost at the first synapse in the neuraxis.

18.8 Advanced reptiles and birds

These animals have a full range of equipment for transmitting specific afferent modalities to the highest parts of the brain: they have specialized receptor endings peripherally, and well-defined ascending tracts in the spinal cord and brainstem. However, it is not clear whereabouts in the brain the discrimination between the conscious modalities is carried out. Since a true neocortex is probably not present in reptiles and birds, some other region or regions must be responsible for this function. The answer may lie in the possibility (outlined in Section 18.5) that the external striatum in these vertebrates includes a population of neurons which are homologous to the neurons of the mammalian neocortex. If so, the external striatum in reptiles and birds is probably the site of conscious discrimination between the various modalities.

18.9 Mammals

Mammals possess differentiated receptors, long collaterals projecting specific modalities to the brain, and a neocortex capable of precise discrimination.

Special features of the avian brain

18.10 Size of the brain

Although proportionately smaller than the brain of mammals, the avian brain is much larger (5 to 10 times) than the brain of reptiles of comparable body weight.

18.11 Poor development of the cerebral cortex

The **dorsal** part of the old original grey matter around the lateral ventricle has scarcely been exploited. In mammals this dorsal part of the grey matter has formed the archicortex, paleocortex, and neocortex. Only the first two of these are present in birds; they lie on the surface as a true cortex, but they are small in area and are not convoluted at all. So poorly exploited is this dorsal wall of the original ventricle that the lateral ventricles of the bird lie only just beneath the dorsomedial surface of the forebrain (Fig. 18.6).

There is no primary motor area of cerebral cortex, no pyramidal tract, and no medullary pyramid. The corticospinal pathway is unique to mammals, being presumably one of the benefits of the rise of the true neocortex in mammals. On the other hand, birds do have a cerebrobulbar and cerebrospinal pathway of long fibres which descend from the external

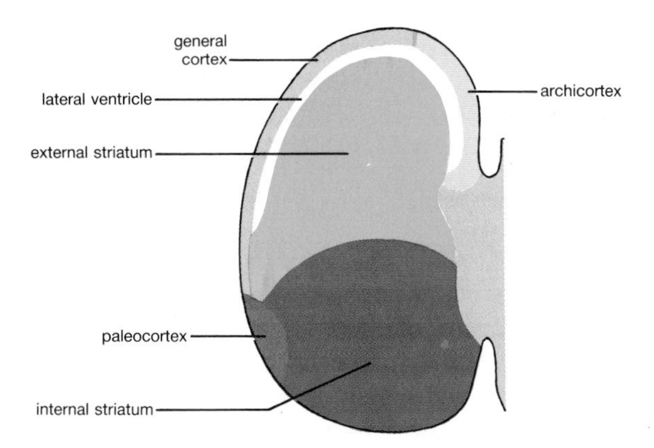

FIG. 18.6. Schematic transverse section through the forebrain of a contemporary bird. The archicortex is medial, forming the hippocampus. The paleocortex is ventrolateral, forming the olfactory cortex. Dorsolaterally the surface layer consists of a relatively thin undifferentiated layer of general cortex, unshaded. Of the huge central zone of the forebrain, the massive dorsolateral component is the external striatum which may contain a population of neurons homologous to those of the mammalian neocortex. The green colour of the external striatum indicates the probable inclusion of this 'neocortical' neuronal population. The more ventral component of the forebrain, known as the internal striatum, is probably homologous to the basal nuclei (corpus striatum) of mammals. The external and internal striatum together were previously regarded as the avian basal nuclei (corpus striatum). The lateral ventricle is very near the dorsal surface.

striatum and project to the medulla oblongata and at least to the cervical spinal cord. This avian pathway could be homologous to the pyramidal system of mammals.

18.12 External striatum

The external striatum is evidently a major new development in birds, paralleling the development of the neocortex in mammals (see Section 18.5).

18.13 Colliculi: 'optic lobe'

The rostral colliculus is enormously enlarged in all birds, forming the so-called optic lobe (Fig. 18.7(a), (b)). Its position is more lateral than the rostral colliculus of the mammal, and this appears to be because it has inadequate space to develop to its full size underneath (ventral to) the

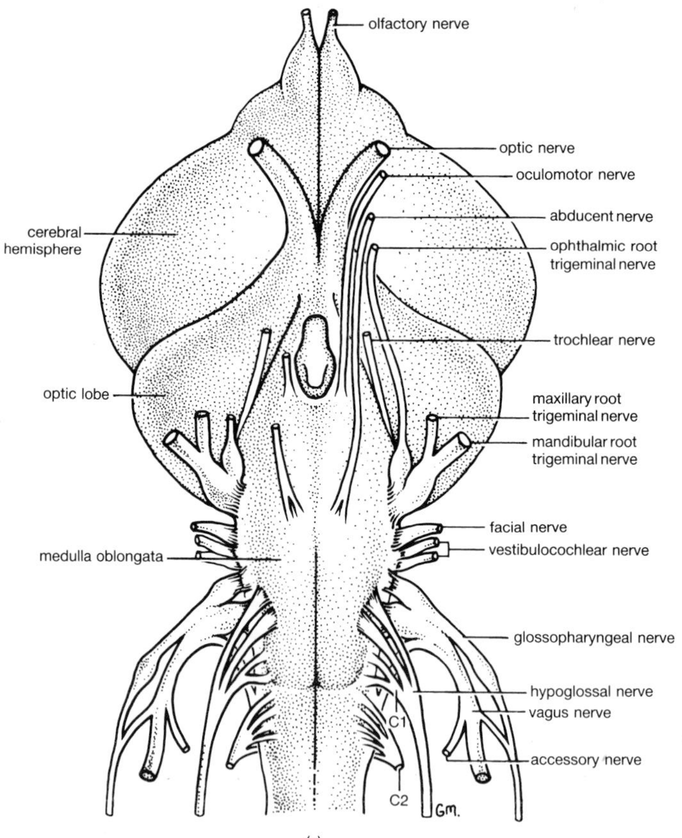

FIG. 18.7. (a) **The brain of a bird.** Ventral view of the brain.

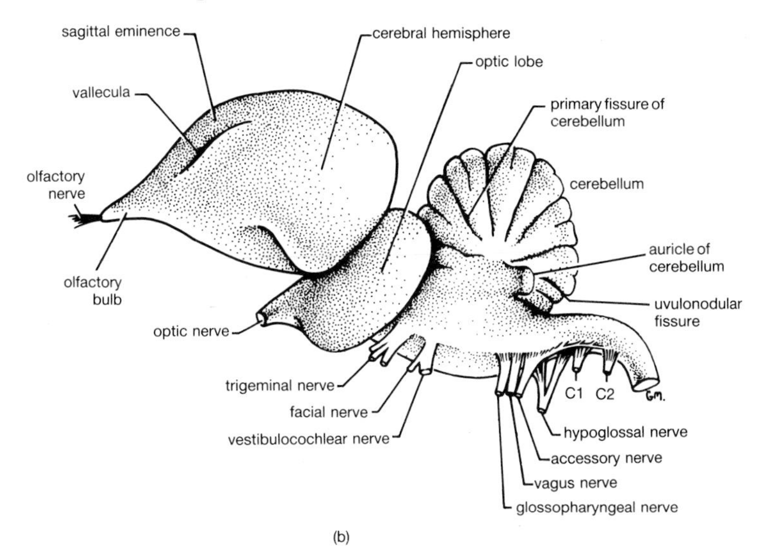

FIG. 18.7. (b) The brain of a bird. Lateral view of the brain.

hemisphere of the forebrain. The optic lobe is also notable in having an outer coat of multilayered cortical grey matter, and this makes its complexity similar to that of the mammalian cerebral cortex or cerebellar cortex.

The huge enlargement of the rostral colliculus is associated with the great development of vision in birds. The avian eyeball is vast in size, that of the ostrich being absolutely the largest among contemporary vertebrates; the cross-sectional area of the optic nerves, if the left and right nerves are added together, exceeds the cross-sectional area of the cervical spinal cord. In the lower vertebrates, including birds, amphibia, and possibly reptiles, extensive analysis of visual impulses occurs in the rostral colliculus. In birds there are also visual projections to the external striatum, thus strengthening the view that the external striatum contains neurons which are homologous to those of the mammalian neocortex.

There is no caudal colliculus, as such, in birds. The nucleus mesencephalicus lateralis, pars dorsalis, in the mesencephalic tectum contains neurons which are homologous to those of the caudal colliculus in mammals, but this nucleus is not large enough in birds to form a distinct round projection like the caudal colliculus of the mammal.

18.14 Olfactory areas

These are relatively reduced in birds generally, though the degree of reduction varies considerably with the species. An airborne existence is not, on the whole, conducive to a well-developed sense of smell. On the other hand, some terrestrial birds like the kiwis do have good olfactory equipment, and there are other species including some marine birds and vultures which seem to have substantial olfactory powers.

18.15 Cerebellum

The main part of the avian cerebellum corresponds to the midline vermis of the mammalian cerebellum. There is also a small cerebellar hemisphere on each side of the vermis. The avian cerebellum exerts essentially the same regulating influence over locomotion as in mammals.

The cerebellum connects to the dorsal aspect of the brainstem by rostral and caudal cerebellar peduncles. Because the pons is almost negligible it is difficult to identify a middle cerebellar peduncle, but this structure has been described in the avian brain.

18.16 Spinocerebellar pathways

These are one of the first of the ascending spinal pathways to become a recognizable entity in vertebrates, being present in the cranial end of the spinal cord of fishes. In birds the spinocerebellar system is much increased and is present at all levels of the spinal cord. One tract in the birds is comparable to the dorsal spinocerebellar tract of mammals, but in the bird it serves the wing whereas in mammals it relates to the hindlimb and trunk. Another appears to be homologous to the ventral spinocerebellar tract; in both birds and mammals this tract serves the hindlimb.

18.17 Cuneate and gracile fascicles

In general these pathways (in the dorsal funiculus) show progressive development through the vertebrate classes from amphibians onwards. This progression is reversed, however, in birds where these fascicles (and consequently the dorsal funiculus) are actually relatively reduced. This implies that peripheral sensibility is less well-developed in birds than in reptiles and mammals. Possibly the feathers largely isolate the animal from contact with surrounding objects and so reduce the need for fine tactile discrimination.

The dorsal roots appear to be relatively thinner than in other higher vertebrates. They probably contain fewer somatic afferent neurons, but this is not enough to account for the small size of the dorsal funiculus. Experiments show that many ascending afferent axons end in the grey matter within a few segments of their entrance into the cord. Thus the main anatomical reason for the reduction of the dorsal funiculus may be not a reduction in the total number of somatic afferent axons entering the cord, but a reduction in the number of long ascending collaterals. This in turn suggests that birds have greatly developed their spinal reflex arcs. The ability of birds to run about after decapitation is consistent with this development.

18.18 Motor spinal pathways

There are indications that rubrospinal, tectospinal, vestibulospinal, and reticulospinal pathways are present, but little is known about their position and termination in the spinal cord. There is also evidence for a cerebrospinal pathway (the occipitomesencephalic tract), originating in the external striatum; this supports the concept that the external striatum contains neurons homologous to those of the mammalian cortex (see Section 18.5).

19. Clinical neurological tests: diagnostic exercises

Clinical neurology is an aspect of clinical studies. However, the following notes summarize the tests and observations which can be carried out under clinical conditions in order to examine the central and peripheral pathways according to anatomical principles.

Clinical examination of the cranial nerves

19.1 Olfactory nerve (nerve I)

Loss of reaction to strong odours.

19.2 Optic nerve (nerve II)

Tests

Pupillary reflex. Constriction of pupil in response to light. Afferent pathway, nerve II; efferent pathway, nerve III. Reflex dilation in response to reduced illumination, obtained by covering the eyes for about 30 seconds: afferent pathway, nerve II; efferent pathway, reticulospinal tract and cervical sympathetic pathway through cranial cervical ganglion.

Visual placing response. See Section 19.15.

Menace response. The animal blinks in response to a threatening movement of the hand. Avoid stimulation of the cornea by air movement. Afferent pathway, cranial nerve II; efferent pathway, cranial nerve VII: also requires a normal visual cortex. It has been reported that the menace reflex is lost in domestic animals if there is a lesion in the cerebellar cortex. If the cerebellar lesion is unilateral, the menace reflex is lost on the same side. The neuronal pathways through the cerebellum are not known.

Fixating response. This test requires reflex movements of eyes, head, and neck, following a moving object, e.g. a handkerchief or pieces of cotton wool thrown in the air and allowed to fall to the ground. Afferent pathway, nerve II; efferent pathway, nerves III, IV, VI.

Deficiency of either the temporal or the nasal visual field (**hemianopia**) can be recognized in man relatively easily. It can also be detected in domestic mammals by carefully testing the menace and fixating responses of each eye in turn, after blindfolding the other eye.

Observations
Observe the subject's movements, especially in dim light. Ophthalmoscopic examination may reveal retinal changes.

19.3 Oculomotor nerve (nerve III)

Tests

Pupillary reflex. Afferent pathway, nerve II; efferent pathway, nerve III. Constriction of the pupil in the same eye as that which receives the light is known as the **direct response**. Constriction of the other pupil is called the **consensual response**.

Observations

Strabismus. Downward and outward deviation of the eye, with inability to converge. The oculomotor nerve innervates the following orbital muscles: ventral oblique, dorsal rectus, medial rectus, and ventral rectus.

Ptosis of the upper lid: nerve III supplies the levator palpebrae superioris muscle. (Ptosis is drooping of the upper eyelid.) See also nerve VII. Ptosis of the upper lid can also follow a sympathetic lesion (due to denervation of smooth muscle in the upper eyelid).

19.4 Trochlear nerve (nerve IV)

Observations

Strabismus. An up-and-in deviation of the eye. The trochlear nerve innervates the dorsal oblique muscle. The axons of this cranial nerve are unique in crossing over before they emerge from the brainstem. Therefore an up-and-in strabismus of the **left** eye indicates a lesion of the motor nucleus of the trochlear nerve on the **right** side (if the lesion is central and not peripheral).

19.5 Trigeminal nerve (nerve V)

Observations, mandibular nerve

A unilateral deficit gives weakness of chewing muscles when biting and wasting of chewing muscles. A bilateral deficit leads to drooping of the lower jaw. Weak closure by the facial muscles is possible.

Tests, ophthalmic nerve

Palpebral reflex. Touching the upper eyelid causes reflex closing of the eyelids. Afferent pathway, ophthalmic nerve; efferent pathway, nerve VII. The facial nerve innervates the orbicularis oculi muscle, which closes the lids.

Corneal reflex. Stimulation of the cornea causes reflex closure of the eyelids. As for the pupillary reflex, the response shown by the stimulated eye is called the direct corneal reflex and the response by the other eye is the consensual reflex. Afferent pathway, ophthalmic nerve; efferent pathway, nerve VII.

Tests, maxillary nerve

Palpebral reflex. Reflex closure of the eyelids following stimulation of the lower eyelid. Afferent pathway, maxillary nerve; efferent pathway, nerve VII.

19.6 Abducent nerve (nerve VI)

Tests

Nictitating reflex. When the lids are immobilized, stimulation of the cornea causes the third eyelid to sweep across the eye. Afferent pathway, ophthalmic nerve; efferent pathway, nerve VI. Nerve VI supplies the retractor oculi muscle. Contraction of the retractor oculi displaces orbital fat and connective tissue, and this displacement has an indirect mechanical effect on the third eyelid causing it to sweep down across the eye. Since the four recti muscles can also combine to retract the eyeball, the third (oculomotor) nerve may also be involved in the nictitating reflex. In the dog and cat, smooth muscle restores the nictitating membrane to its resting position, the motor nerve supply being sympathetic.

Observations

Strabismus. Medial deviation of the eye with inability to gaze laterally. The abducent nerve supplies only the lateral rectus muscle.

19.7 Facial nerve (nerve VII)

Tests

Menace reflex. See optic nerve.

Palpebral reflex. See trigeminal nerve.

Corneal reflex. See trigeminal nerve.

Handclap reflex. The ear pinna is directed towards the source of the sound. Afferent pathway, nerve VIII; efferent pathway, nerve VII innervates muscles which move the pinna.

Observations
Signs of **facial paralysis**. Drooping of the ear on the affected side, with no response to stimulation. The lip is retracted toward the sound side; drooping of the lip on the affected side. There may be dribbling.

Ptosis of the upper and lower eyelid: the facial nerve innervates the levator anguli oculi medialis muscle (previously called the corrugator supercilii m.), which lifts the upper lid. See also oculomotor nerve. There may be pouching of food in the cheek on the affected side (the facial nerve innervates the buccinator muscle). Inability to dilate the nostril on the affected side in the horse (the facial nerve innervates the dilator of the nostril).

19.8 Vestibulocochlear nerve (nerve VIII)

Tests, cochlear nerve

Handclap reflex. The ears, and/or the head as a whole, turn towards the sound. Afferent pathway, cochlear nerve; efferent pathway, nerve VII, for movement of the ear pinna; or efferent pathway, tectospinal tract, for movement of the head and neck.

Observations, vestibular nerve

Vestibular lesions. The head is **tilted**, the ear going down on the side of the lesion. There is a tendency to roll over. Spontaneous **nystagmus** may occur (the slow component usually going to the same side as the lesion). Abnormal positional nystagmus (see Section 19.13) may occur.

19.9 Glossopharyngeal nerve (nerve IX)

Tests

Gag reflex. Touching the pharyngeal wall of the dog induces retching. The reflex is lost in bilateral lesions. Afferent pathway, nerve IX; efferent pathway, nerve X. Nerve IX supplies a large sensory area within the pharynx and on the back of the tongue; the vagus nerve innervates the pharyngeal muscles.

Observations

Dysphagia (difficulty in swallowing) may occur owing to reduced afferent information from the tongue and pharynx.

19.10 Vagus nerve (nerve X)

Tests

No retching response to the **gag reflex** in bilateral lesions (see glossopharyngeal nerve).

Observations

Dysphagia may arise because of loss of afferent vagal pathways from the laryngeal aditus, and of motor vagal pathways to the pharyngeal muscles. Deficiency of the gag reflex and difficulty in swallowing will be only partial in a unilateral lesion of the glossopharyngeal or vagus nerves, and will be difficult to detect clinically.

Tachycardia, altered depth of **respiration**, and **digestive disturbances** may be evident.

Lesions of the vagus nerve can lead to **laryngeal paralysis**. The loss of **abduction** of the vocal and vestibular folds is usually more obvious than the loss of adduction. In the horse the loss of abduction of the folds can be observed endoscopically, and causes characteristic sounds at exercise during the inspiratory phase (**roaring** in horses). Loss of barking may occur in dogs. In man there may be hoarseness if the lesion is unilateral, and severe stridor and loss of voice if bilateral. The recurrent laryngeal nerve (strictly, its caudal laryngeal branch) innervates all of the laryngeal muscles except the cricothyroid muscle, and therefore laryngeal paralysis can arise from lesions of this nerve as well as the main trunk of the vagus.

19.11 Spinal division of accessory nerve (nerve XI)

Observations

Paralysis of the trapezius muscle and the other muscles which this innervates is of doubtful clinical significance in domestic animals, but can be a serious handicap in man.

19.12 Hypoglossal nerve (nerve XII)

Observations

In the dog the tongue may protrude from the mouth and deviate towards the side of the lesion in a unilateral lesion; atrophy of the intrinsic tongue muscles is evident on the affected side, if the lesion is unilateral. Minor dysphagia may arise.

Testing postural and locomotor responses

Predictable responses to these tests should occur in all normal subjects, but considerable variability may be encountered within the normal range.

19.13 Tonic neck and eye responses

When the nose is elevated the forelimbs extend and the hindlimbs flex in the normal dog. With a motor or kinaesthetic deficit the forepaws knuckle over.

Positional nystagmus is induced by turning the head dorsally or laterally. This indicates vestibular disease. The normal animal shows no nystagmus.

19.14 Proprioceptive positioning responses

Abduct one limb from its normal position, or cross one limb in front of the other, or flex the carpometacarpal or tarsometatarsal joints and rest the limb on the dorsum of the paw ('knuckling' the paw); the normal animal will immediately return the displaced limb to its normal position. These are tests of the modality of kinaesthesia (joint awareness), and some of them (e.g. 'knuckling' the paw) also test the sense of touch. In the dog and cat the response to abduction or adduction of the paw can be tested by standing the foot on a piece of paper and pulling the paper slowly in a lateral or medial direction.

19.15 Placing responses

These are divided into visual and tactile responses. (i) The **visual placing response** is tested by carrying a dog or a cat towards a table top. On approaching the surface the animal will reach out to support itself on the table. This reaction requires intact visual and motor pathways. (ii) To test the **tactile placing response**, carry a **blindfolded** dog towards a table. Allow the dorsum of one forepaw to touch the edge of the table. The normal animal will immediately place both forefeet on the table to support its weight. Repeat the test with the other forepaw. Normal touch and motor pathways are needed.

19.16 Extensor postural thrust

Lift the animal by the thorax and lower its hindlimbs towards the ground. The normal animal will extend its limbs to support its weight. (If the animal is blindfolded, the tactile pathways will be tested without the visual pathways.) This test requires intact afferent and efferent spinal pathways, and intact higher motor centres and cerebellum. By moving the animal backwards, forwards, and to the side, a very mild deficit, sometimes on only one side, may be detected. This test on the hindlimbs is essentialy the same as the wheelbarrow test on the forelimbs.

19.17 Hopping

Hold an animal with three of its limbs off the ground. Shift its centre of gravity over the fourth limb which is supporting the animal. Move the body in all directions, and particularly sideways; the normal animal will hop to keep the supporting limb under its body. This is a good way of identifying one slightly paretic limb. A mild deficit may be revealed, even when the gait is apparently normal.

19.18 Wheelbarrow test

While supporting the abdomen, make the animal walk on its forelimbs only. An abnormal animal may stumble or knuckle on to the dorsum of the paw. This is a valuable test for detecting a mild deficit in the forelimbs of an animal with an apparently normal gait.

19.19 Hemiwalking

While lifting the fore- and hindlimbs on one side, make the animal walk on the limbs of the other side. This is another very valuable test which can reveal a slight paresis on one side.

19.20 Righting

A normal animal rights itself when released from lateral recumbency. Another test is to suspend the animal by its hips and pelvis and turn its body from side to side. The normal animal will carry and hold its head in the normal position in relation to the ground. Blindfolding is advisable, to eliminate visual compensation.

19.21 Blindfolding

Covering both eyes with a blindfold will greatly accentuate either a bilateral or a unilateral motor deficit, kinaesthetic deficit, cerebellar ataxia, or vestibular disturbance. (Blindfolding eliminates visual compensation for these deficits).

Blindness in one eye can be made obvious by blindfolding the other eye, thus rendering the animal totally unsighted.

Examination of spinal reflexes

19.22 Withdrawal reflex

Pinching the pad or the web between the digits, or the hoof in small farm animals, induces reflex withdrawal by flexing all the joints of the limb. Signs of pain, such as turning the head or crying out, also indicate that the pain pathways are intact all the way from the peripheral ending to the cerebral cortex. Differentiation between voluntary withdrawal to avoid the painful stimulus, and reflex withdrawal, may be difficult. This is also called the **pedal** or **flexor** reflex. The peripheral pathways of this reflex require intact afferent, spinal, and efferent pathways. (i) The afferent pathway in the **forelimbs** involves the median, ulnar, and radial nerves; the efferent pathway utilizes the median, ulnar, musculocutaneous, and axillary nerves. Spinal nerves C6 to T1 inclusive contribute to these nerves. (ii) The **hindlimb** withdrawal reflex utilizes the peroneal and tibial nerves as the afferent pathway; the efferent pathway requires the sciatic, peroneal, and tibial nerves. These nerves are formed by spinal nerves L6 to S1 inclusive in the dog.

19.23 Patellar tendon reflex

Support the hindlimb so that it is relaxed, the animal lying on its other side, and tap the patellar ligament. The stifle joint is reflexly extended. Both the afferent and the efferent pathways lie in the femoral nerve. Spinal nerves L4 and 5 (and sometimes also L3 and L6) form the femoral nerve in the dog.

19.24 Triceps tendon reflex

With the animal lying on its side, tap the triceps tendon. Slight reflex extension of the elbow may be detected, but the response is difficult to elicit in the normal dog. Both the afferent and the efferent pathways are in the radial nerve. Spinal nerves C7 to T2 form the radial nerve.

19.25 Biceps tendon reflex

Slight contraction of the biceps muscle can sometimes be elicited by tapping the biceps tendon, the limb being in a relaxed position. Again, this reflex is difficult to elicit in the normal dog. Together with the triceps tendon reflex, it may be useful to detect hyperreflexia, as in upper motor neuron disease. The afferent and efferent pathways pass in the musculocutaneous nerve. Spinal nerves mainly C7, and occasionally also C6 and C8 form the musculocutaneous nerve.

19.26 Cutaneous (panniculus) muscle reflex

In the normal dog, lightly tweaking the skin of the flank with forceps between vertebrae T1 and L4 induces an ipsilateral cutaneous twitch. This reflex can be elicited in the normal cat by simply touching the fur of the flank, and in the horse by gently pricking the skin. The afferent pathway enters the spinal cord through spinal nerves T1 to L3. The ascending pathway within the spinal cord is probably the propriospinal system. The efferent pathway is via spinal nerves C8 and T1 only. The afferent and efferent peripheral pathways, and the ascending pathways in the spinal cord, must all be intact. This reflex is useful to determine the level of a severe injury to the spinal cord. For example, in the dog virtual transection of the spinal cord by a severe disc protrusion at T13–L1 is followed by loss of the panniculus reflex caudal to this level and persistence of the reflex cranial to this site.

19.27 Perineal reflex

Touching the anus, particularly with a cold metal instrument or a cotton wool bud soaked in surgical spirit, may induce contraction of the external anal sphincter and flexion of the tail over the anus. This reflex is commonly observed in the normal animal when taking the temperature per rectum. The afferent and efferent pathways are spinal nerves S1 to S3, and there are efferent fibres also in coccygeal (caudal) nerves for tail movement.

19.28 Crossed extensor reflex

When the withdrawal reflex is tested, with the animal on its side, the contralateral limb extends. This typically indicates an extensive lesion of the spinal cord. The explanation of this abnormal reflex is that all inhibitory control of the lower neurons by the higher centres has been lost. It does not occur in the normal animal.

19.29 Babinski reflex

With the dog lying on its side, the hindlimb is supported above the hock, the hock and digits being slightly flexed. The tip of a blunt probe strokes the plantar surface of the paw from the hock to the toes. The response in a normal animal is a slight flexion of the toes, or there may be no response at all. The abnormal response is a fan-like extension of the digits, and this indicates an upper motor neuron lesion. The value of this test is that it may enable an upper motor neuron lesion to be distinguished from ataxia of afferent or cerebellar origin.

Other tests

19.30 Assessment of muscle tone

Muscle tone can be assessed in small animals by manipulating the limbs with the animal lying on its side. Hypotonia can be detected by shaking the limb, which then looks completely floppy. It is difficult in domestic mammals to distinguish hypertonia from a voluntary increase in tone.

19.31 Testing conscious pain responses

Pinprick pain responses are commonly tested in the dog by using a sterile hypodermic needle with a blunt tip. If the animal feels the pain of the prick, it responds by turning the head or crying out; the afferent pathway is uncertain, but may be via either the spinothalamic tract or the propriospinal system, or both of these routes. True pain can be tested in the dog by pinching the web between the toes, or nipping the toenail with artery forceps; the afferent pathway is spinoreticular. (See also Section 19.22.)

19.32 Special techniques

More sophisticated techniques for a neurological examination are available. These include: (i) examination of cerebrospinal fluid; (ii) radiographic techniques, namely vertebral column and skull radiography, spinal cord myelography, ventriculography, pneumoencephalography, cerebral angiography, and tomography; (iii) electroencephalography; (iv) electromyography; (v) radioisotope 'scintiscan' techniques; and (vi) sonar.

Case sheet

A case sheet is a useful aid to making a systematic neurological examination, without leaving anything out. The following is an example:

Case sheet for a veterinary neurological examination

Case no:	Owner's name: ....
Date:	Breed/species: ....
Sex:	Age: ....

1. History

2. General observations

Overall impression (including mental status of an animal, risk of handling, owner's influence).

Musculature

Appearance (symmetry or asymmetry, wasting) of right and left, fore- and hindlimbs.
Degree of muscle tone in fore- and hindlimbs.

Gait

Postural abnormalities.
Orthopaedic complications.

3. Cranial nerves

I Olfactory nerve

II Optic nerve

Pupillary reflex
- Direct
- Consensual

Menace response
Visual placing response
Fixating response
Ophthalmoscopic examination

III Oculomotor nerve

Pupillary reflex
- Direct
- Consensual

Strabismus
Ptosis of upper lid

IV Trochlear nerve

Strabismus

V Trigeminal nerve

Mandibular V—Jaw movements
Ophthalmic V
- Palpebral reflex (upper lid)
- Corneal reflex
  - Direct
  - Consensual

Maxillary V—Palpebral reflex (lower lid)

VI Abducent nerve

Strabismus
Nictitating reflex

VII Facial nerve

Menace reflex
Palpebral reflex
Corneal reflex
  Direct
  Consensual
Handclap reflex
Facial muscles—eyelids, ears, lips, nostrils, cheeks

VIII Cochlear division

Handclap reflex

VIII Vestibular division

Head tilt
Rolling
Spontaneous nystagmus
Positional nystagmus

IX Glossopharyngeal nerve

Gag reflex
Dysphagia

X Vagus nerve

Gag reflex
Dysphagia
Laryngeal paralysis
Tachycardia

XI Accessory nerve

Weakness and wasting of appropriate muscles of forelimb

XII Hypoglossal nerve

Position and mobility of tongue
Muscle atrophy

4. Postural reactions

Tonic neck and eye responses
  Neck
  Positional nystagmus
Proprioceptive positioning
  Forelimb
  Hindlimb
Placing responses
  Visual
  Tactile
Extensor postural thrust
Hopping
Wheelbarrow test
Hemiwalking
Righting
  From lateral recumbency
  Suspended from hips
Blindfolding

5. Spinal reflexes

Withdrawal reflex
  Left fore Right fore
  Left hind Right hind
Patellar reflex
  Left
  Right
Triceps tendon reflex
  Left
  Right
Biceps tendon reflex
  Left
  Right
Panniculus reflex
  Left
  Right
Perineal reflex
Crossed extensor reflex
  Fore
  Hind
Babinski reflex
  Left
  Right

6. Conscious pain response

7. Muscle tone

Left fore	Right fore
Left hind	Right hind

8. Special tests

Diagnostic exercises

Most of the following cases could be accounted for by a single lesion, small or large, within the brain or spinal cord. Some of the cases could be explained by peripheral lesions, but only central lesions are considered in the Solutions (except for visual pathways).

The cases are arranged in related groups to enable the effects of similar kinds of lesions to be compared. The Solutions are given after the Exercises.

Case 1

A dog:

1. Loss of proprioceptive positioning and tactile placing responses from both forelimbs.
2. Proprioceptive positioning and tactile placing responses normal in both hindlimbs.
3. Paralysis of the tongue with bilateral atrophy of the intrinsic tongue muscles.

Case 2

A cat:

1. Complete loss of responses to touch and proprioceptive positioning in the left forelimb and left hindlimb, and loss of responses to touch on the left side of the head.
2. Adduction of the right eyeball.

Case 3

A cat:

1. Complete loss of responses to touch and proprioceptive positioning in the right forelimb and right hindlimb, and loss of responses to touch on the right side of the head.

2. Visual deficit in the nasal field of the left eye and the temporal field of the right eye.

Case 4

A dog:
1. Loss of the withdrawal reflex from the right forelimb.
2. Loss of proprioceptive positioning from right forelimb.
3. Loss of panniculus reflex from all of the right side.
4. Proprioceptive positioning is normal in the left hindlimb, but the right hindlimb shows a slight proprioceptive deficit.
5. Withdrawal reflexes normal in both hindlimbs.
6. Distal to the right shoulder, no response to pain, except for the lateral aspect of the brachium and the medial aspect of the antebrachium.

Case 5

A dog:
1. In ordinary daylight the left and right pupils are of normal diameter.
2. There is no strabismus.
3. Light directed into the right eye produces both a direct and a consensual response.
4. Light directed into the left eye produces neither a direct nor a consensual response.
5. Menace and visual placing responses normal for right eye, negative for left eye.

Case 6

A dog:
1. In ordinary daylight the right pupil is abnormally dilated.
2. There is no strabismus.
3. Light directed into the right eye produces a consensual response but not a direct response.
4. Light directed into the left eye produces a direct response but not a consensual response.
5. Visual placing, menace, and fixating responses normal in both eyes.

Case 7

A dog:
1. The owner reports that the animal seems to be able to see only what is directly in front of it.

2. Examination shows that the dog is blind, except for a small part of both the left and the right nasal fields immediately adjacent to the midline.
3. Direct and consensual pupillary reflexes normal in both eyes.

Case 8

A horse:

1. Nystagmus, slow phase being towards the left.
2. Tendency to fall to the left.
3. Left facial paralysis.
4. Head tilted, the left ear going down.

Case 9

A cat:

1. A visual deficit is observed in the temporal field of the left eye and in the nasal field of the right eye.
2. Direct and consensual pupillary reflexes are normal in both eyes.
3. There is a slight impairment of balance, without clear emphasis on either the left or right side of the body.
4. The owner reports a general reduction in hearing.

Case 10

A dog:

1. The menace response is negative for both the left and the right eyes.
2. The visual placing response is negative for both eyes.
3. Light directed into the left eye produces a direct and consensual pupillary response. So does light directed into the right eye.
4. There is hypertonus of the limbs on both sides.

Case 11

A cat:

1. Difficulty in swallowing.
2. Loss of gag reflex on left side.

Case 12

A cat:

1. The tongue tends to protrude from the mouth, deviating to the right.
2. Wasting of the right side of the tongue.

Case 13

A dog:
1. Difficulty in swallowing.
2. Change of voice.
3. Right vocal fold permanently adducted.

Case 14

A horse:
1. Left eye normal.
2. Up-and-in strabismus of right eye.

Case 15

A dog:
1. Right eye normal.
2. Downward and outward strabismus of left eye.

Case 16

A horse:
1. Lips drawn towards the left side.
2. Right lower eyelid drooping.
3. Right ear drooping.

Case 17

A horse:
1. Drying of right cornea.
2. Drying of nasal mucosa on right side.
3. Reduced salivation when feeding.

Case 18

A dog:
1. Diabetes insipidus.
2. Extremities and ears feel cold.
3. Periodic shivering.
4. Loss of appetite.

Case 19

A dog:
1. Flaccid paralysis of left forelimb.
2. Total loss of withdrawal reflex from left forelimb.
3. Some wasting of left triceps muscle.
4. Left forelimb warmer than the right.
5. Right forelimb normal.
6. Slight paresis of left hindlimb with exaggerated reflexes.
7. Right hindlimb normal.
8. Slow heart rate.

Case 20

A sheep:
1. Continuous circling to the left.
2. Intermittent aversion of head, the ear going down on the right side.
3. Blindfolding of the left eye causes no change in posture or locomotion, but blindfolding of the right eye causes the animal to stagger and fall repeatedly to its left side.
4. Hemiwalking tests indicate a locomotory deficit in the left limbs.

Case 21

A cat:
1. Circling to the left.
2. Occasional falling and rolling over to the left side.
3. The left side of the face is constantly tilted downwards.
4. Vision is entirely normal in both eyes.

Case 22

A dog:
1. General insecurity of posture, with the limbs wide apart to maintain balance when standing.
2. Hypertonia of all limbs.
3. Tremor of limbs, accentuated during fine movements like placing the paw on an object to restrain it.

Case 23

A dog:
1. Hypertonus of right forelimb, the other limbs being normal.

2. Intermittent tremor of right forelimb, not increased during movement and disappearing during sleep.
3. Occasional spontaneous exaggerated and incoordinated movements of the right forelimb.
4. All afferent responses normal.

Case 24

A horse:
1. Paresis (weakness) of all four limbs.
2. Hyperreflexia of all four limbs.
3. Stiff gait, suggestive of hypertonus.
4. Defective proprioceptive positioning responses from the hindlimbs on both sides.
5. Distressed breathing (dyspnoea).
6. All cranial nerve responses normal.

Case 25

A dog:
1. Retention and dribbling of urine and retention of faeces.
2. Perineal reflex and tail response lost.
3. The hindlimbs show severe paresis, incoordination, and hypotonus.
4. Withdrawal reflex almost lost in both hindlimbs.
5. Some wasting of parts of the hamstring muscles of both legs.
6. Loss of pain response from the skin distal to the strifle, except the medial aspect of leg and paw, of both hindlimbs.
7. Normal pain responses from the cranial region of left and right thigh.
8. Patellar reflex normal in both limbs.

Case 26

A dog:
1. Paresis of both hindlimbs.
2. Some wasting of vastus muscles of both limbs.
3. Patellar reflex absent in both limbs.
4. Normal pain responses from the skin of both hindlimbs, including the cranial and medial aspects of the thigh.
5. Withdrawal reflex exaggerated in both hindlimbs.
6. Normal control of urination and defecation.
7. Perineal reflex normal.

Solutions to diagnostic exercises

Case 1

The bilateral loss of kinaesthesia from the forelimbs but not the hindlimbs is consistent with a lesion in the midline of the medulla oblongata in the dorsal part of the medial lemniscal system. The bilateral paralysis of the tongue indicates destruction of the motor nucleus of the hypoglossal nerve on both sides and confirms that the level of the lesion is in the caudal part of the medulla oblongata. The lesion could be in the region of Fig. 6.1.

Case 2

The whole of the right medial lemniscal system is interrupted. The level is at the motor nucleus of the abducent nerve, at the rostral end of the medulla oblongata, just caudal to the middle diagram in Fig. 7.3.

Case 3

A lesion in the left cerebral cortex (Fig. 8.2(a)).

Case 4

A lesion has affected the right dorsal roots and/or the right dorsal horn at segments C8 to T2 of the spinal cord. The hindlimb proprioceptive deficit suggests that the lesion may also have affected the gracile fascicle in the right dorsal funiculus. The loss of the panniculus reflex could be explained by involvement of the dorsal horn at C8 and T1, since the final ascending projections of this reflex must enter the grey matter at this level. The loss of the withdrawal reflex and the absence of pain response from the paw indicate that spinal nerves C8, T1, and T2 are involved, since the median, ulnar, and radial nerves arise almost entirely from these spinal nerves. The presence of a pain response from the lateral aspect of the brachium and from the medial aspect of the antebrachium indicates that C7 is not involved, since this nerve is the main source of the axillary and musculocutaneous nerves. The lesion is unlikely to be peripheral, since there is both a hind- and forelimb deficit.

Case 5

The animal is blind in the left eye only, so the lesion is in the left optic nerve (or left retina) (Fig. 8.1).

Case 6

The optic nerve and optic tracts are normal, but the right parasympathetic oculomotor nucleus is destroyed. The lesion is in the midbrain, approximately at the level of the second diagram (down) in Fig. 11.1 (but the motor nucleus of the oculomotor nerve is normal, since there is no strabismus).

Case 7

The loss of vision in one-half of each visual field is called hemianopia. In this dog the temporal half of the visual field of each eye is lost, constituting bitemporal hemianopia. Also lost is the adjoining part of the nasal half of the visual field of each eye, leaving intact only the most medial part of the nasal field. Since the animal can only see what is directly in front of it, the deficit can be described as 'tunnel vision'. Only the decussating fibres have been damaged; in the dog these fibres account for 75 per cent of the total optic pathway. A midline lesion of the optic chiasma, such as a pituitary tumour, could account for these visual defects (Fig. 8.1). The pupillary reflexes are normal in both eyes, because the most lateral part of the retina of both eyes still responds normally to the light stimulus.

Case 8

A lesion in the left side of the medulla oblongata, involving the left vestibular nuclei and left motor nucleus of the facial nerve. Lesions in the middle and inner ear also could readily account for these signs. So also could compression of the roots of the left VIIth and VIIIth nerves by a tumour, as they arise from the brainstem.

Case 9

A lesion in the right lateral and medial geniculate bodies (Fig. 8.1).

Case 10

The optic nerves, optic tracts, and rostral colliculi are normal on both sides, since the pupillary reflexes are fully functional (Fig. 8.1). But there is complete blindness. This could mean destruction of the visual cortex on both sides. Alternatively, both lateral geniculate nuclei could be destroyed. The hypertonus of all limbs is consistent with destruction of motor areas of the cerebral cortex (see Section 11.6), and confirms the presence of widespread damage to the left and right cerebral cortex. Lesions and clinical signs of this type, termed cerebrocortical necrosis, have been recognized in the cat following cardiac arrest during anaesthesia and in ruminants suffering from thiamine deficiency.

Case 11

Destruction of the left nucleus of the solitary tract could account for this case, since there is loss of all afferent information from the left side of the pharyngeal mucosa (Fig. 6.4) travelling mainly in nerve IX and also in nerve X.

Case 12

Destruction of the right motor nucleus of the hypoglossal nerve (Fig. 6.1). This amounts to a lower motor neuron lesion, as shown by muscle wasting.

Case 13

This combination suggests a deficit in the motor innervation of the pharynx and larynx (vagal branches), and could be due to a lesion of the right nucleus ambiguus (Fig. 6.1).

Case 14

Up-and-in strabismus of the right eye indicates failure of the right dorsal oblique muscle, and hence of the motor nucleus of the trochlear nerve. Since the axons of this nucleus decussate, this lesion is in the **left** nucleus.

Case 15

The left dorsal oblique and lateral rectus muscles are still in a state of tone, unopposed by the paralysed ventral oblique and medial rectus muscles. The lesion is in the left motor nucleus of the oculomotor nerve.

Case 16

A lesion of the right motor nucleus of the facial nerve.

Case 17

A lesion of the right parasympathetic nucleus of the facial nerve, which distributes parasympathetic motor pathways through the trigeminal nerve to the lacrimal gland, glands of the nasal mucosa, and the submandibular and sublingual salivary glands (Fig. 6.4).

Case 18

A lesion in the hypothalamus, disturbing various autonomic functions (Section 16.9).

Case 19

A lesion of the left side of the spinal cord at segments C6 to T2. Flaccid paralysis and muscle wasting are explicit signs of a lower motor neuron lesion. The total loss of all response to the withdrawal reflex means that all the flexor muscle groups and their nerves are involved, i.e. the axillary, musculocutaneous, median, and ulnar nerves; the wasting of triceps shows that the radial nerve is also affected. Therefore, the lesion has taken out the left ventral horn of segments C6, C7, C8, T1, and T2 (forming the brachial plexus). The autonomic effects indicate involvement of the left lateral horns of segments T1 to T5. The lesion has also disturbed the white matter in the left lateral funiculus (e.g. it may have interrupted some or all of the left lateral corticospinal, left rubrospinal, and left reticulospinal tracts), causing an upper motor neuron disorder of the left hind limb (hyperreflexia).

Case 20

These signs suggest a unilateral cerebral cortical lesion (caused for example by a Coenurus cyst), but the question is, on which side is the lesion? The direction of circling and head aversion are not reliable indicators of the side. The relatively dramatic effect of blindfolding the right eye suggests that the animal has vision with this eye; the left eye is probably blind, and this is consistent with a right cerebral cortical lesion. The locomotory deficit which was revealed in the left limbs, by hemiwalking, strongly supports a right cerebral cortical lesion. The postural–locomotory deficit of the left limbs could be largely compensated by vision in the right eye. But blindfolding the right eye renders the animal virtually blind; the postural–locomotory deficit of the left limbs then becomes unmanageable, so the animal falls to the left side. Conclusion: a large unilateral lesion in the right cerebral cortex.

Case 21

These signs indicate a destructive lesion of the left vestibular nuclei (see Section 8.3). The normal vision argues against an extensive lesion of the cerebral hemispheres.

Case 22

These signs suggest a midline cerebellar lesion. The difficulty in maintaining balance indicates involvement of the vestibulocerebellum. The hypertonia suggests that the spinocerebellum is affected. The tremor reveals pontocerebellar involvement. Thus all three cerebellar components are affected. There is no indication of unilateral deficits (see Section 15.13).

Case 23

Hypertonus, tremor, and particularly the spontaneous exaggerated movements on the right side suggest a lesion somewhere in the left basal nuclei, the effects of such lesions being usually contralateral (Section 13.5).

Case 24

The paresis, hyperreflexia, and hypertonus are consistent with an upper motor neuron disorder resulting from a lesion of the white matter of the spinal cord (see Section 13.6). The proprioceptive deficit also suggests damage to the white matter of the spinal cord. Clinical signs like these can occur from very widespread and scattered lesions of the white matter on both sides of the cervical spinal cord, somewhere between C3 and C7 (as in the wobbler syndrome in the horse and dog). The greater involvement of the hindlimbs rather than the forelimbs is typical of the wobbler syndrome. The dorsal funiculus is only mildly affected in wobbler horses, so that a joint proprioceptive deficit is not obvious in the forelimb. But many of the proprioceptor pathways from the joints of the hindlimb have probably transferred from the gracile fascicle to the dorsal spinocerebellar tract in the dorsolateral part of the lateral funiculus (see Section 7.4). In wobblers degeneration affects the dorsolateral part of the lateral funiculus, thereby involving both the joint proprioceptors from the hindlimbs and also the dorsal spinocerebellar tract from the hindlimb. Spinocerebellar fibres are now known to contribute to the sense of kinaesthesia (see Section 9.5). The dyspnoea may arise from disturbance of reticulospinal pathways, leading to erratic and inadequate operation of the diaphragm and intercostal muscles.

Case 25

This could be a lesion which has extensively damaged both the grey and white matter of the spinal cord in segments L6 to S3 inclusive. Some of the lower motor neurons in the sciatic nerve (spinal nerves L6, L7, S1), as well as preganglionic neurons of the pelvic nerve (see Section 16.9), are involved (see Fig. 20.6). Since the patellar reflex is normal the femoral nerve (including the saphenous nerve which is sensory to the medial aspect of the lower leg and paw) has escaped. The femoral nerve is formed mainly by spinal nerves L4 and 5. The cause could be a protrusion of the intervertebral disc at L4–5, affecting the spinal cord segments L6 to S3. Alternatively, there could be a massive compressive lesion (abscess, tumour, fracture) within the vertebral canal of vertebra L7, damaging the relevant nerves within the cauda equina (L7, S1, S2, S3).

Case 26

The absence of the patellar reflex and the presence of muscle wasting in the vastus muscles indicate a lower motor neuron lesion, in this instance affecting the ventral horn on both sides of segments L4 and 5 of the spinal cord (see Fig. 20.6), thus destroying the motor outflow of the left and right femoral nerves. The presence of pain responses from the skin of the cranial and medial aspects of the thigh (conducted by the femoral nerve) shows that there is a central lesion, rather than peripheral damage to the femoral nerves. The exaggerated withdrawal reflex indicates that the white matter of the lateral and/or ventral funiculi has become involved, giving an upper motor neuron lesion of segments L6 and L7 of the spinal cord (sciatic nerve is L6, L7, and S1). However, pain pathways through the spinoreticular tracts in the lateral column have survived. Normal urination, defecation, and perineal reflex indicate that the reticulospinal tracts and segments S1, S2, S3 of the spinal cord have escaped. The cause could be a disc protrusion at disc L3–4, damaging spinal cord segments L3, L4, and L5.

20. Radiographic anatomy of the head and vertebral column in the diagnosis of disorders of the central nervous system

G. C. SKERRITT
Department of Veterinary Anatomy, University of Liverpool

General considerations

20.1 Species

Radiographs of the head and vertebral column are commonly taken in veterinary practice. Dogs and cats are the most frequent subjects but, increasingly, radiography of the head and neck is being carried out in the large animals, especially in the horse. This account is restricted to the small animals and is concerned only with the radiographic anatomy of those structures that may be involved in disturbances of the central nervous system.

20.2 Objectives of radiography in clinical neurology

Usually the purpose of radiography of these regions is to aid in the diagnosis of skeletal abnormalities. The brain and the spinal cord are not themselves sufficiently radio-dense to be seen on radiographs. Even space-occupying lesions of the CNS such as tumours, cysts, and abscesses rarely show radiocontrast with the surrounding tissues. Due to the close topographical relationship between the vertebrae and the spinal cord, injuries and diseases of the vertebrae often result in neurological disturbance. A tentative diagnosis can be made in most cases of spinal cord disease on the basis of a complete case history and careful physical examination. Focal lesions can quite often be accurately localized by means of a neurological examination, so that the value of radiography lies mainly in **confirmation** of findings, especially in cases where surgery is in prospect. It follows that radiography is not a substitute for a careful physical and neurological examination; indeed it is often necessary to use neurological tests in order to know where to point the X-ray beam.

Symptomless abnormalities of the vertebral column may sometimes be identified on films taken for other conditions, since the vertebrae are usually visible in thoracic and abdominal radiographs.

Radiography can be used in one of three ways to aid in the diagnosis of CNS disorders: (i) to identify skeletal damage of traumatic origin, such as fractures, dislocations, protruding intervertebral discs; (ii) to reveal slowly progressive soft tissue lesions which have invaded or caused rarefaction of adjacent bone; (iii) by means of injected contrast media to help to visualize a soft tissue lesion.

Intracranial structures

20.3 Positioning of the head

The tissues of the head represent a wide range of radiodensity and are located at varying depths. It is therefore vitally important to position the animal properly if a confusing mass of superimposed images is to be avoided. It is often difficult, if not impossible, to imobilize and accurately position the conscious patient for radiography of the head, so that sedation or anaesthesia is usually necessary. A single radiograph of the head never shows all the structures that can be demonstrated, and therefore various views and degrees of contrast must be studied if an accurate survey of the radiographic anatomy is to be made. On the other hand, diagnosis of a clinical condition may only require a single radiograph taken in the correct projection.

The most commonly used projections in radiography of the head are as follows: (i) **Dorsoventral**. The saggital plane should be perpendicular to the film so that the X-ray beam enters the dorsal aspect of the head. (ii) **Ventro-dorsal**. The beam enters the ventral aspect of the head. The saggital plane should again be perpendicular to the film, but for this projection the animal is positioned in dorsal recumbency with the head and neck extended. (iii) **Lateral**. The saggital plane should be parallel to the film in order to achieve exact superimposition of the images of one side upon those of the other side.

In addition to these projections, oblique and frontal views, and intraoral placement of film are all techniques that are indicated in certain clinical conditions.

20.4 Interpretation of plain radiographs of the head

The head is difficult to study radiologically. Its bone structure is complex and there is considerable variation in shape between breeds and individuals. In dogs, three types of head shape are defined: dolicocephalic (long-nosed, e.g. Rough Collie); mesaticephalic (intermediate shape, e.g. German Shepherd Dog); and brachycephalic (short-nosed, e.g. Pekingese). These variations affect the spacing and number of the teeth, the extent of the frontal sinus, and the relative sizes of the nasal and cranial areas.

Radiographs of the heads of very young animals can be difficult to interpret. The suture lines between the skull bones do not fuse until a few weeks after birth and appear as radiolucent lines that should not be mistaken for fractures. In some small breeds (e.g. Chi-hua-hua, Maltese Terrier) the suture lines may never close. The large patches of unossified tissue which occur at the junction of the sutures between the parietal bone and the frontal, sphenoid, and temporal bones are called **fontanelles**. In young puppies most of the bony cranium is very thin, making it difficult to obtain good radiographic contrast and detail of hard structures. The shadows of the developing permanent teeth alongside those of the deciduous teeth present a further complication in the interpretation of radiographs of the heads of young animals.

The wide range of variation in head shape between the different breeds

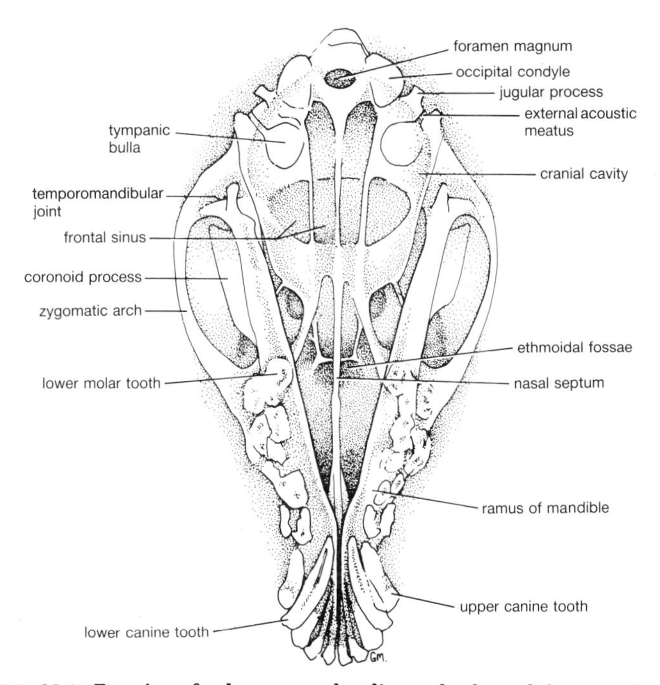

FIG. 20.1. Drawing of a dorsoventral radiograph of an adult canine head.

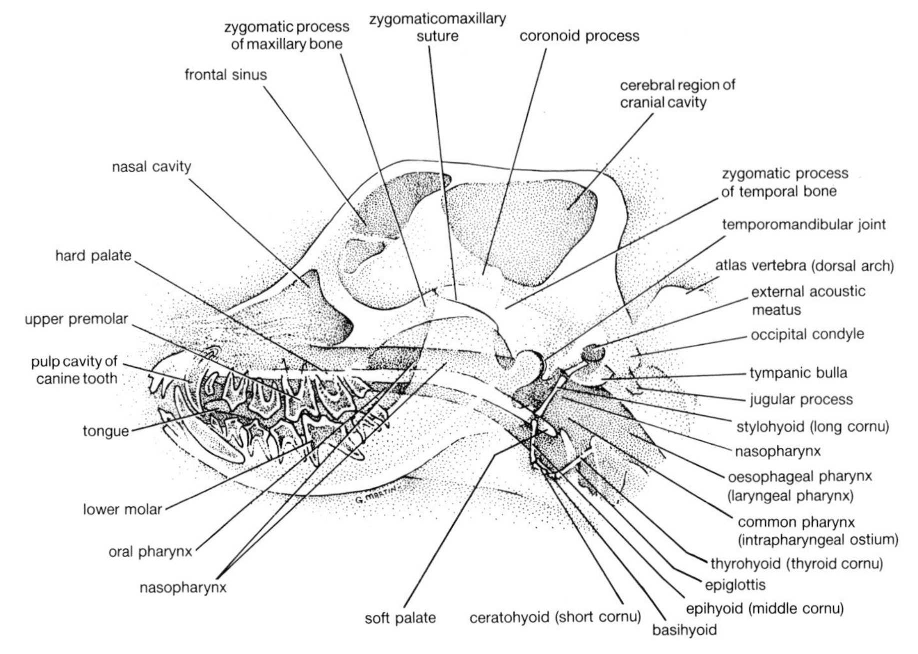

FIG. 20.2. Drawing of a lateral radiograph of an adult canine head.

and ages of dog makes the preparation of **reference films** essential in veterinary practice.

The main landmarks which are visible in dorsoventral and lateral radiographs of the normal adult canine head are shown in Figs. 20.1 and 20.2.

20.5 Contrast radiography of intracranial structures

Contrast techniques are used in radiography to allow the visualization of structures which are not normally visible on plain radiographs. Thus contrast media can be introduced into the bloodstream or a cavity, which then appears either more or less radiopaque than the surrounding tissues. The fluid contrast media used in radiographic studies of the CNS are water-soluble, non-toxic, iodine preparations, e.g. iopamidol (Niopam 200, 300, and 370, Merck). Contrast techniques are not widely used in the diagnosis of brain disorders of small animals, largely because of the difficulty of interpretation. However, they can provide a helpful diagnostic aid and may well become more popular as experience accumulates.

Cerebral arteriography

The injection of a suitable contrast medium into the cerebral arterial circulation is the procedure of choice for the radiodiagnosis of tumours in the forebrain or diencephalon. If the injection is made into the common carotid artery both the intra- and extracranial vessels are filled, resulting in a confusing superimposition of vascular shadows.

Demonstration of the intracranial circulation alone is achieved by the injection of contrast medium into the **internal carotid artery**. Two principal techniques are available for this procedure. After surgical exposure of the carotid bifurcation a polythene catheter is inserted into the common carotid artery and threaded up to the region of the internal carotid artery; the external carotid artery is then temporarily occluded to prevent reflux of contrast medium. This method is not entirely satisfactory, however, since some contrast medium still enters the extracranial circulation via anastomotic vessels, due to the fall in pressure in the extracranial vessels. A better method is to use image-intensified fluoroscopy to introduce a catheter into the common carotid artery, and thence into the internal carotid artery. Although more complicated than the first method this technique does have the advantage that surgical dissection is minimal. After taking the films the common carotid artery must either be sutured or ligated.

Interpretation of **cerebral arteriograms** is always difficult, even for the experienced radiologist. The difficulties are lessened if the head is held with its saggital plane either exactly parallel or at right angles to the film. The abnormalities that can be revealed include tumours, thromboses, ruptured arteries, and dilations of arteries. These radiologic abnormalities are indicated by displacement of the main arteries, by discrete areas showing excessive vascularity, or by zones showing abnormally low vascularity.

Within 3–4 seconds of the initial intra-arterial injection, a **venogram** is obtained as the contrast medium transfers to the venous circulation. Very vascular lesions are characterized by marked retention of the contrast medium.

Cavernous sinus venography

In this procedure the contrast medium is injected into the **angularis oculi vein**, an anastomotic valveless vein that connects the facial vein with the cavernous sinus via the dorsal external ophthalmic vein and ophthalmic plexus (see Section 3.5). Blood can flow in either direction in the angularis oculi vein and can reach the cavernous sinus. The injection is best made by exposing the angularis oculi vein through a skin incision 1 cm long just rostral to the medial canthus of the eye. Temporary occlusion of the external jugular veins by placing a tape around the neck ensures that the contrast medium flows in the direction of the cranial venous sinuses. In the normal dog the contrast medium is seen mainly on the injected side. Bilateral filling is essential for visualization of the anastomoses crossing the midline, and if this is required both sides must be injected simultaneously.

The lesions revealed by cavernous sinus venography are restricted to the orbit and the floor of the cranial cavity, e.g. space-occupying lesions involving the pituitary gland or the optic chiasma. The procedure is relatively simple and seems to be free from adverse side-effects.

Ventriculography

The introduction of a suitable contrast medium into the ventricular system of the brain allows it to be visualized radiographically (Fig. 20.3). The contrast medium may be either air (or oxygen) to give negative (radio-transparent) contrast, or a non-toxic, water-soluble, iodine preparation, e.g. iopamidol (Niopam, Merck) to give positive contrast. Ventriculography is considered fairly safe and consistently effective in the dog.

The contrast medium is injected directly into a lateral ventricle after trephining the skull under general anaesthesia. The trephine hole is sited about 1 cm from the midline at a point halfway between the level of the lateral canthi and the external occipital protuberance. A spinal needle is inserted through the cerebral hemisphere, parallel to the falx cerebri (Fig. 20.4), to a depth of 1–2 cm. Cerebrospinal fluid trickles from the needle when the lateral ventricle is entered. A small quantity (5–10 ml) of fluid is removed before injecting either air (5–10 ml) or an iodine preparation (1–5 ml) into the ventricle. A better distribution of air between the two ventricles is achieved by trephining and injecting on both sides.

The interpretation of a **ventriculogram** requires experience of the rather complex normal radiographic anatomy of the ventricles. Any pathological conditions that alter the normal outline of the ventricles, e.g. tumours, abscesses, or hydrocephalus, can be revealed by ventriculography.

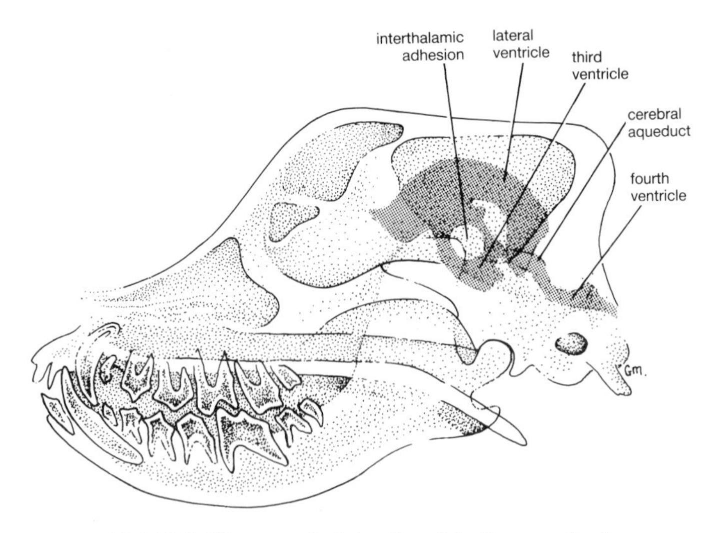

FIG. 20.3. Diagram of a lateral ventriculogram of a dog.

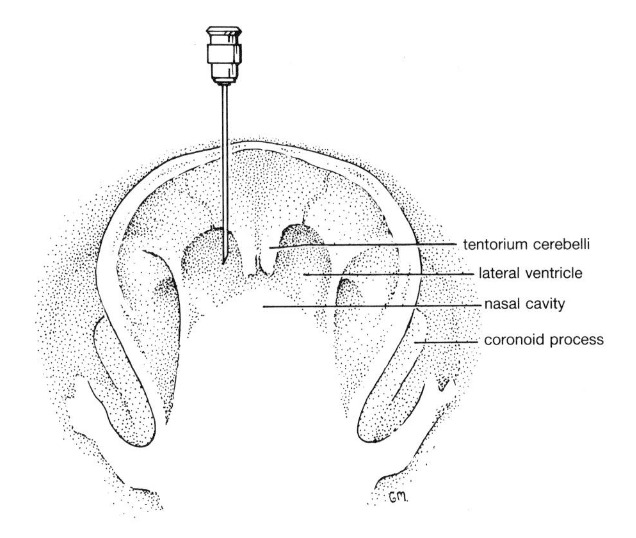

FIG. 20.4. Diagram of a rostrocaudal radiograph of the canine head. The needle has been introduced into the lateral ventricle, and air has been injected to produce a pneumoventriculogram.

Vertebral column

20.6 Positioning of the animal

Because of the complex form of the vertebrae exact positioning of the animal for spinal radiography is essential. Sedation or general anaesthesia are usually necessary so that there is absolutely no movement during exposure. Radiographs of the lateral view of the vertebral column may be obtained with the animal in lateral recumbency and supported with foam rubber cushions to ensure that the saggital plane of the vertebral column is parallel to the film. Alternatively a lateral view may be taken by repositioning the tube and film, and placing the animal in sternal recumbency. The ventrodorsal view is obtained by supporting the animal in dorsal recumbency so that the saggital plane of the vertebral column is perpendicular to the film.

20.7 Interpretation of plain radiographs of the vertebral column

The vertebrae are extremely complex bones varying in shape from the distinctive atlas to the fused bones of the sacrum. The accurate interpretation of radiographs of the vertebral column depends on exact positioning of the animal and a sound knowledge of the anatomy of these bones. In particular it is important to be fully familiar with the normal characteristics of vertebral bodies, transverse and spinous processes, intervertebral disc spaces, and intervertebral foramina.

The vertebral column of the dog and cat consists of about 50 vertebrae morphologically arranged in groups, viz. cervical (7), thoracic (13), lumbar (7), sacral (3 fused together), and caudal (approximately 20) also known as coccygeal. The first cervical vertebra, or atlas, possesses wing-like transverse processes, each perforated by a transverse foramen. The second cervical vertebra, or axis, is charcterized by a long blade-like dorsal spinous process, and the dens (odontoid process) which extends cranially from the body to project into the ventral part of the vertebral foramen of the atlas. The remaining five cervical vertebrae possess prominent and irregularly shaped transverse processes. The thoracic vertebrae all carry elongated dorsal spinous processes which are directed caudally until the eleventh (anticlinal) vertebra. The dorsal spinous processes of the lumbar vertebrae are directed cranially and the prominent transverse processes point craniolaterally. The sacrum of the dog and cat comprises three fused vertebrae.

Radiographically visible lesions of the vertebrae do not necessarily result in neurological signs. Similarly lesions involving the spinal cord and meninges alone may not be apparent radiographically, except perhaps with the use of a contrast medium. The following clinical conditions are examples of disorders of the vertebral column that may be responsible for neurological signs.

Atlanto-axial subluxation

This condition occurs in toy and miniature breeds, and may result from a congenital malformation or traumatic fracture of the dens, or a traumatic tearing of the transverse atlantal ligament (see Section 4.13). In consequence either the dorsally displaced dens or the body of the axis compresses the spinal cord causing cervical pain, neck stiffness, and upper motor neuron quadriplegia.

Cervical spondylopathy

This condition is also known as cervical stenosis, cervical spondylolisthesis, or **wobbler syndrome**. It is characterized by subluxation of the more caudal cervical vertebrae (C3–C7), and occurs in the larger breeds of dog, especially the Great Dane and Doberman Pinscher (see Section 4.11), and also in young Thoroughbred horses. Affected animals generally show a progressive development of neurological deficits, beginning with proprioceptive losses of the hindlimbs (see Section 7.4) and progressing to a spastic quadriplegia. The severity of the clinical signs reflects the degree of vertebral canal stenosis.

Spina bifida

Radiography may be used to confirm that there has been incomplete fusion of the dorsal vertebral arches. Spina bifida usually affects thoracic and/or lumbar vertebrae and is seen particularly in the Manx cat and the Bulldog. In some affected animals there are no neurological deficits, but commonly there is a mild pelvic limb ataxia and paresis together with faecal and urinary incontinence.

Spondylosis

This is a common finding during unrelated radiographic examinations. Spurs or bridges of bone are seen to cross the intervertebral spaces ventrally. Unless very extensive and involving spinal nerves, this condition does not result in neurological deficits. An extensive spondylosis of the cervical vertebrae is occasionally seen in cats that have been fed a diet predominantly of liver. The condition is believed to be due to excessive levels of vitamin A.

Discospondylitis

In this condition there is an infection of an intervertebral disc which usually spreads to involve the bodies of the adjacent vertebrae (usually lumbo-sacral). Radiographically there is lysis of bone and the production of new bone. Affected animals are pyrexic, depressed, and show varying degrees of ataxia/paresis of the pelvic limbs and vertebral pain.

Lesions of intervertebral discs

Overextension, overflexion, compression, or torsion of the vertebral column can result in protrusion of an intervertebral disc towards or into the spinal canal (see Section 4.9). The case history, neurological examination, and radiographic examination are used to differentiate disc protrusions from other causes of compression of the spinal cord. Figure 20.5 shows the necessity for using a narrow, vertical X-ray beam to view only two or three intervertebral spaces on any one film.

The following are listed by Hoerlein (1978) as the possible radiographic findings: (i) Calcified nucleus pulposus (increased opacity) with little or no involvement of the anulus fibrosus. (ii) Calcified nucleus with involvement of the anulus fibrosus. (iii) Narrowing of the disc space. (iv) Narrowing of the intervertebral foramina, abnormally spaced articular processes, or both. (v) Opaque mass in the vertebral canal above a normal, narrow or calcified disc.

The list is a useful guide but its limitations need consideration. The first two features can occur without protrusion being present, and the third and fourth depend on using a perfectly vertical X-ray beam and a perfectly positioned vertebral column. Only the fifth item is direct evidence of a protrusion, but even then that particular protrusion may not be clinically

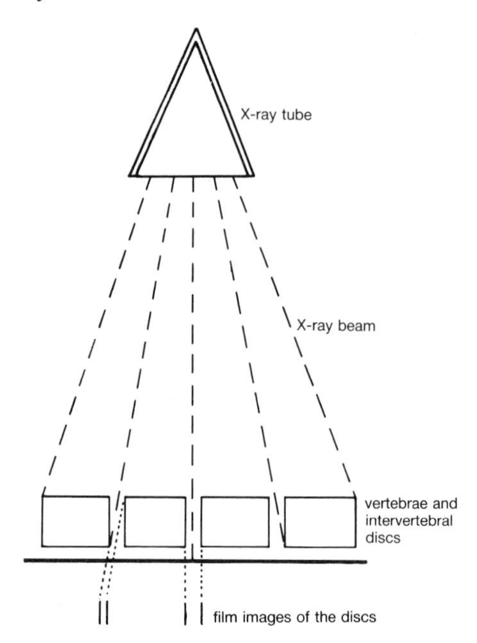

FIG. 20.5. Diagram to show that only perpendicular X-rays will produce an accurate image of intervertebral discs. Rays at an angle result in an apparent narrowing of the disc space on the film.

significant; furthermore, the clinically important, explosive, type 1 protrusions are often not calcified and can be far too small to be visible radiographically. Consequently, despite the most careful radiography, there are cases where protrusions cannot be demonstrated by plain radiography.

20.8 Contrast radiography of the vertebral column

Although lesions of the vertebral canal and spinal cord such as haemorrhage, disc protrusion, or neoplasia may result in neurological deficits suggestive of cord compression, they are rarely visible by plain radiography. Contrast techniques are needed, therefore, to identify the lesion more precisely and to indicate its extent and location. Two techniques are available—myelography and epidurography.

Myelography

This is the introduction of a non-toxic, water-soluble radiographic contrast medium (e.g. iopamidol) into the **subarachnoid space** around the spinal cord. The contrast medium may be injected into either the cerebellomedullary cistern or in the lumbar region, depending on the region of the vertebral canal to be investigated. The site for **cisternal injection** is between the occiput and the atlas (Fig. 2.4). After injection in sternal recumbency, the head and neck are elevated for a few minutes to promote caudal flow of the contrast medium. Radiographs may then be taken in several planes. The

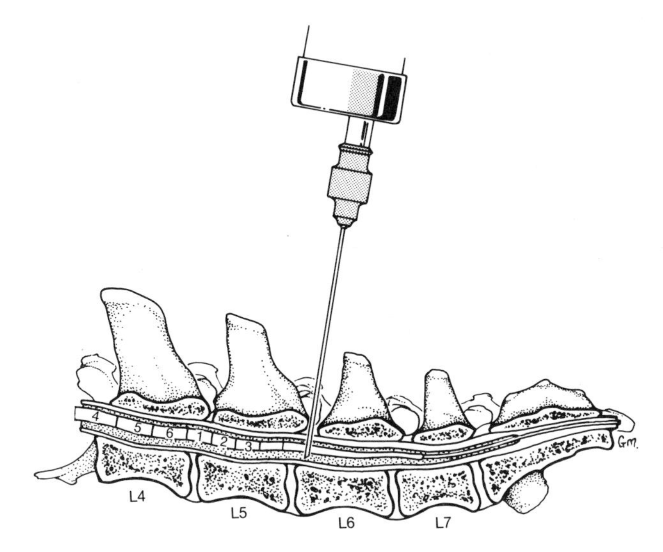

FIG. 20.6. Diagram of the vertebral canal of the dog showing a site for injecting contrast medium into the subarachnoid space (stippled). The diagram also shows the segments of the spinal cord and their relations to the intervertebral discs: 4, 5, 6, 7, lumbar segments of the spinal cord; 1, 2, 3, sacral segments of the cord; remainder (unlabelled), coccygeal segments. Thus a dorsal protrusion of disc L4–5 will strike segment L7 of the spinal cord.

site for **lumbar injection** of contrast medium is usually through the dorsal intervertebral spaces of either L5/L6 or L4/L5 with the dog in lateral recumbency (Fig. 20.6). The recommended procedure is to pass the needle right through the cord to strike the ventral floor of the vertebral canal, so that the injection can be made into the ventral subarachnoid space. The reason for injecting ventral to the spinal cord is that the dorsal subarachnoid space is smaller. However, the difficulty in accurately entering the dorsal subarachnoid space seems strange justification for destroying a needle tract of spinal cord and risking injury to its vertical arteries (Fig. 1.6). Perhaps it is fortunate for the radiographer that most patients are already suffering from paraplegia before being subjected to this technique of myelography.

Epidurography

The introduction of a suitable contrast medium into the epidural space may give an alternative means of evaluating a lumbosacral lesion. The injection may be made between the last lumbar vertebra and the sacrum. The technique is seldom used.

21. Topographical anatomy of the central nervous system

D. A. HOGG
Department of Anatomy, University of Glasgow

Spinal cord

21.1 Regions of the spinal cord

The spinal cord arises at the level of the foramen magnum; the position of its caudal end in relation to the vertebral column varies with the species and the maturity of the animal (see Section 4.2). It is divided into regions according to the groups of spinal nerves which attach to it, i.e. cervical, thoracic, lumbar, sacral, and coccygeal. The caudal part of the cervical region and the lumbar region are thickened to form respectively the **cervical** and **lumbar enlargements**, which give rise to the brachial and lumbosacral plexuses. In the dog and cat, the cervical enlargement includes spinal cord segments C6 to T1; the lumbar enlargement includes segments L5 to S1. These enlargements are due to the increased number of neurons associated with the innervation of the limbs. The thoracic region on the other hand is particularly slender, being less than the diameter of a pencil in a medium-sized dog. Caudal to the lumbar enlargement the spinal cord tapers to form the **conus medullaris**, which contains the remaining sacral and caudal (coccygeal) segments of the spinal cord. At about the level of the last lumbar and first sacral vertebrae, the spinal cord ends as the **filum terminale**, which is a fine cord consisting of glial and ependymal cells. The **dural tube** continues for a short distance after the end of the spinal cord, containing the filum terminale and the first part of the cauda equina; the tube ends between sacral vertebral segments S1 and S4, depending on the species (see Section 4.3). Caudal to the end of the dural tube, the filum combines with the dura mater to form the **filum of the dura mater**. The sacral and caudal (coccygeal) nerves form a leash around the filum terminale and conus medullaris known as the **cauda equina** (L. horse's tail).

The spinal and cranial **meninges** are surveyed in Chapters 2 and 4.

21.2 Segments of spinal cord and their relations to vertebrae

The portion of the spinal cord which gives rise to one pair of spinal nerves is defined as one **segment of spinal cord**. The segments of the spinal cord are not precisely related topographically to their corresponding vertebrae, each segment of the spinal cord usually being somewhat cranial in relation to its corresponding vertebra. Two factors contribute to this relationship. First, the segment of spinal cord is truly segmental, but the vertebra is intersegmental because it arises developmentally by fusion of the cranial and caudal halves of two adjacent segmental sclerotomes. Second, the vertebral column grows longer than the spinal cord, so that the spinal cord in adult domestic mammals ends at about the last lumbar or first sacral vertebra, depending on the species. In man, the spinal cord is very much shorter than the vertebral column, ending at about vertebra L1. The length of the spinal cord segments relative to the length of the corresponding vertebrae varies between species. In the horse, for example, it is only the cervical segments and the last segments of the spinal cord (L5, and the segments that follow it) that are cranial to their corresponding vertebrae. In contrast, the majority of the segments of the spinal cord of the dog lie within the vertebra immediately cranial to the expected level. For instance spinal cord segment T6 lies entirely within vertebra T5. The only segments of the canine spinal cord which lie mainly within the vertebra of the same number are the first one or two cervical segments, the last two thoracic segments, and the first two lumbar segments. In veterinary clinical practice the relatively cranial position of the segments of the spinal cord in relation to the vertebrae is particularly important for the last lumbar and the three sacral segments of the spinal cord of the dog; these segments are situated several vertebrae cranial to the expected level. Thus in the dog lumbar segments 5 and 6 of the spinal cord lie within the fourth lumbar vertebra; lumbar segment 7 lies over the intervertebral disc between vertebrae L4 and 5; and the three sacral segments of the spinal cord lie within the fifth **lumbar** vertebra (Fig. 20.6). In other words, no fewer than six segments of the canine spinal cord (from L5 to S3 inclusive) are crowded within two vertebrae (L4 and 5). Hence a severe midline disc protrusion between vertebrae L4 and 5 could damage segments L6, 7, S1, and S2 of the spinal cord. The result could be a lower motor neuron lesion affecting the sciatic and pelvic nerves, whereas the femoral nerve may escape even though it arises partly from fibres of the fourth lumbar spinal nerve which emerges between vertebrae L4 and 5.

21.3 General organization of grey and white matter

Centrally placed within the spinal cord lies the small **central canal** (Fig. 21.1(a)), lined by ependymal cells. The canal contains cerebrospinal fluid and is continuous with the ventricular system of the brain (Fig. 21.17). The

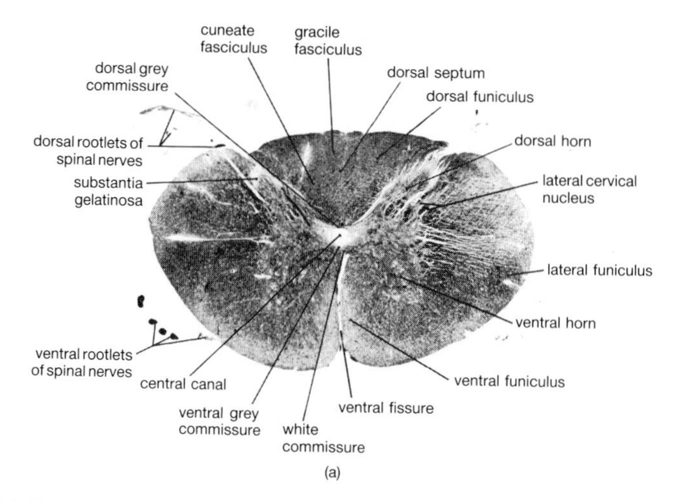

FIG. 21.1 **(a) Transverse section of the spinal cord of the dog at about C2.** The dorsal funiculus attains its maximum width here, since the gracile fascicle and the much larger cuneate fascicle are virtually complete. The ventral horn in this segment is relatively small compared with that of the cervical and lumbar enlargements. The lateral cervical nucleus is the cell location of the third neuron in the spinocervical tract; it occurs in the lateral funiculus in segments C1 and C2. In this and the other stained slices the white matter is darker than the grey. See Fig. 21.4 (b) for site of the slice.

internal structure of the spinal cord is divided into grey and white matter. In a transverse section of the spinal cord, the **grey matter** is a centrally placed, butterfly-shaped area composed mainly of nerve cell bodies and axons which are unmyelinated or only very lightly myelinated. The **white matter** is peripheral in position and consists mainly of myelinated axons but no nerve cell bodies. These regions are so named because of their colour in the fresh state, which depends on the extent to which their nerve fibres are myelinated. Both the grey and the white matter contain numerous neuroglial cells.

21.4 Dorsal, lateral, and ventral horns of grey matter

The **dorsal horn** occurs at all levels of the spinal cord and contains neurons of the somatic afferent and visceral afferent cell columns, which receive projections from the primary afferent neurons of the spinal nerves. The tip of the dorsal horn is termed the **substantia gelatinosa** (Figs. 21.1(a), (b)). In the first few cervical segments a small isolated area of grey matter lies in the lateral funiculus. This is the **lateral cervical nucleus** (Fig. 21.1(a)), a relay point on the spinocervical tract. The **ventral horn** is also present at all levels (Fig. 21.1(a)), and consists of the cell bodies of the somatic efferent neurons which supply the striated muscles of the body. In the thoracic and cranial lumbar segments there is a **lateral horn**, which represents the general visceral efferent cell column. This column comprises the cell bodies of preganglionic sympathetic neurons; in the sacral segments the corresponding lateral region of the grey matter consists of preganglionic parasympathetic neurons which distribute their axons in the sacral spinal nerves. Throughout the spinal cord two very narrow bands of grey matter extend across the midline dorsally and ventrally in relation to the central canal, forming the **dorsal** and **ventral grey commissures** (Fig. 21.1(a)).

21.5 Laminae of grey matter

The grey matter can be divided into 10 layers (**Rexed's laminae**) which have differing cytological and functional properties (Fig. 21.1(b)). The majority of the primary afferent fibres which arrive from the dorsal roots enter laminae I to VI inclusive. Laminae I, II, and III receive the small unmyelinated 'C' fibres which transmit true pain; lamina II, the **substantia gelatinosa**, is particularly involved in this pathway (see Section 9.9).

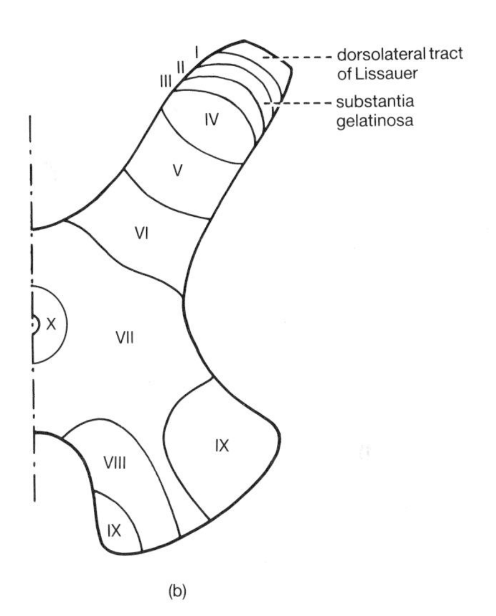

FIG. 21.1. (b) Diagrammatic transverse section of the grey matter of the spinal cord. The grey matter can be divided into 10 laminae (of Rexed) with different cytological and functional characteristics. Lamina II is the substantia gelatinosa.

Laminae IV, V, and VI receive afferent projections from progressively deeper structures. Thus lamina IV, which receives the greatest afferent input of all the first six layers, obtains its projections from cutaneous mechanoreceptors. Lamina V derives its input from deep mechanoreceptors, as well as from thermal cutaneous receptors which are thought to be a source of spinothalamic fibres. Lamina VI (**Clarke's column**) acquires an input from muscle spindles and Golgi tendon organs, with onward transmission through the dorsal and ventral spinocerebellar tracts respectively. The above summary shows that some sensory modalities can be allocated to individual laminae, but the relationships of laminae to modalities are not precise. Laminae IV, V, and VI also contain interneurons which receive projections from the lateral corticospinal and rubrospinal tracts; these interneurons then project on the motoneurons in lamina IX. Lamina VII should be interpreted as the spinal component of the reticular formation; it receives the terminals of descending reticulospinal fibres as well as a large input from the more dorsal laminae, and forms spinoreticular fibres which ascend to the brainstem. The neurons of the lateral horn lie in lamina VII. Reticulospinal fibres also end in lamina VIII, together with vestibulospinal fibres. Lamina VIII consists of a mass of short neurons which project to adjacent laminae both ipsilaterally and contralaterally, among them being many propriospinal neurons. Lamina IX contains alpha and gamma motoneurons. Lamina X consists of descussating neurons which influence laminae VII and VIII.

21.6 Funiculi of white matter

The white matter of the spinal cord is divided into three **funiculi** or columns (Fig. 12.3). The dorsal funiculus lies between the **dorsal sulcus**, a groove on the surface of the spinal cord in the dorsal midline, and the line of emergence of the dorsal rootlets of the spinal nerves. The lateral funiculus extends from the dorsal rootlets to the ventral rootlets. The ventral funiculus lies between the ventral rootlets and the **ventral median fissure**. The dorsal sulcus is largely filled by a blend of pia mater and neuroglia, thus forming the **dorsal septum**. The ventral fissure remains open, but is occupied by vertical arteries arising from the ventral spinal artery (Fig. 1.6).

The left and right **dorsal funiculi** are completely separated in the dorsal midline by the dorsal septum. Each is partly subdivided by an additional septum into a medial part, the **gracile fasciculus**, and a more lateral **cuneate fasciculus** (Fig. 21.1(a)). These two great tracts are among the few which are recognizable anatomically without special procedures. The left and right ventral funiculi are not completely separated by the intervening **ventral median fissure** (Fig. 21.1(a)), since the fissure does not extend into the white matter which lies immediately ventral to the grey matter; this narrow transverse band of white matter is the **white commissure** (Fig. 21.1(a)), and forms an important region for fibres crossing from one side of the spinal cord to the other (see Section 10.13).

The absolute quantity of white matter increases fairly uniformly in a cranial direction due to the increasing number of myelinated fibres in the various ascending and descending tract systems.

21.7 Tracts of the white matter

The approximate positions of the tracts of the spinal cord are shown diagrammatically in Fig. 12.3. This diagram suggests that the tracts are sharply defined, but actually they overlap and blend with each other.

Dorsal funiculus

The dorsal funiculus consists essentially of **ascending tracts**. In the cervical region it contains: (i) the gracile fascicle; (ii) the cuneate fascicle; (iii) the spinal part of the spinocuneocerebellar pathway; and (iv) the dorsal corticospinal tract in ungulates. These are afferent (ascending) pathways, except for the last named tract.

Lateral funiculus

Both **ascending** and **descending** tracts are contained within the lateral funiculus. In dorsoventral sequence, the **ascending tracts** in the lateral funiculus of the cervical spinal cord are: (i) the dorsolateral tract (of Lissauer); (ii) the dorsal spinocerebellar tract; (iii) the spinocervical (spinocervicothalamic) tract; (iv) the ventral spinocerebellar tract; (v) the cranial spinocerebellar tract; (vi) the spinothalamic tract; (vii) the spinoreticular tract; (viii) the spinotectal tract; and (ix) the spino-olivary tract.

The **descending tracts** in the lateral funiculus in dorsoventral order are: (i) the raphe spinal tract; (ii) the lateral corticospinal tract; (iii) the rubrospinal tract; and (iv) the medullary (lateral) reticulospinal tract.

Ventral funiculus

This contains only **descending tracts**. In the cervical spinal cord in dorsoventral order, these are: (i) the medial longitudinal fasciculus (bundle); (ii) the pontine reticulospinal tract; (iii) the ventral corticospinal tract; (iv) the vestibulospinal tract; and (v) the tectospinal tract.

All three funiculi contain short ascending and descending intersegmental fibres of the fasciculus proprius (Fig. 12.3).

Medulla oblongata

21.8 Gross structure

The medulla oblongata lies ventral to the cerebellum. Caudally it continues into the spinal cord, the boundary between the two usually being regarded as the most cranial rootlet of the first cervical spinal nerve. In the caudal third of the medulla oblongata, the dorsal septum and ventral fissure are still present as in the spinal cord. Rostrally the medulla oblongata is sharply divided from the pons by the transverse fibres which form the caudal border of the pons. Dorsally the medulla oblongata is attached to the cerebellum by the **caudal cerebellar peduncles** (Figs. 7.4, 21.12).

On the ventral surface of the medulla oblongata are the **pyramids**, two prominent parallel ridges (Fig. 21.2). At the rostral end of the ventral surface, on each side of the pyramids, there is a conspicuous transverse elevation, the **trapezoid body** (Fig. 21.2). On each side of the dorsal surface, at the caudal end of the medulla oblongata, the **cuneate fascicle** and the much smaller **gracile fascicle** are visible to the naked eye; so also is the relatively large **spinal tract of the trigeminal nerve** (Fig. 21.12).

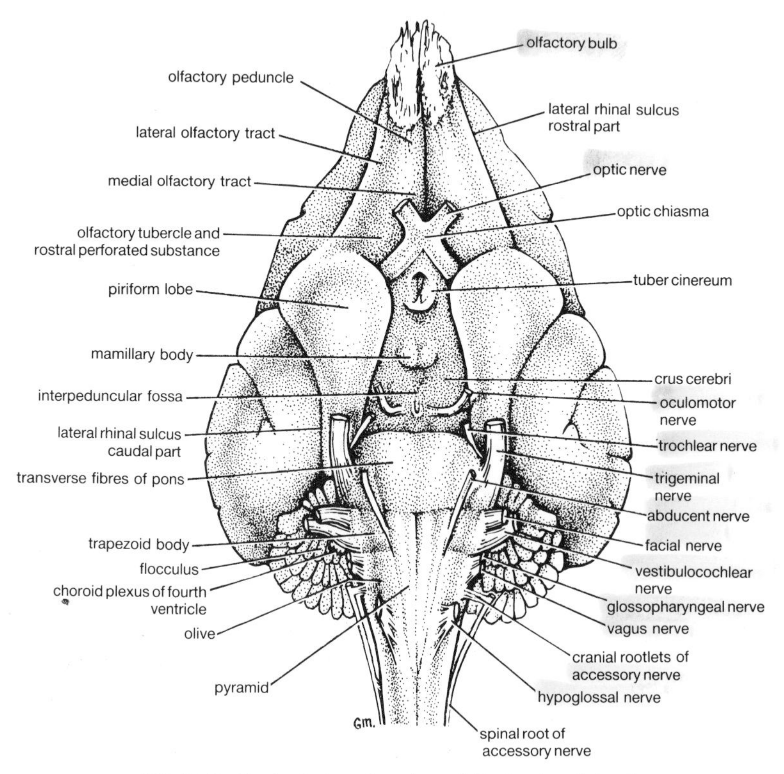

FIG. 21.2. Ventral view drawing of the brain of the dog.

21.9 Cranial nerves

The cranial nerves from the VIth to the XIIth inclusive are attached to the medulla oblongata (Figs. 21.2, 21.3). Nerves VI (**abducent nerve**) and XII (**hypoglossal nerve**), with nerve III (**oculomotor nerve**), form a series which arise from the ventral aspect of the brainstem, along a line running parallel with the midline (Fig. 21.2). Nerves III, IV, VI, and XII contain somatic efferent axons; nerve III also contains autonomic efferent fibres. These four

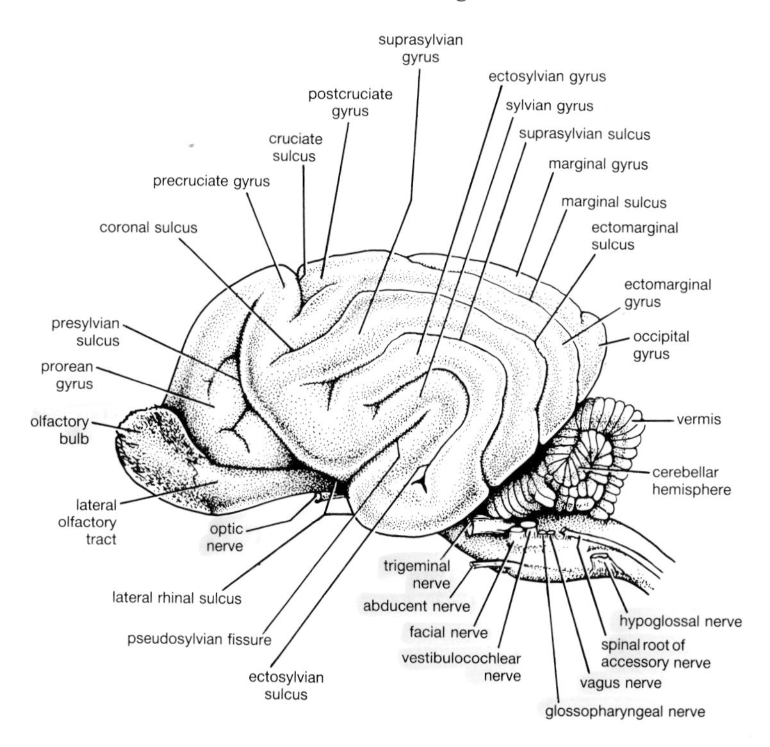

FIG. 21.3. Lateral view drawing of the brain of the dog.

cranial nerves form a series which is comparable to the ventral roots of the spinal nerves. Cranial nerves VII (**facial nerve**), IX (**glossopharyngeal nerve**), X (**vagus nerve**), and the cranial root of XI (**accessory nerve**), with cranial nerve V (**trigeminal nerve**) which comes from the pons, arise in a more lateral position (Fig. 21.2). These nerves contain special visceral efferent components supplying the muscles derived from the embryonic pharyngeal arches (see Section 6.9). Nerves VII, IX, and X also contain visceral (autonomic) efferent, visceral afferent, and somatic afferent fibres. Cranial nerve XI is attached to the medulla oblongata only by the rootlets of its cranial component; its larger spinal component arises from the cervical segments of the spinal cord and passes rostrally through the foramen magnum to unite with the cranial component. (The cranial component immediately transfers to the vagus nerve, and subsequently forms the pharyngeal and laryngeal branches of the vagus nerve; see Fig. 6.4.) The VIIIth (**vestibulocochlear**) nerve arises immediately caudal to nerve VII (Fig. 21.2).

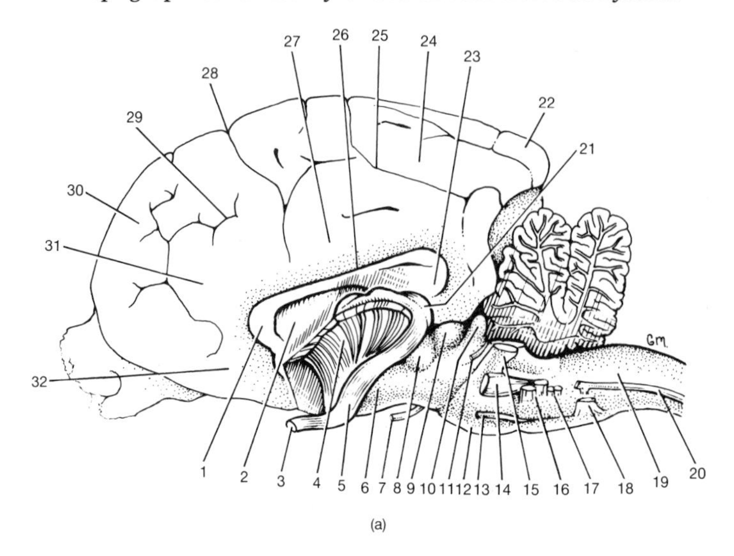

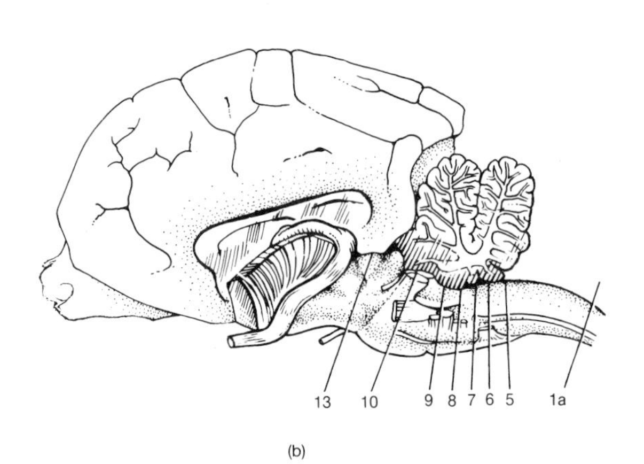

FIG. 21.4 (a) Drawing of the brain of the dog after removal of the left cerebral hemisphere and left half of the cerebellum. The left lateral aspect of the brainstem has been exposed. 1, genu of corpus callosum; 2, septum pellucidum; 3, left optic nerve; 4, left internal capsule; 5, left optic tract; 6, left cerebral crus; 7, left oculomotor nerve; 8, left medial geniculate body; 9, left rostral colliculus; 10, left caudal colliculus; 11, left trochlear nerve; 12, pons; 13, left abducent nerve; 14, left trigeminal nerve; 15, left middle cerebellar peduncle; 16, left facial and vestibulocochlear nerves; 17, left glossopharyngeal and vagus nerves; 18, left hypoglossal nerve; 19, medulla oblongata; 20, left accessory nerve; 21, left lateral geniculate body; 22, occipital gyrus; 23, splenium; 24, splenial gyrus; 25, splenial sulcus; 26, callosal sulcus; 27, cingulate gyrus; 28, cruciate sulcus; 29, genual sulcus; 30, prorean sulcus; 31, genual gyrus; 32, paraterminal gyrus.

(b) Key to the slices through the spinal cord and brainstem. Photographs of the slices are shown in Figs, 21.1 (a), 21.5–21.10 and 21.13. The slices were cut as indicted here (1a, 5–10, and 13 indicate corresponding Figures in this chapter). The white matter is stained more darkly than the grey. For the general anatomy of this aspect of the brain, see Fig. 21.4 (a).

21.10 Ventricular system

The central canal continues rostrally into the medulla oblongata from the spinal cord (Figs. 21.5, 21.6, 21.17). More rostrally still it opens out to form the much larger, diamond-shaped cavity of the fourth ventricle, which occupies a position dorsal to the medulla oblongata and pons (Figs. 21.12, 21.14, 21.17). The fourth ventricle like the central canal is lined by ependymal cells, the thin tent-like roof being formed only by ependymal epithelium with a thin partial covering of nervous tissue. The apex of the roof (the pole of the tent) projects dorsally into an indentation of the cerebellum. The rostral sloping surface of the tent is termed the **rostral medullary velum** (Figs. 21.10, 21.14(b)) and the caudal surface is the **caudal medullary velum** (Fig. 21.14(b)). The pia mater covering the caudal surface

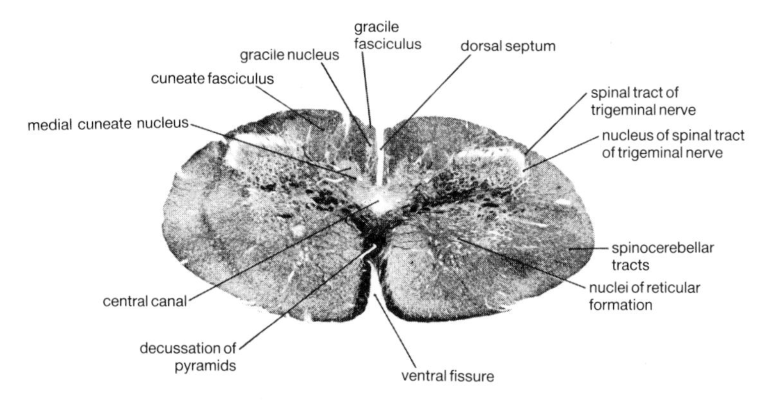

FIG. 21.5. Transverse section of the caudal end of the medulla oblongata of the dog. See Fig. 21.4. (b) for the site of the section.

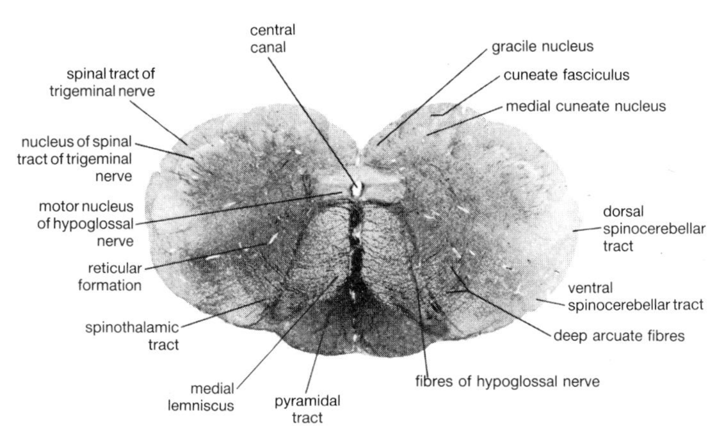

FIG. 21.6. Transverse section of the caudal quarter of the medulla oblongata of the dog. See Fig. 21.4 (b) for the site of the section.

of the cerebellum and the dorsal surface of the medulla oblongata projects rostrally to contact the ependymal epithelium of the caudal medullary velum. The apposition of the vascular pia mater to the ependyma in this and other situations within the brain produces a **choroid plexus** (Fig. 2.6), in this case the **choroid plexus of the fourth ventricle** (Figs. 21.2, 21.14(b)). The role of the choroid plexus in the production of cerebrospinal fluid is described in Section 2.10. In the most lateral parts of the roof of the fourth ventricle the small left and right **lateral apertures** open into the subarachnoid space. A small tuft of the choroid plexus of the fourth ventricle protrudes through each of these apertures (Fig. 21.14(b)). The lateral aperture lies near the **lateral recess** of the fourth ventricle (Fig. 21.17). The single median aperture of man is generally considered to be absent in the domestic mammals.

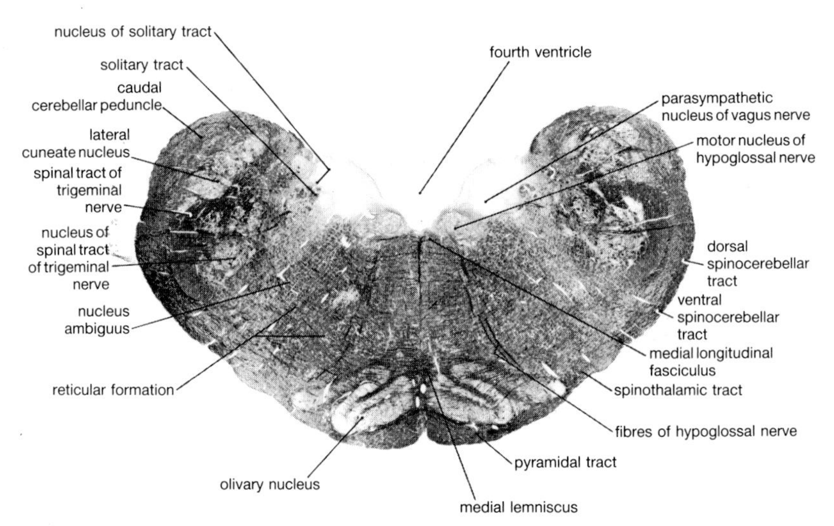

FIG. 21.7. Transverse section through the caudal half of the medulla oblongata of the dog. See Fig. 21.4 (b) for the site of the section.

21.11 Internal structure

Pyramids

The pyramids are formed by the descending fibres of the **pyramidal tracts** (Figs. 21.2, 21.6–21.9); these include the **corticospinal fibres** projecting into the spinal cord, and **corticonuclear** (corticobulbar) **fibres** which project to the cranial nerve motor nuclei within the medulla oblongata. Caudally the pyramids reduce in size and appear to merge due to the crossing over of a large proportion of their fibres in the **decussation of the pyramids** (Fig. 21.5). Caudal to the decussation, the corticospinal fibres which have crossed over leave the ventral aspect of the brainstem and take up a more dorsal position, in preparation for continuing in the lateral funiculus of the spinal cord as the lateral corticospinal tract (Fig. 12.3).

Gracile and cuneate fascicles

These two major tracts continue rostrally from the dorsal funiculus of the spinal cord to occupy a similar dorsal position in the caudal region of the medulla, before ending in the **gracile nucleus** and **medial cuneate nucleus** (Figs. 21.5, 21.6). The axons from these nuclei sweep ventromedially as the **deep arcuate fibres** (Fig. 21.6) and decussate to join the contralateral **medial lemniscus**. This bundle, which is initially vertical and adjacent to the midline in position, ascends through the medulla oblongata (Figs. 21.7–21.9) to higher levels of the brainstem. Immediately rostral to the gracile nucleus lies the small group of cells known as **nucleus Z**, which forms part of the pathway from muscle proprioceptors of the hindlimb to the cerebral cortex (see Section 9.5).

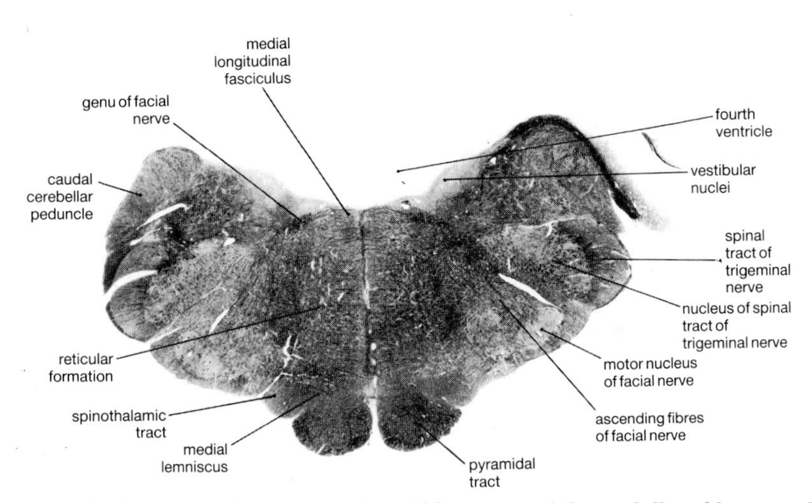

FIG. 21.8. Transverse section through the middle region of the medullar oblongata of the dog. See Fig. 21.4 (b) for the site of the section.

Spinothalamic tract

The spinothalamic tract ascends from the spinal cord to the thalamus and occupies a ventrolateral position throughout the medulla oblongata (Figs. 21.6–21.9).

Spinal tract of the trigeminal nerve

This tract (Figs. 21.5–21.9) is formed by central processes of primary afferent neurons with their cell bodies in the trigeminal ganglion (see Section 6.6). These fibres descend caudally from the pontine level in the dorsolateral region of the medulla, and conduct principally the somatic afferent sense of pain and temperature from the head. They end by synapsing in the medially adjacent **nucleus of the spinal tract of the trigeminal nerve**, an elongated structure which extends rostrally throughout the medulla oblongata (Figs. 21.5–21.9) and caudally into the first few cervical segments of the spinal cord, hence the name 'spinal'.

Dorsal and ventral spinocerebellar tracts

These convey muscle proprioceptive fibres from the hindlimb into the medulla oblongata (Figs. 21.5–21.7). The ventral tract continues rostrally, but the fibres of the dorsal tract abruptly turn dorsally to enter the ipsilateral caudal cerebellar peduncle, through which they ascend to the cerebellar cortex (Fig. 9.1). The caudal cerebellar peduncle also contains muscle proprioceptive fibres belonging to the **spinocuneocerebellar pathway**; these arise from the **lateral cuneate nucleus** (Figs. 9.1, 21.7), and continue to the caudal peduncle as the (dorsal) **superficial arcuate fibres. Nucleus X** which is apparently a relay station on the pathway of forelimb muscle proprioceptors to the cerebral cortex (see Section 9.5) lies near the rostral pole of the lateral cuneate nucleus. The **olivary nucleus** (previously known as the **inferior olive**) is a large folded mass of grey matter situated in the ventrolateral region of the medulla oblongata (Fig. 21.7) and forming a surface elevation termed the **olive** (Fig. 21.2). This nucleus receives feedback impulses from higher centres within the extrapyramidal system (see Sections 13.1, 13.2), and also an input of muscle proprioceptive and cutaneous impulses from spinal levels via the **spino-olivary tract**; the olivary nucleus projects to the contralateral cerebellar cortex, via the caudal cerebellar peduncle. The **arcuate nucleus** lies on the ventral surface of the pyramid; it appears to be a caudally displaced fragment of the pontine nuclei, projecting pyramidal feedback pathways into the cerebellar cortex by means of the (ventral) superficial arcuate fibres which pass through the caudal cerebellar peduncle.

Cochlear nuclei

In the rostral part of the medulla oblongata lie the laterally placed **dorsal and ventral cochlear nuclei** (Fig. 21.9). These are large structures in the domestic species, especially the cat and dog, so much so that the dorsal cochlear nucleus bulges laterally from the brainstem to form the **acoustic tubercle** (particularly prominent in the cat) to which the VIIIth cranial nerve attaches. Fibres from both cochlear nuclei ascend to higher centres through the **lateral lemniscus** (Figs. 21.9, 21.10), both ipsilaterally and contralaterally (Fig. 8.1). From the ventral cochlear nucleus many fibres decussate ventrally (though passing dorsally over the pyramids) to form the **trapezoid body**, a particularly conspicuous feature in the domestic species

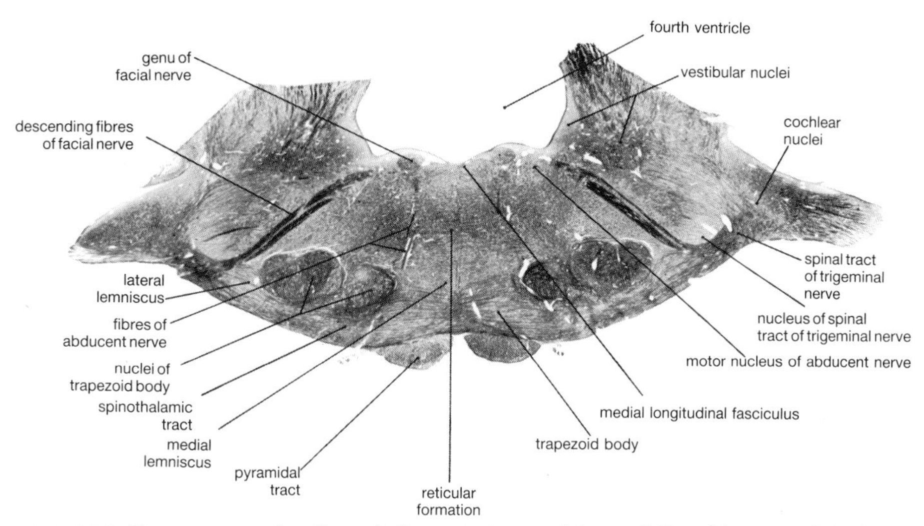

FIG. 21.9. Transverse section through the rostral part of the medullar oblongata of the dog. See Fig. 21.4 (b) for the site of the section.

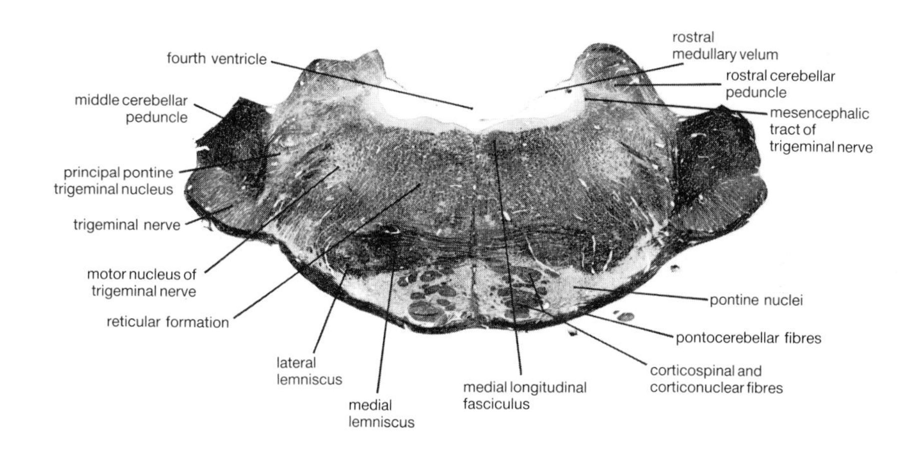

FIG. 21.10. Transverse section through the pons of the dog. See Fig. 21.4 (b) for the site of the section.

(Figs. 21.2, 21.9, 21.11). Some of these fibres are believed to synapse in the **dorsal and ventral nuclei of the trapezoid body** (Fig. 21.9), leading from there into the auditory pathway to the cerebral cortex (see Section 8.2).

Vestibular nuclei

Lying slightly medial to the cochlear nuclei are the rostral, caudal, lateral, and medial **vestibular nuclei** (Figs. 21.8, 21.9), which receive impulses from the vestibular division of the VIIIth cranial nerve. The **vestibulospinal tract** arises from the lateral vestibular nucleus and projects caudally through the medulla oblongata to the spinal cord. The vestibular nuclei have many connections with the cerebellum through the caudal cerebellar peduncle. All of the vestibular nuclei also project via the **rostral part** of the **medial longitudinal fasciculus** (Figs. 21.7–21.10, 21.13) to the motor nuclei of the cranial nerves that supply the extrinsic muscles of the eye (Fig. 8.1). Furthermore, the medial vestibular nucleus projects uncrossed fibres which descend caudally, within the ventral funiculus of the spinal cord, as the vestibulospinal part of the medial longitudinal fasciculus; these reach as far caudally as the most cranial thoracic segments of the spinal cord.

Solitary tract

The solitary tract (Fig. 21.7) is a prominent bundle in the dorsal region of the medulla oblongata, formed by the central processes of primary afferent neurons of nerves VII, IX, and X transmitting visceral afferent modalities. These fibres end by synapsing in the **nucleus of the solitary tract** (Fig. 21.7) which represents the visceral afferent cell column in the medulla oblongata (Fig. 6.2); it lies medial to the somatic afferent column, represented by the nucleus of the spinal tract of the trigeminal nerve (Fig. 21.7).

Motor nuclei of cranial nerves VI, VII, IX, X, and XII

The **motor nuclei** of the **abducent** (VIth) and **hypoglossal** (XIIth) **nerves** lie in the medulla oblongata close to the midline and immediately ventral to the fourth ventricle, and constitute a portion of the somatic efferent cell column (Fig. 6.2). The **parasympathetic nucleus of the vagus nerve** is situated immediately lateral to the **hypoglossal motor nucleus** (Figs. 6.1, 21.7); it forms part of the general visceral (autonomic) efferent column (Fig. 6.2), and contains the cell bodies of the preganglionic parasympathetic neurons of the vagus nerve. Further rostrally lie two other nuclei of the autonomic

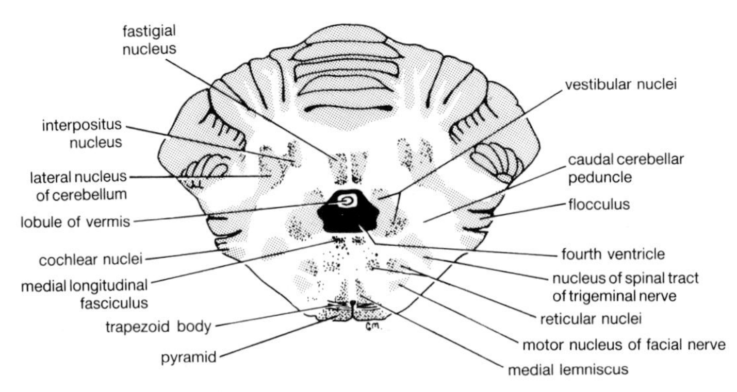

FIG. 21.11. Semischematic drawing of a transverse section through the hindbrain of a dog. Among the main landmarks are the trapezoid body and cerebellar nuclei.

efferent column (Section 6.8), i.e. the **parasympathetic nucleus of the glossopharyngeal nerve** (caudal salivatory nucleus), and the **parasympathetic nucleus of the facial nerve** (rostral salivatory nucleus). These two small nuclei are difficult to distinguish, being more or less continuous with each other and with the vagal parasympathetic nucleus. The special visceral efferent column, which developmentally arises between the two preceding columns, has been 'squeezed out' during development and has come to lie in a more ventrolateral position (Figs. 6.1, 6.2). In the medulla oblongata it is represented by the **motor nucleus of the facial nerve** (Fig. 21.8), and by the **nucleus ambiguus** (Fig. 21.7). The axons from the neurons of the facial motor nucleus reflect this migration from the original location of the nucleus, since they turn dorsomedially to loop around the motor nucleus of the abducent nerve as the **genu of the facial nerve** (Fig. 21.9), before they pass ventrolaterally to escape from the brainstem in the root of the facial nerve. The nucleus ambiguus is a combined special visceral motor nucleus for nerves IX, X, and XI.

Spinotectal, tectospinal, rubrospinal tracts

In addition to those already mentioned, several other ascending or descending tracts travel in the medulla oblongata to or from higher centres. These include the spinotectal, tectospinal, and rubrospinal tracts.

Reticular formation

The remaining substance of the medulla oblongata is made up of the **reticular formation** (Figs. 21.6–21.9), a network of grey and white matter which extends throughout the brainstem. Certain nuclear masses are recognizable within this network. One group of nuclei (the **precerebellar reticular nuclei**) projects to the cerebellum. Another group (the **noncerebellar reticular nuclei**) is subdivided into the following three longitudinal and parallel series of nuclei, extending between the medulla oblongata and the midbrain. (i) The **raphe nuclei** lie in the midline, ventral to the fourth ventricle and mesencephalic aqueduct. Of these, the **nucleus raphe magnus** (in the medulla oblongata) is of particular interest since, like the central grey matter (**periaqueductal grey**) of the midbrain, it is involved in mechanisms controlling pain (see Section 9.9). The neurons of the nucleus raphe magnus contain **serotonin**, and project serotoninergic axons down the spinal cord in the **raphe spinal tract**. (ii) The **central group** of nuclei (e.g. the **gigantocellular reticular nucleus**) forms a line on each side of the midline. It includes the **medial medullary motor reticular centres**, which form the **medullary reticulospinal tract**. It also includes inspiratory components of the **medullary 'respiratory centre'**, and cardiac depressor components of the **medullary 'cardiovascular centre'** (see Section 16.6). (iii) The **lateral group** (e.g. the **parvicellular reticular nucleus**) lies lateral to and parallel with the central group. It includes the **lateral medullary motor reticular centres**, which project inhibitory fibres medially into the central group of nuclei, especially into the medullary motor reticular centres. It also contains expiratory neurons of the respiratory centre and cardiac accelerator neurons of the cardiovascular centre.

Pons

21.12 Gross structure

On its ventral surface the pons is sharply demarcated from the midbrain and medulla oblongata by the rostral and caudal borders of the **transverse fibres of the pons** (Fig. 21.2), a raised mass of fibres forming a 'bridge' which gives its name to the whole region (L. *pons*, a bridge). This mass of fibres shows considerable variation in size in different species. In man it is particularly large and extends much further caudally, overlapping other structures including the **trapezoid body** which is then regarded as being a constituent of the pons. Dorsally the pons attaches to the cerebellum by the **middle cerebellar peduncle**. The pons is divided into a **ventral part** consisting of pyramidal fibres and descending pyramidal feedback pathways and nuclei, and a **dorsal part** (the **tegmentum of the pons**) containing the remaining nuclei and tracts.

21.13 Cranial nerves

The sensory and motor roots of the Vth (**trigeminal**) nerve attach to the ventrolateral aspect of the pons (Figs. 21.2–21.4), the point of attachment forming an arbitrary boundary between the transverse fibres of the pons and the middle cerebellar peduncle. The attachment of nerve V is in line with the **lateral** series of cranial nerves (see Section 21.9).

21.14 Ventricular system

The pons forms the rostral part of the floor of the fourth ventricle (Figs. 21.10, 21.14(a)).

21.15 Internal structure

Corticospinal, corticonuclear, corticopontocerebellar pathways

Descending tracts which pass through the **ventral part of the pons** include the corticospinal and corticonuclear tracts. Fibres in the **corticopontine**

tract descend from the cerebral cortex and end in the ventral part of the pons by forming synapses in the scattered **pontine nuclei** (Fig. 21.10). Axons from the pontine nuclei run in the **transverse fibres of the pons** and ascend to the cerebellar cortex through the contralateral **middle cerebellar peduncle** (Figs. 20.10, 20.12), thus completing the **corticoponto-cerebellar pathway**. The large size of the middle peduncle even in domestic species indicates the importance of this pyramidal feedback pathway from the cerebral hemisphere to the cerebellar cortex of the opposite side (see Section 11.2).

Ascending and descending tracts in the tegmentum of the pons

Ascending pathways in the more dorsal part of the pons include fibres of the **medial lemniscus** (Fig. 21.10), which is now horizontal rather than vertical in position. The more laterally placed **lateral lemniscus** (Fig. 21.10) forms part of the ascending auditory relay (Fig. 8.1). The rostral part of the **medial longitudinal fasciculus** (Fig. 21.10) lies in the midline under the floor of the fourth ventricle, and carries ascending axons from the vestibular nuclei to the motor nuclei of cranial nerves III and IV, and also VI (Fig. 8.1). The **rubrospinal** and **tectospinal** tracts descend through the dorsal part of the pons.

Motor and sensory trigeminal nuclei

The **motor nucleus of the trigeminal nerve** lies in the dorsal part of the pons (Fig. 21.10), continuing the special visceral efferent column (Fig. 6.2). On its lateral aspect lies the **principal** (or pontine) **sensory nucleus of the trigeminal nerve** (Fig. 21.10), which receives its input from the pathways of touch and pressure from the face.

Reticular formation

The reticular formation continues throughout the dorsal part of the pons (Fig. 21.10). It includes the **pontine motor reticular centres**, which give rise to the **pontine reticulospinal tract**. It also contains the **pneumotaxic centre**. The **locus ceruleus** is a small pigmented component of the reticular formation lying adjacent to the **mesencephalic tract of the trigeminal nerve** (Fig. 20.10) in the rostral and lateral part of the floor of the fourth ventricle, partly in the pons and partly in the midbrain. Despite its small size, the locus ceruleus has noradrenergic projections to extremely widespread parts of the neuraxis, including the entire cerebral cortex, the thalamus, hypothalamus, brainstem, cerebellum, and spinal cord; afferent projections are received from the raphe nuclei, hypothalamus, amygdaloid body, and cingulate gyrus. It may be involved in mechanisms which switch on episodes of paradoxical sleep (see Section 16.3).

Midbrain

21.16 Gross structure

This is the smallest division of the brainstem, interposed between the pons caudally and the diencephalon rostrally. It consists of a dorsal component, the tectum (L. roof), and a ventral component, the cerebral peduncle. The **tectum** (strictly, the mesencephalic tectum) is formed by the rounded **rostral** and **caudal colliculi** (Figs. 7.4, 21.4, 21.12) and their brachia. The rostral colliculus connects to the lateral geniculate body by the **brachium of the rostral colliculus**, and the caudal colliculus is joined by the **brachium of the caudal colliculus** to the medial geniculate body. The rostral boundary of the tectum is formed by the **caudal commissure** (Fig. 21.14(b)); the caudal border lies at the beginning of the fourth ventricle. Caudodorsally the midbrain is linked to the cerebellum by the paired **rostral cerebellar peduncles** (Fig. 21.12).

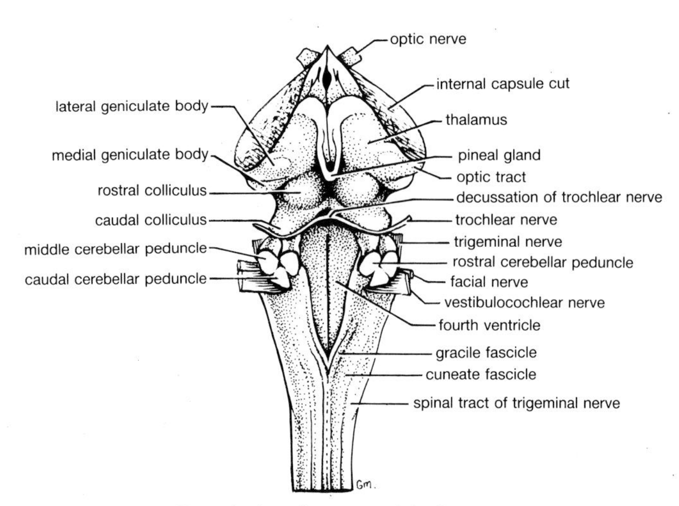

FIG. 21.12. Dorsal view drawing of the brainstem of a dog.

Ventrally, the left and right **cerebral peduncles** form a pair of prominent columns which diverge rostrally forming the **interpeduncular fossa** between them (Fig. 21.2). In the depth of the fossa lies the **caudal perforated substance** formed by the penetration of blood vessels. Each cerebral peduncle consists of a dorsal component, the **tegmentum** (L. *tegmen*, a covering) which constitutes the central core of the midbrain, and a ventral component, the **cerebral crus** (Figs. 21.2, 21.4); between them lies an intermediate component, the **substantia nigra** (Fig. 21.13).

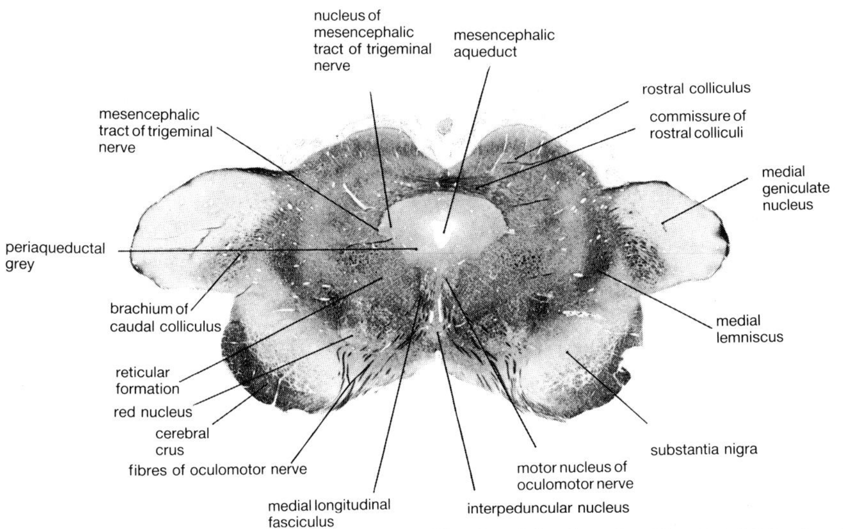

FIG. 21.13. Transverse section through the midbrain of the dog. See Fig. 21.4 (b) for the site of the section.

21.17 Cranial nerves

Cranial nerves III (oculomotor nerve) and IV (trochlear nerve) arise from the midbrain. The **oculomotor nerve** attaches ventrally in the interpeduncular fossa (Fig. 21.2), and topographically is in line with nerves VI and XII. All of this series of nerves contain somatic efferent neurons. Although the **trochlear nerve** is also a somatic efferent nerve its site of emergence from the brainstem is not in line with nerves III, VI, and XII; instead it arises from the **dorsal** surface of the brainstem, immediately caudal to the caudal colliculus (Fig. 21.4(a)). The reason for this is that its fibres descussate dorsally in the rostral medullary velum to form the **decussation of the trochlear nerve** (Fig. 21.12), a slender but distinct transverse band in the roof of the fourth ventricle.

21.18 Ventricular system

The part of the ventricular system associated with the midbrain is the **mesencephalic aqueduct** (cerebral aqueduct), a small, centrally located channel. It expands caudally at pontine level into the fourth ventricle (Figs. 21.13, 21.14(a), (b), 21.17). Rostrally it continues into the third ventricle.

21.19 Internal structure

Tectum

The caudal colliculus lies on the dorsal aspect of the midbrain (Figs.

21.4(a), 21.12). It serves as a relay station and reflex centre on the auditory pathways (Figs. 8.1, 8.5). Fibres leave this colliculus via the **brachium of the caudal colliculus** and pass to the **medial geniculate body** of the thalamus. Just rostral to the caudal colliculus lies the **rostral colliculus** (Figs. 21.4, 21.12, 21.13), which functions as a reflex centre for the visual pathways (Figs. 8.1, 8.4). The rostral colliculus connects by the **brachium of the rostral colliculus** to the **lateral geniculate body** (Fig. 7.4). The left and right rostral colliculi are linked by the **commissure of the rostral colliculi** (Fig. 21.13); a similar **commissure of the caudal colliculi** joins the caudal colliculi. The **spinotectal tract** projects proprioceptive signals into the tectum from spinal segments, which help to integrate spinovisual reflexes causing the head and eyes to turn towards the source of visual or auditory stimulation. The rostral border of the tectum is crossed by the **caudal commissure** which lies just ventral to the pineal body and dorsal to the mesencephalic aqueduct (Fig. 21.14(b)). It connects the right and left sides of several midbrain and diencephalic structures, but is of very uncertain function.

Tegmentum of the midbrain

The tegmentum contains the following fibre pathways and nuclei.

(i) Ascending fibres of the **medial lemniscal system** lie dorsal to the substantia nigra (Fig. 21.13). The **medial longitudinal fasciculus** (Fig. 21.13), which has already projected upon the motor nuclei of the abducent and trochlear nerves, now ends in the motor nucleus of the oculomotor nerve. The **lateral lemniscus** abruptly diverges dorsally to reach the **caudal colliculus**. At the level of the caudal colliculus a bundle of transverse fibres traverses the tegmentum ventral to the medial longitudinal fasciculus; this is the so-called **decussation of the rostral cerebellar peduncles** and consists of pyramidal and extrapyramidal feedback fibres which are decussating as they return from the cerebellum to the higher centres.

(ii) The **nucleus of the mesencephalic tract of the trigeminal nerve** is located in the lateral part of the central grey matter of the midbrain (Fig. 21.13), lateral to the mesencephalic aqueduct, and is closely associated with the **mesencephalic tract of the trigeminal nerve** (Figs. 21.10, 21.13). This nucleus consists of the cell bodies of primary afferent proprioceptive neurons of the Vth nerve, which have forsaken the characteristic ganglionic location of primary afferent cell bodies and are situated within the brainstem itself (see Section 6.6).

(iii) The **motor nuclei of the oculomotor** (Fig. 21.13) and **trochlear nerves** are located close to the midline, in line with the somatic efferent cell column (Fig. 6.2). However, as already stated, the axons from the trochlear nucleus decussate over the roof of the fourth ventricle (Fig. 21.12). The **parasympathetic nucleus of the oculomotor nerve** (Edinger–Westphal nucleus) belongs to the general visceral efferent (parasympathetic) series

(Fig. 6.2) and lies at the rostrodorsal aspect of the motor nucleus of the oculomotor nerve.

(iv) The **red nucleus** is a large nucleus located ventral to the oculomotor nucleus and within the reticular formation of the midbrain (Fig. 21.13). It is so named because its vascularity gives it a red colour in the fresh state. It gives rise to the **rubrospinal tract**, a major pathway of the extrapyramidal system and of particular importance in domestic species. The **tectospinal tract** arises from the tectum and descends to the cervical spinal cord, its function being to turn the head and neck in response to visual and auditory impulses (Figs. 8.4, 8.5). The rubrospinal and tectospinal tracts decussate in the midbrain forming the **ventral** and **dorsal tegmental decussations** respectively.

(v) The **interpeduncular nucleus** lies, as its name suggests, between the cerebral peduncles in the ventral midline region of the midbrain (Fig. 21.13) ventral to the periaqueductal grey. It has connections (Fig. 21.18) with the rhinencephalon, mamillary body, hypothalamus, habenular nuclei, and reticular formation in the pons and medulla; it forms a relay station on pathways linking the olfactory and limbic parts of the rhinencephalon with autonomic nuclei in the medulla oblongata and with autonomic reticulo-spinal pathways (see habenular nuclei for further details, Section 21.23).

(vi) There is extensive **reticular formation** in the tegmentum of the midbrain. One of its components, the **central grey matter (periaqueductal grey)** (Fig. 21.13) is of particular interest since it contains enkephalinergic nerve endings which are involved in the control of pain. (See also Section 21.11, the **Reticular formation** of the medulla oblongata.)

Cerebral crus

The cerebral crus (Fig. 21.13) is composed of descending fibres belonging mainly to the pyramidal system, but it also contains the **corticopontine fibres**. Therefore the cerebral crus is much larger than the pyramid since it includes the corticopontine pyramidal feedback fibres, whereas the pyramid contains only fibres which either end in the medulla oblongata or descend into the spinal cord.

Substantia nigra

Dorsal to each cerebral crus lies the pigmented grey area known as the **substantia nigra** (Fig. 21.13), which is associated with the extrapyramidal system. It becomes prominent in mammals and is best developed in man. It projects to many regions of the brain, including the cerebral cortex, basal nuclei, thalamus, tectum, amygdaloid body, and hypothalamus, the functional significance of all these connections being obscure. However, it is clearly known to be involved with the basal nuclei, and especially in inhibiting the globus pallidus. It shares these general activities with the

subthalamic nucleus, to which it is closely related rostrally. Lesions in both of these components can cause severe hyperkinesia. The cells of the substantia nigra normally contain **dopamine**, but in **Parkinson's disease** this substance is virtually absent. It appears likely that the substantia nigra regulates the basal nuclei through the action of dopamine as a neurotransmitter.

Diencephalon

21.20 Gross structure

The diencephalon is a midline structure forming the most rostral part of the brainstem. It consists essentially of the structures that enclose the third ventricle, and is itself entirely surrounded by the cerebral hemisphere except ventrally in the hypothalamic region. Caudally it joins the midbrain at the level of the caudal commissure and the caudal border of the mamillary bodies. Rostrally it reaches to the rostral commissure of the rhinencephalon. Laterally it is extensively attached to the medial surface of

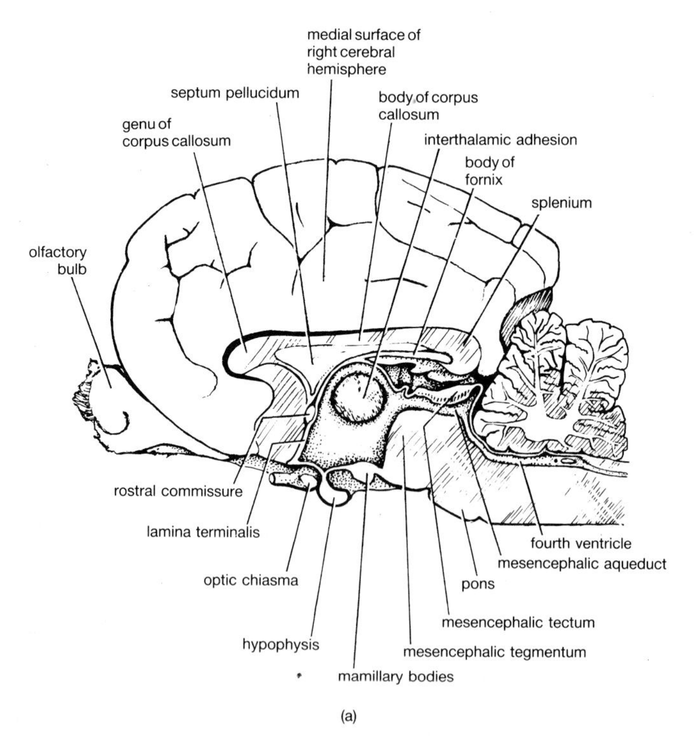

FIG. 21.14 (a) Drawing of a midline sagittal section through the brain of a dog.

the cerebral hemisphere by the fibres of the caudal limb of the internal capsule (Fig. 21.12) and the optic and acoustic radiations.

Epithalamus

This is a small, and most dorsal component, of the diencephalon, situated at the caudal end of the third ventricle. As its name implies, it lies 'above' the thalamus. From its dorsal surface projects the **pineal gland**.

Thalamus

The thalamus is the largest part of the diencephalon. It lies on either side of the third ventricle, the left thalamus being joined across the midline to the right thalamus (Fig. 21.14(a), (b)) by the **interthalamic adhesion** (see Section 21.22). The thalamus is shaped like an egg, with the pointed end directed cranially (Figs. 21.12, 21.19).

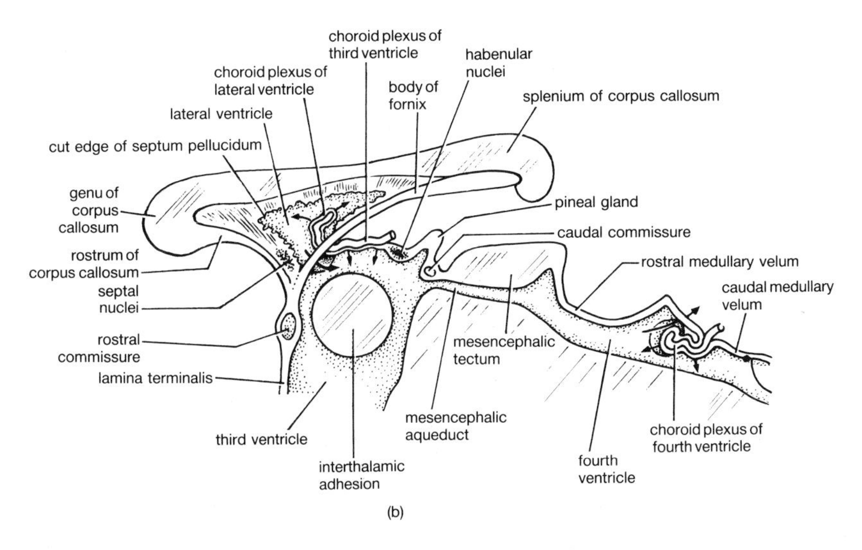

FIG. 21.14 (b) **Diagram of the ventricles of the brain of the dog.** The diagram is an enlarged view of part of Fig. 21.14 (a). Arrows indicate the formation of cerebrospinal fluid from the choroid plexuses of the lateral ventricle and third ventricle, and from that of the fourth ventricle. Formation of cerebrospinal fluid by the choroid plexuses of the lateral, third, and fourth ventricles is indicated by short arrows directed into the ventricles. A curved arrow passes through the interventricular foramen (at the rostral end of the diagram), and through the lateral aperture of the fourth ventricle (at the caudal end of the diagram), indicating the direction of flow of cerebrospinal fluid. The diagram shows that the choroid plexus of the third ventricle is continuous with that of the lateral ventricle. (Based on Fig. 16.16 of Evans and Christensen (1979), *Miller's anatomy of the dog*, by courtesy of Dr H. E. Evans and Cornell University.)

Metathalamus

The metathalamus comprises the **medial** and **lateral geniculate bodies** (Figs. 21.4, 21.12). These are caudolateral projections which somewhat overlie the midbrain laterally.

Hypothalamus

This component of the diencephalon lies 'below', i.e. ventral to, the thalamus on either side of the midline. In rostrocaudal order its superficial components are the **optic chiasma, tuber cinereum,** and **mamillary bodies**. The tuber cinereum is the region bordered by the optic chiasma, the beginning of the optic tracts, and the mamillary bodies. The **optic tract** extends from the optic chiasma to the lateral geniculate body (Figs. 21.4, 21.15(a)). The **hypophysis** is attached to the infundibulum, which in turn is attached to the tuber cinereum. The mamillary bodies are so named because they resemble two small breasts (Fig. 21.2). The right and left hypothalami are separated by the third ventricle (Figs. 16.1, 21.16).

Subthalamus

This is a small ventral region at the junction of diencephalon and midbrain, caudolateral to the hypothalamus and ventral to the thalamus.

21.21 Cranial nerves

The **optic nerve** (cranial nerve II) arises from the ventral surface of the hypothalamus at the **optic chiasma** (Figs. 21.2, 21.4).

21.22 Ventricular system

The **third ventricle** is a vertical slit within the diencephalon (Figs. 21.14(a), (b), 21.15–21.17). The slit is converted to a ring-like space by the solid cylindrical **interthalamic adhesion**, which joins the left thalamus to the right thalamus (Figs. 21.14(a), (b), 21.15(a)); at least in man, only a few axons actually connect the right and left thalamus via the adhesion, most of it being formed of neuroglial tissue. In man the adhesion is occasionally absent. The third ventricle has a thin membranous roof from which the **choroid plexus of the third ventricle** protrudes into the lumen (Figs. 21.14(b), 21.15(b)). The choroid plexus of the third ventricle is continuous through the interventricular foramen with the choroid plexus of the lateral ventricle (Fig. 21.14(b)). The floor of the third ventricle is formed by the optic chiasma, infundibulum, tuber cinereum, and the mamillary bodies. The funnel-shaped **infundibular recess** (Fig. 21.17) extends the ventricle ventrally into the infundibulum, which is the stalk of the hypophysis. The rostral wall is formed by the **lamina terminalis**, a thin layer of grey matter extending from the rostrum of the corpus callosum to the dorsal surface of the optic chiasma (Fig. 21.14). At the junction of the roof with the rostrolateral wall is the **interventricular foramen** (Figs. 21.14(b), 21.17). The caudal boundary of the third ventricle is formed by the pineal gland, the caudal commissure, and the mesencephalic aqueduct (Fig. 21.14). The **pineal recess** enters the pineal gland (Fig. 21.17).

21.23 Internal structure

Epithalamus

The epithalamus consists of the **pineal gland** (Figs. 21.12, 21.14(b)), and the **habenular nuclei** (Figs. 21.14(b), 21.15(a), (b)) and their connections (L. *habenula*, a small strap). The afferent fibres of the habenular nuclei, which are partly olfactory and partly limbic, arrive from the septal nuclei (Fig. 21.18) through the **habenular stria** (the **stria habenularis thalami**, also known as the **stria medullaris thalami**). The efferent fibres leave the habenular nuclei via the **fasciculus retroflexus** (Fig. 21.15(a), 21.18), which then projects to the interpeduncular nucleus (Fig. 21.18). The latter connects (indirectly) with the pontine and medullary reticular formation; this pathway runs from the interpeduncular nucleus to the **periventricular fibres** of the hypothalamus, and then through the **dorsal longitudinal fasciculus** to the pontine and medullary reticular formation (see Section 16.5 for functional considerations). The **caudal commissure** is a stout bundle of fibres which crosses the midline near the pineal gland (Fig. 21.14(b)), but is of uncertain distribution and function.

Thalamus

The thalamus consists of many large nuclear masses and their connections. The basic characteristic of the thalamic nuclei is that most of them project to the cerebral cortex (by thalamocortical fibres). Various classifications of the thalamic nuclei have been attempted, according to either topographical, embryological, or physiological criteria. However, much remains to be done before fully rational subdivisions can be achieved. (The extent of this problem is revealed by the almost total lack of agreement between the *Nomina Anatomica Veterinaria* and the *Nomina Anatomica.*) The nuclei are named here topographically, after Bowsher (1975) and Meyer (1979). Their afferent and efferent connections are discussed in Sections 17.17–17.22.

(i) The **ventral nuclei** are separated into topographical subgroups. The **caudal subgroup** is associated with the major specific sensory pathways ascending to conscious levels of the cerebral cortex. Thus the **ventrocaudal nucleus, lateral part**, (often known as **nucleus VPL**), is associated with projections from the body via the medial lemniscal system (Fig. 21.15(a)); the **ventrocaudal nucleus, medial part (nucleus VPM)** (Fig. 21.15(a)), serves a similar function for the head region. The **ventrolateral nucleus**

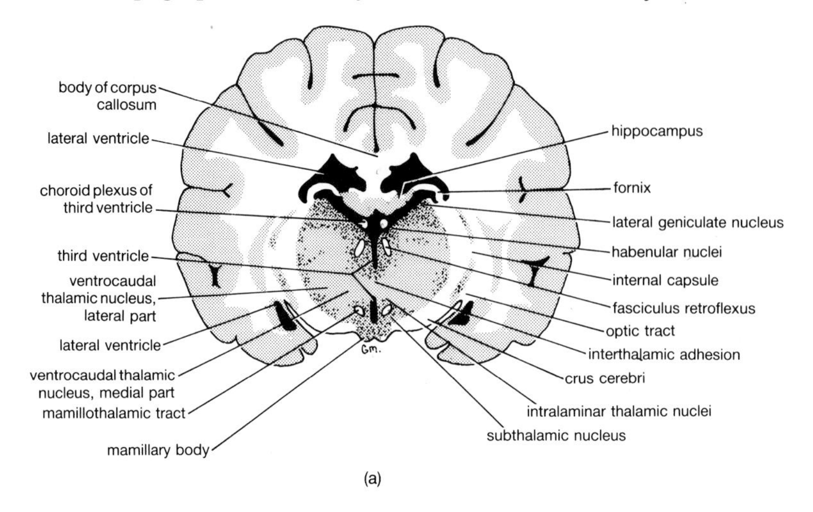

FIG. 21.15. (a) Semischematic drawing of a transverse section through the forebrain of a dog. The section was cut in the region of the mamillary bodies and thalamus.

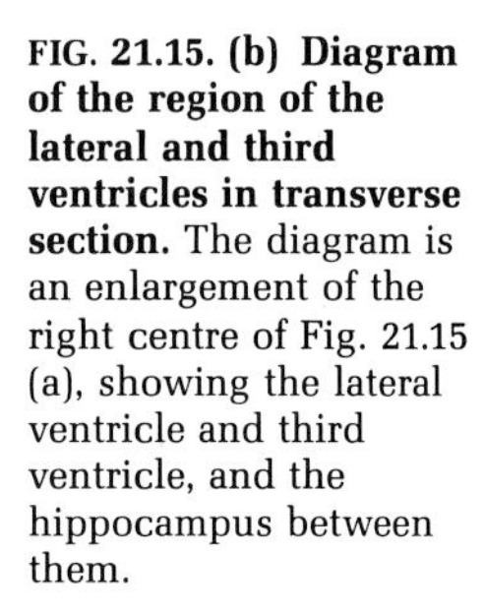

FIG. 21.15. (b) Diagram of the region of the lateral and third ventricles in transverse section. The diagram is an enlargement of the right centre of Fig. 21.15 (a), showing the lateral ventricle and third ventricle, and the hippocampus between them.

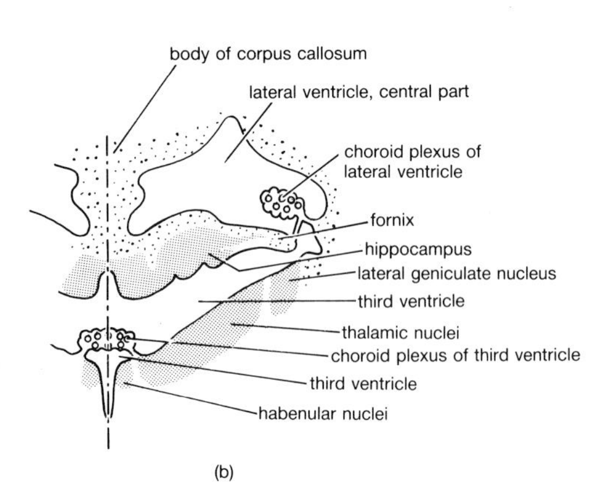

projects to the pyramidal and extrapyramidal cortex and is therefore the final relay for cerebellar–pyramidal, and cerebellar–extrapyramidal feedback pathways (Figs. 11.2, 13.1).

(ii) The **intralaminar nuclei** include five nuclei, among them the **nucleus centralis thalami** (also known as the **nucleus centrum medianum**). In Section 17.19 the intralaminar nuclei are called the **central group** of thalamic nuclei. They lie adjacent to the third ventricle (Figs. 21.15(a), 21.16). Representing the rostral end of the ascending reticular formation, they receive non-specific information arriving through the ascending reticular formation, and project diffusely all over the cerebral cortex (Fig. 9.3), thus providing for the arousal function of the ascending reticular formation. They also project to the lateral, dorsomedial, and rostral nuclei,

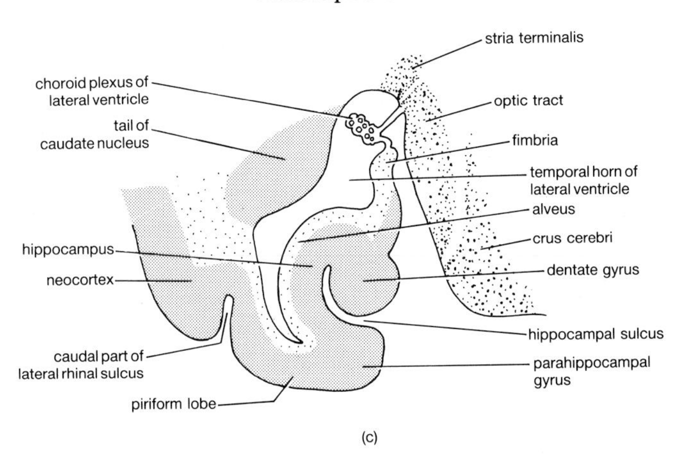

FIG. 21.15. (c) Diagrammatic transverse section of the lateral ventricle in the region of the piriform lobe. The field shown is an enlargement of the lower left part of Fig. 21.15 (a), showing the temporal horn of the lateral ventricle, the nuclei and gyri which bound it, and the choroid plexus within it. (Based on Abb.45 of *Lehrbuch der Anatomie der Haustiere,* by Nickel, Schummer, and Seiferle, Vol. IV (1975) by courtesy of Verlag Paul Parey.)

which in turn project to widespread regions of the cortex. This entire system, orchestrated by the intralaminar nuclei, is the anatomical basis for the consciousness maintained by the ascending reticular formation. It also accounts for much of the cerebral activity in the electroencephalogram.

(iii) There are also lateral, dorsomedial, and rostral nuclei. The **lateral thalamic nuclei** receive projections from the ventral nuclei and project to the cognitive cerebral cortex (Fig. 17.3). The **dorsomedial thalamic nuclei** receive projections from the **periventricular fibres** of the hypothalamus and project to the frontal lobe of the cerebral cortex (Fig. 17.3), this being part of the two-way circuitry which places the hypothalamus under the control of the cortex. The **rostral thalamic nuclei** receive an indirect afferent input from the hippocampus via the mamillary body and mamillothalamic tract (Fig. 21.18), and project to the cingulate gyrus of the cerebral cortex; these thalamic nuclei are considered to be a part of the **limbic system** (see Section 8.7).

Metathalamus

This part of the diencephalon consists of the **lateral** and **medial geniculate nuclei** (Figs. 21.13, 21.15(a)), which project from the surface of the brain as the lateral and medial geniculate bodies (Figs. 21.4, 21.12). The lateral geniculate nucleus is an important visual relay centre, and the medial geniculate nucleus plays a similar role in the auditory pathways (Fig. 8.1). The axons from the lateral geniculate nucleus pass caudally to the visual area in the occipital lobe of the cerebral cortex (Fig. 8.2(a)) as the **optic**

radiation. These are the final fibres in the optic pathway. Axons from the medial geniculate nucleus project to the auditory area of the temporal lobe as the **auditory radiation**.

Hypothalamus

The hypothalamus can be subdivided into many nuclei. The *Nomina Anatomica Veterinaria* lists 22, and organizes them into a rostral area lying above the optic chiasma, an intermediate area placed over the tuber cinereum, and a caudal area related to the mamillary bodies. The well-known **supraoptic** and **paraventricular nuclei** lie in the rostral area of the hypothalamus. The hypothalamus can also be divided into a **medial** and **lateral zone**, by the postcommissural fornix. The medial zone contains most of the hypothalamic nuclei, including the supraoptic and paraventricular nuclei (Figs. 16.1, 16.2, 21.16). They project to the neurohypophysis by the **supraoptico-** and **paraventriculohypophyseal tracts**, respectively. These two nuclei are involved in osmoregulation.

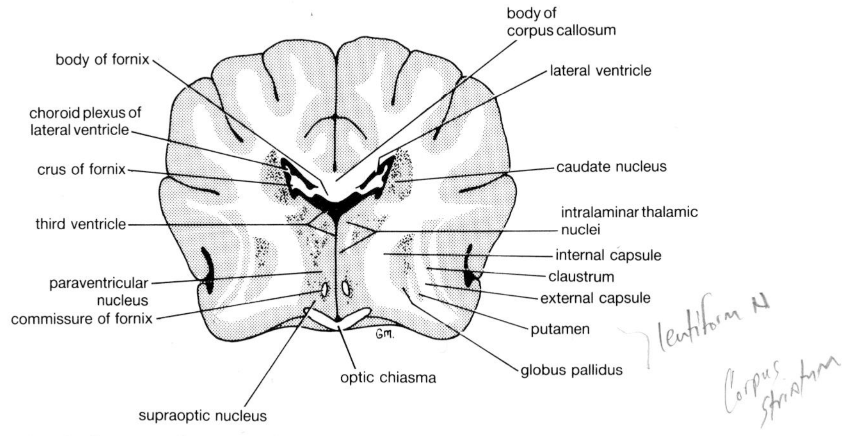

FIG. 21.16. Semischematic drawing of a transverse section through the forebrain of a dog. The section was cut in the region of the optic chiasma, hypothalamus, and basal nuclei.

The **mamillary body** contains hypothalamic nuclei linking the hippocampus and midbrain reticular formation. The mamillary body projects to the rostral thalamic nuclei by the **mamillothalamic tract** (Figs. 21.15(a), 21.18). Although the contribution of the mamillary body to the structural basis of emotion is uncertain, it is commonly included in the **limbic system**, as are the rostral thalamic nuclei (see Section 8.7).

The **stria terminalis** and **fornix** are two conspicuous fibre bundles which project into the hypothalamus. The stria terminalis consists of efferent fibres from the amygdaloid body to the septal nuclei (Fig. 21.18) and hypothalamus (Fig. 21.19(d)); the fornix (Fig. 21.19(e)) comprises two-way connections between the hippocampus on the one hand and the hypothalamus and mamillary body on the other (Fig. 21.18). The stria terminalis and fornix run parallel to each other (Figs. 21.18, 21.20), separated by the lateral ventricle. The stria terminalis lies on the floor of the lateral ventricle in a groove between the caudate nucleus and thalamus (Fig. 21.20). The body of the fornix is in the midline, immediately ventral to the corpus callosum (Figs. 21.14(a), (b), 21.16). For further details of the fornix see the end of Section 21.29.

The **medial forebrain bundle** carries a considerable proportion of the efferent and afferent fibres of the hypothalamus. Essentially it links the hypothalamus in both directions, with more rostral components such as the septal nuclei and olfactory bulb (Fig. 21.18), and with more caudal structures such as the reticular formation of the hindbrain and the parasympathetic nuclei of the cranial nerves (see Section 8.5).

The **periventricular fibres** of the hypothalamus are a link in the pathway from the habenular nuclei to the hindbrain reticular centres (see Section 21.23, **Epithalamus**) and in the connections between the hypothalamus and dorsomedial thalamic nuclei (see Section 21.23, **Thalamus**).

Subthalamus

This small region is composed of several nuclei and fibre bundles which connect with the extrapyramidal system. The **subthalamic nucleus** is of clinical interest in man because of its association with hemiballismus (see Section 13.4).

Cerebellum

21.24 Gross structure

The cerebellum (L. little brain) is separated from the cerebral hemispheres by the **transverse cerebral fissure** in which lies the **tentorium cerebelli** (Fig. 2.4). It lies dorsal to the pons and medulla oblongata and is separated from the fourth ventricle only by the thin, tent-like roof of the ventricle, i.e. the rostral and caudal medullary velum. The cerebellum is attached to the brainstem by the three paired **cerebellar peduncles**.

The cerebellum consists of two laterally placed **hemispheres** and a relatively narrow midline **vermis** (L. a worm) (Figs. 15.3, 21.3). Two deep fissures divide the cerebellum into three main lobes, i.e. the **primary fissure** separating the **rostral lobe** from the **caudal lobe**, and the **uvulonodular** (caudolateral) **fissure** dividing the flocculonodular lobe from the caudal lobe (Fig. 15.3); the **flocculonodular lobe** consists of the midline **nodulus** of the vermis (Fig. 15.3) and the paired **flocculus** (L. a little tuft) (Figs. 21.2,

21.11) of the hemispheres. The vermis of the rostral lobe is divided transversely into five lobules, I to V. The vermis of the caudal lobe is divided into lobules VI to IX. The nodulus is lobule X. The small part of the rostral lobe that contributes to the hemisphere is divided into lobules HI to HV, and the lateral part of the caudal lobe forms lobules HVI to HIX. The flocculus is lobule HX. The rostral and caudal extremities of the cerebellum are tucked ventrally under the middle part (Figs. 21.2, 21.14(a)), so that the flocculonodular lobe is almost completely hidden under the caudoventral aspect of the cerebellum. Functional aspects of the main components of the cerebellum are considered in Sections 15.9–15.12.

21.25 Internal structure

Cerebellar cortex

This is the outermost layer consisting of grey matter and characterized by folds which increase its surface area. The folds are termed the **cerebellar folia** (L. *folium*, a leaf; the plural, *folia*, leaves of a book) and the intervening grooves are the **cerebellar sulci**. For details of the functional histology of the cortex see the end of Chapter 15.

White matter

The white matter forms a core of nerve fibres responsible for relaying impulses between different parts of the cerebellum, and to and from other parts of the neuraxis. Due to the folds of the cortex, projections of the white matter extend within the substance of the folia in tree-like fashion and are collectively termed the **arbor vitae** (L. tree of life).

Nuclei of the cerebellum

These are located within the white matter of the cerebellum. They receive projections from the cortex and give rise to fibres which pass from the cerebellum to various centres in the brainstem. The three main, paired, nuclei are from lateral to medial, the **lateral** (dentate), **interpositus**, and **fastigial nuclei** (Fig. 21.11). Of these, the fastigial nucleus (sometimes known as the 'roof' nucleus, from L. *fastigium*, a roof) lies in the cerebellum immediately dorsal to the fourth ventricle. The position of the lateral nucleus speaks for itself; as its name implies, the interpositus nucleus is placed between the other two, though connected to the lateral nucleus by a small bridge. (In man the interpositus nucleus is represented by two nuclei, the **globose** and **emboliform nuclei**.)

21.26 Cerebellar peduncles

The cerebellar peduncles are the three paired stalks which attach the cerebellum to the brainstem (Fig. 7.4) and convey nerve fibres to and from the cerebellum (see Sections 15.3–15.5). The **rostral peduncle** connects to the midbrain (Figs. 21.10, 21.12), the **middle peduncle** to the pons (Figs. 21.10, 21.12), and the **caudal peduncle** to the medulla oblongata (Figs. 20.7, 20.8, 20.11, 20.12). On the medial aspect of the caudal peduncle lies a small separate bundle, the **juxtarestiform body**, which carries the afferent and efferent vestibulocerebellar fibres; this structure is believed to be involved in the **paradoxical vestibular syndrome** (see Section 8.3, **Vestibular disease**). The peduncles rearrange their positions as they pass dorsally to join the cerebellum, so that they become mediolateral in position rather than rostrocaudal, the rostral peduncle being the most medial and the middle the most lateral in position (Fig. 21.12).

Cerebral hemispheres

It is generally (though not universally) agreed that the cerebrum comprises the two cerebral hemispheres, and that it constitutes the telencephalon. According to these definitions, the terms cerebrum, cerebral hemispheres, and telencephalon are therefore simply different names for the same structure, and this usage is adopted here.

21.27 Gross structure

The paired cerebral hemispheres are the largest component of the brain. In the dorsal midline they are separated by the **longitudinal cerebral fissure** which contains a fold of dura mater, the **falx cerebri** (see Section 2.7). They are connected across the midline by the **commissural fibres**, which lie in the depths of the longitudinal fissure.

The size of the cerebral hemisphere in relation to body weight varies greatly among the mammalian species, being by far the largest in man. The evolution of the cerebral hemisphere is considered in Chapter 18 (see Sections 18.1–18.5). The uneven growth of the telencephalon which occurred in the course of evolution led to a great development of the frontal lobe in primates. It also produced a greatly enlarged temporal lobe, and this creates a hemisphere which is roughly C-shaped when viewed laterally, especially in man (Fig. 17.2(a)) but also in other animals including the dog (Fig. 21.19(a)). This C-shaped curvature is repeated again and again by a whole series of structures deep within the hemisphere, such as the lateral ventricle, caudate nucleus, stria terminalis, hippocampus, and fornix (Figs. 21.18, 21.19). The midline thalamus enlarges as the brain evolves, and pushes its way laterally into the centre of these repeated C-shaped cerebral configurations, which therefore encircle the thalamus (Fig. 21.19).

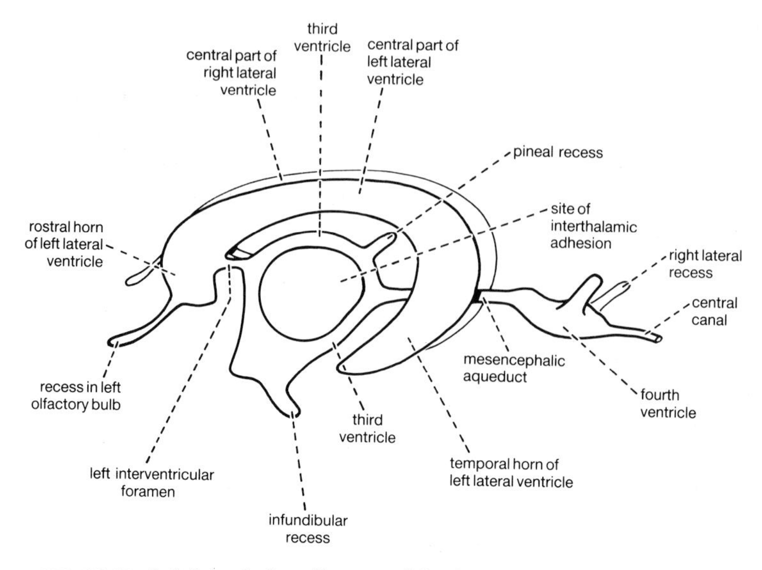

FIG. 21.17. Left lateral view diagram of the four ventricles of the brain.

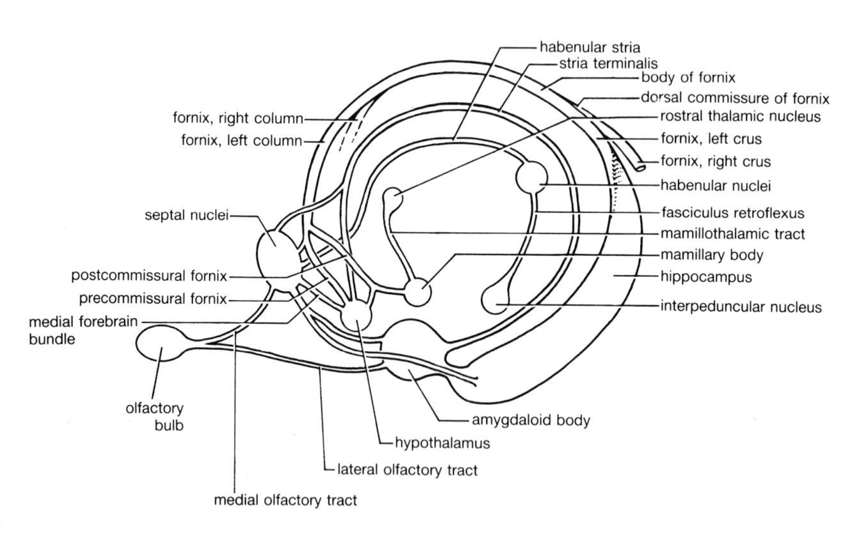

FIG. 21.18. Lateral view diagram of some of the main components of the left limbic system. The components of the limbic system form a series of ring-like structures around the thalamus. The olfactory bulb and olfactory tracts are not parts of the limbic system but are included because they project into it. Some major components of the limbic system are not shown, and these include the parahippocampal gyrus, cingulate gyrus, and cingulum. Several components are only incompletely shown, for example the medial forebrain bundle. The constituent parts are not drawn to scale.

Lobes

The cerebral hemisphere (excluding the ventrally situated olfactory components of the rhinencephalon) can be divided into the frontal, parietal, occipital, and temporal lobes (Figs. 17.2(a), (b)), which roughly correspond to the overlying skull bones. However, in contrast to the human hemisphere, that of the carnivore lacks specific sulci or other boundaries which delineate the lobes, and consequently there is no precise agreement about the lobar boundaries. (For discussion of the association areas within each lobe see Chapter 17, Sections 17.7–17.12.) Like that of the cerebellar cortex, the surfaces of the lobes are much folded in higher mammals to increase their surface area, the greatest folding occuring in man. The folds are termed **gyri**, and the intervening grooves **sulci**. The pattern of folding varies considerably between species, and also within species so that the sulci and gyri may not be identical in, for example, a series of dogs. However, the principal gyri and sulci in the dog are shown in Figs, 21.3, 21.4.

Olfactory components of the rhinencephalon

The ventral regions of the cerebral hemisphere are formed by the olfactory components of the rhinencephalon. The most rostral of these components is the **olfactory bulb** (Figs. 21.2, 21.3). The olfactory peduncle joins the olfactory bulb to the **lateral** and **medial olfactory tracts**, which diverge caudally (Fig. 8.6). Between the tracts lie the **olfactory tubercle** and **rostral perforated substance** (Fig. 21.2), the latter being formed by the penetration of numerous blood vessels to supply the basal nuclei. The most caudal olfactory component is the **piriform lobe** (Fig. 21.2). Along its whole length the rhinencephalon is separated from the rest of the cerebral hemisphere by the rostral and caudal parts of the **lateral rhinal sulcus** (Figs. 21.2, 21.15(c)). This sulcus is formed during the evolution of the brain by the pushing of the developing neocortex ventrally upon the paleocortex (Fig. 18.5), and it therefore constitutes the boundary between these two cortical areas.

21.28 Ventricular system

The curved C-shaped lateral ventricle lies within the cerebral hemisphere (Figs. 21.15(a), (b), (c), 21.16, 21.17, 21.19 (a), (b)). Its **central part** communicates medially with the third ventricle through the **interventricular foramen** (Figs. 21.14(b), 21.17, 21.19(b)). Rostral to this, within the frontal lobe, is the **rostral horn** of the lateral ventricle, which lies close to the midline being separated by only the thin **septum pellucidum** from the rostral horn of the opposite side (Fig. 21.14(b)). From the rostral horn a narrow recess may extend rostroventrally into the olfactory bulb (Fig. 21.17). Caudally the **temporal horn** (Figs. 21.17, 21.19(b)) curves laterally

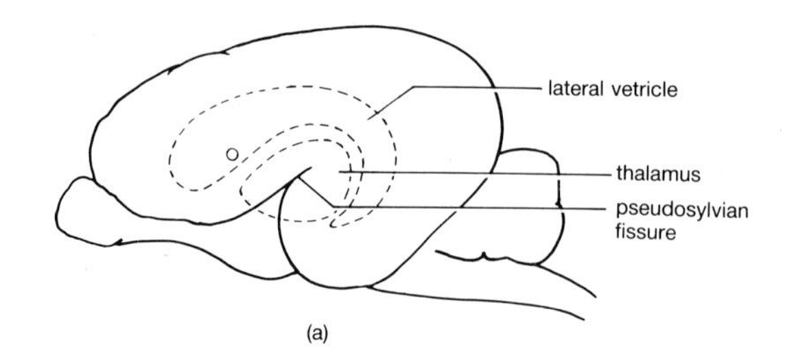

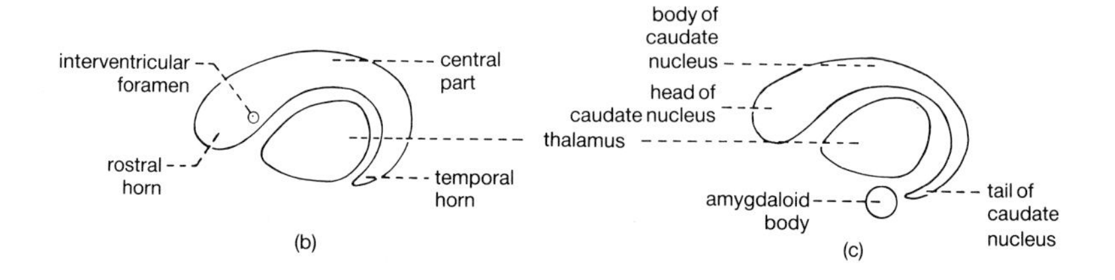

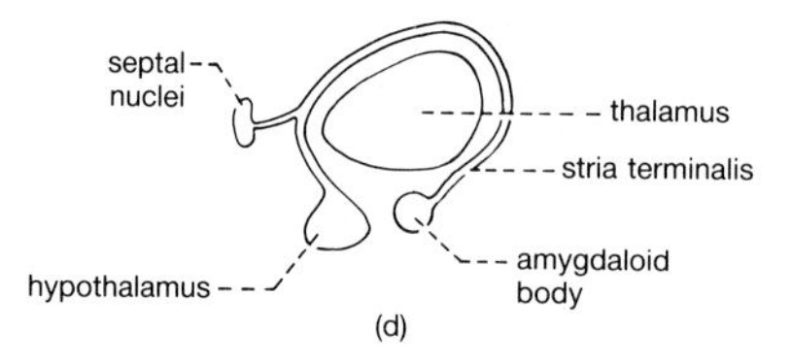

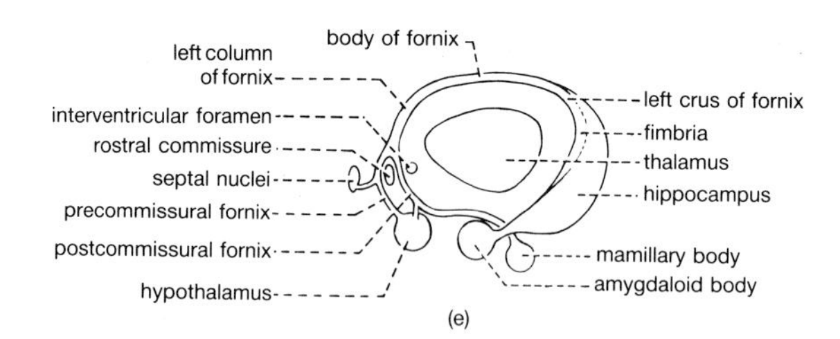

FIG. 21.19. Lateral view diagrams of C-shaped and circular components of the left cerebral hemisphere. All these components are shown in relation to the left thalamus.

(a) showing the essentially C-shaped form of the left cerebral hemisphere as a whole, and the outline of the left lateral ventricle and left thalamus within it.

(b) Showing the left lateral ventricle encircling the left thalamus.

(c) Showing the left caudate nucleus winding round the left thalamus.

(d) Showing the left stria terminalis, projecting from the left amygdaloid body to the left septal nuclei and left hypothalamus and girding the left thalamus.

(e) Showing the left hippocampus and left fornix and their projections, encompassing the left thalamus.

and ventrally into the temporal lobe of the hemisphere, accompanied by the tail of the caudate nucleus (Fig. 21.15(c)). Like the other ventricles of the brain, the paired lateral ventricle has a **choroid plexus** protruding from its floor along the fimbria and fornix (Figs. 21.15(b), (c), 21.16, 21.20); the choroid plexus of the lateral ventricle is continuous, via the interventricular foramen, with the choroid plexus of the third ventricle (Fig. 21.14(b)).

21.29 Internal structure

White matter

The white matter of the cerebral hemisphere is composed of association fibres, projection fibres, and commissural fibres.

(i) The **association fibres** comprise a vast network of interneurons joining different areas of the same hemisphere (see Section 17.1). Associaton fibres are classified into short and long fibres. The short fibres connect adjacent or neighbouring gyri by curving under the sulci, thus forming U-shaped loops known as **arcuate fibres**. The **long association fibres** form bundles connecting the lobes and projection areas within the same hemisphere. Although the form and function of these bundles have been controversial, it is now apparent that their organization is precise and very complex. They seem to be largely reciprocal, e.g. parieto–occipital and occipito–parietal as in Fig. 17.3. The **superior longitudinal fasciculus** connects the frontal lobe with the parietal, temporal, and occipital lobes. The **inferior longitudinal fasciculus** passes between the occipital and temporal lobes. The **cingulum** is a particularly massive bundle, interconnecting the hippocampus and other cortical parts of the limbic system, and is itself therefore included in the **limbic system** (see Section 8.7). As its name implies (L. *cingulum*, a girdle), this is yet another circular structure, curving dorsally over the lateral ventricle. It starts ventral to the rostrum of the corpus callosum, curves dorsally over the body of the corpus callosum, and ends caudally by winding ventrally around the hippocampal region and entering the parahippocampal gyrus. Throughout its course it lies within the similarly curved **cingulate gyrus** (Fig. 21.4).

(ii) Each hemisphere is linked to other parts of the central nervous system by **projection fibres**. They consist of efferent (corticofugal) fibres such as those of the pyramidal system, and afferent (corticopetal) fibres such as those projecting from the thalamus to the neocortex. Immediately beneath the neocortex large numbers of corticofugal and corticopetal fibres form a fan-shaped radiation, the **corona radiata**, as they converge towards the brainstem to form the **internal capsule** (Figs. 21.4, 21.12, 21.15(a)). This massive bundle of fibres (the internal capsule) has a rostral and a caudal crus. The **rostral crus** is sandwiched between the caudate nucleus and the lentiform nucleus; the **caudal crus** lies along the lateral aspect of the thalamus, separating the lateral border of the thalamus from the middle and caudal part of the lentiform nucleus.

The **optic radiation** is a band of fibres which runs from the lateral geniculate nucleus and travels caudally in the caudal crus of the internal capsule to end in the visual cortex on the dorsal, medial, and caudal surface of the occipital lobe (Fig. 8.2(a)). The **auditory radiation** consists of fibres which arise from the medial geniculate nucleus (Fig. 8.1) and project to the auditory area of the temporal lobe (Figs. 8.2(a), 8.3).

(iii) The **commissural fibres** connect the left and right cerebral hemispheres across the midline. By far the largest commissure is the **corpus callosum** (Figs. 21.14–21.16). It forms the most important pathway by which the cortical areas of one hemisphere are connected to those of the other hemisphere (see Section 17.13). The fibres of the middle part, or **body of the corpus callosum**, form the roof of the rostral horn and central part of the lateral ventricle, and are attached ventrally to a midline membrane, the **septum pellucidum**, which bridges the gap between the corpus callosum and the fornix (Fig. 21.14). Most of the septum pellucidum is thin and membranous, but the part nearest the rostral commissure is thickened by the presence of the **septal nuclei** (Fig. 21.14(b), and see below); in this region the septum pellucidum therefore bulges slightly into the lateral ventricle. Caudally the corpus callosum is thickened to form the **splenium** (Gk. *splenion*, a bandage). At its rostral end the corpus callosum is hook-shaped, since it bends sharply ventrally at its **genu** and then rapidly tapers to a point as the **rostrum of the corpus callosum**, becoming continuous with the lamina terminalis (Fig. 21.14). The fibres of the corpus callosum fan out laterally into each hemisphere from its body (Figs. 21.15(a), 21.16), and also rostrally and caudally from its genu and splenium respectively, intersecting the fibres of the internal capsule at the corona radiata. The fibres also turn dorsally and ventrally, so as to reach all parts of the hemisphere.

The **rostral commissure** is a cylinder of fibres embedded in the lamina terminalis (Fig. 21.14). It connects the olfactory components (e.g. the olfactory bulb and piriform lobe) with the limbic components (e.g. the amygdaloid body) of the left and right rhinencephalon.

Grey matter

The grey matter of the cerebral hemisphere includes (i) the cerebral cortex, (ii) the basal nuclei, and (iii) the limbic components of the rhinencephalon.

(i) In mammals the **cerebral cortex** has three general components, the paleocortex, archicortex, and neocortex (see Section 18.4). The term '**-pallium**' is sometimes substituted for '-cortex', but the terms are virtually synonymous (the only difference apparently being that '-pallium' is thought to be more appropriate for the **entire** wall of the hemisphere, including ependyma and pia mater). In mammals the paleocortex becomes the olfactory region of the cerebral hemisphere (olfactory bulb and piriform lobe), and the archicortex becomes the hippocampus; the paleocortex and archicortex together are known as the **allocortex**.

The **neocortex** can be divided into projection areas and association areas. The **projection areas** are the primary motor area, primary somatic sensory area, visual area, and auditory area (Figs. 8.2(a), (b), 8.3). The **association cortex** consists of the cognitive association area of the parietal and occipital lobes, the interpretative association area of the temporal lobe, and the premotor association area of the frontal lobe (Figs. 8.2(a), 17.2(a), (b) 17.3). The functions and interconnections of the projection and association areas are surveyed in Sections 17.1–17.12.

(ii) The terms **basal nuclei** and **corpus striatum** are regarded here as synonymous. It will be considered that they comprise the following six components: (1) the caudate nucleus; (2) the lentiform nucleus; (3) the claustrum; (4) the internal capsule; (5) the external capsule; and (6) the nucleus accumbens.

There is no agreed definition of the terms basal nuclei and corpus striatum. The *Nomina Anatomica* lists the term basal nuclei, but the *Nomina Anatomica Veterinaria* omits it. One of the particular difficulties is disagreement about where the claustrum should be placed; the *Nomina Anatomica Veterinaria* (1983) includes it in the corpus striatum, but the *Nomina Anatomica* (1977) does not. The amygdaloid body is also a problem. The *Nomina Anatomica* and various authors include it in the basal nuclei; the *Nomina Anatomica Veterinaria* and other authors exclude it, mainly because of its largely different connections and functions. This book follows the *Nomina Anatomica* in retaining the term basal nuclei, and follows the *Nomina Anatomica Veterinaria* in placing the amygdaloid body under the rhinencephalon instead of in the basal nuclei.

The **caudate nucleus** is a C-shaped nucleus which winds around the convex dorsal surface of the thalamus (Fig. 21.19(c)). Its enlarged **head** forms the floor of the rostral horn of the lateral ventricle, bulging into the lumen from the lateral aspect (Figs. 21.16, 21.20). Its tapered **tail** forms the roof of the temporal horn of the ventricle, and fades out caudally near the amygdaloid body (Fig. 21.19(c)). The caudate nucleus is separated from the **globus pallidus** and **putamen** by the rostral crus of the internal capsule.

The **lentiform nucleus** is composed of the putamen and the globus pallidus. The **putamen** lies lateral to the **globus pallidus** (Fig. 21.16), as is consistent with the meaning of *putamen* (L. a husk or outer shell, that falls off in trimming). The medial border of the lentiform nucleus adjoins the internal capsule (Fig. 21.16). Laterally the lentiform nucleus is bounded by the **external capsule**, a thin layer of white matter (Fig. 21.16). The external capsule in turn abuts laterally on to a parallel thin lamina of grey matter, the **claustrum** (Fig. 21.16). Thus the corpus striatum, as its name implies,

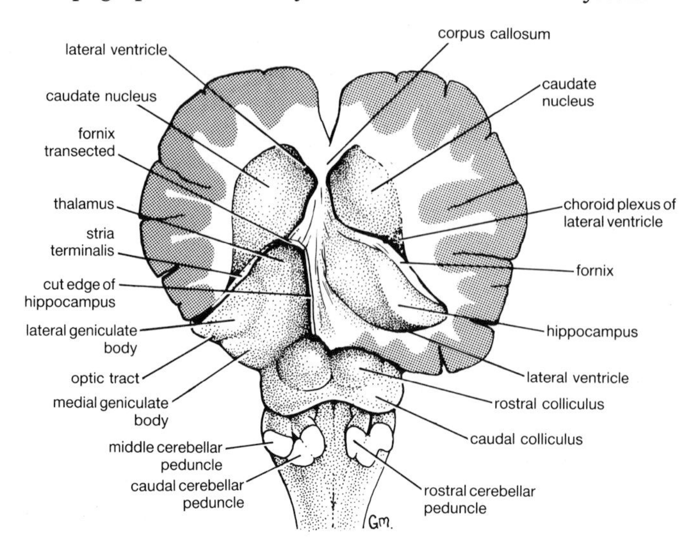

FIG. 21.20. Dorsal view drawing of the deep structures of the cerebral hemispheres. On the right side the dorsal part of the neocortex has been removed to expose the caudate nucleus and hippocampus projecting dorsally from the floor of the lateral ventricle. On the left side the hippocampus has been removed, thus exposing the geniculate bodies and thalamus.

consists of alternating grey and white bands forming a 'striated body'. The caudate and lentiform nuclei are extensively involved in the functions of the extrapyramidal system (see Section 12.2), but the function of the claustrum is not known. (The term **claustrum** is Latin for a barrier, so-called because in some species it forms a barrier between the external capsule and the **insula**, an island of cerebral cortex which is buried in the lateral sulcus.)

The **nucleus accumbens** (L. reclining nucleus) consists of a minor group of cells lying between the head of the caudate nucleus and the septal nuclei. It may be regarded as a fragment of the caudate nucleus.

The putamen and caudate nucleus together are sometimes referred to as the **neostriatum**, and this term is often shortened to **striatum**; the shortened version is then used to name afferent and efferent pathways such as 'nigro-striatal' projections. The term globus pallidus is also often shortened, to **pallidum**, and the shortened term is again applied to pathways, e.g. 'pallidofugal' projections. Furthermore, the globus pallidus is sometimes known as the **paleostriatum**. This is phylogenetically appropriate, since the mammalian globus pallidus appears to be homologous to a part of the avian paleostriatum; in contrast, the term neostriatum is less satisfactory, since the mammalian neostriatum is not homologous to the avian neostriatum (see Section 18.6).

(iii) The **limbic components of the rhinencephalon** are discussed in

Sections 8.6 and 8.7, and the difficulties of defining the limbic system are there considered. By no means all parts of the limbic system belong to the rhinencephalon, but many do and these include: (1) the parahippocampal gyrus; (2) the hippocampus; (3) the fimbria; (4) the fornix; (5) the amygdaloid body; and (6) the septal nuclei.

At its rostral end, the **parahippocampal gyrus** blends with the piriform lobe (Figs. 21.15(c)). It passes caudodorsally on the caudal part of the medial surface of the hemisphere where it becomes continuous with the caudal end of the **cingulate gyrus**, which encircles the rostral, dorsal, and caudal aspects of the corpus callosum (Fig. 21.4). (The cingulate gyrus is a part of the neocortex, not the rhinencephalon.)

The **hippocampus** (Gk. a sea-horse) extends caudally from the piriform lobe, and together with its rostral extension, the fornix, forms another C-shaped structure winding round the thalamus (Fig. 21.19(e)). In horizontal sections of the forebrain it projects dorsally from the caudo-medial region of the floor of the lateral ventricle, like the rounded hump of a whale's back (Figs. 21.15(a), 21.20). Originally an external cerebral gyrus, the hippocampus finds itself in this unusual internal position (Fig. 21.15(c)) through being rolled dorsally into the lateral ventricle during development. Dorsal to the caudal part of the thalamus, the left and right hippocampi converge and meet in the midline.

The **fimbria** (L. a fringe) is a ledge which projects all along the lateral side of the hippocampus (Figs. 21.15(c), 21.19(e)). It represents a major pathway of afferent fibres entering the hippocampus, and is the sole exit from the hippocampus of efferent fibres, which then continue in the fornix. A thin layer of white matter, the **alveus**, (L. *alveus*, a trough, as in alveolus for a tooth) covers the ventricular surface of the hippocampus (Fig. 21.15(c)); this consists of efferent and commissural fibres of the hippocampus, which pass through the fimbria.

The Latin meaning of **fornix**, an arch, well describes the gross form of this structure, which curves dorsally over the thalamus (Figs. 21.18, 21.19(e)). The **crus of the fornix** arises from the fimbria and then meets its partner of the other side in the midline to form the **body of the fornix** (Fig. 21.18). This lies in contact with the ventral surface of the splenium and body of the corpus callosum (Fig. 21.14), and here hippocampal fibres are exchanged between the left and right sides at the **dorsal commissure of the fornix** (Fig. 21.18). At the rostral aspect of the thalamus the left and right fornices diverge (Fig. 21.18), part company with the corpus callosum (Fig. 21.14), and pass ventrally as the left and right **columns of the fornix** (Fig. 21.18). Each column continues ventrally towards the hypothalamus, running down the lamina terminalis (Fig. 21.14). At the rostral commissure the column splits into the **pre-** and **postcommissural fornix**, which project to the septal nuclei and hypothalamus, and to the mamillary body and hypothalamus respectively (Figs. 21.18, 21.19(e)). The fornix provides two-way connections between the hippocampus and hypothalamus, and is therefore a major participant in the control which the hippocampus exerts over the hypothalamus (see Section 16.2).

So-called because of its shape (Gk. *amygdale*, an almond), the **amygdaloid body** lies deep within the piriform lobe and is continuous caudally with the hippocampus (Figs. 21.18, 21.19(e)). It is also associated with the tail of the caudate nucleus (Fig. 21.19(c)). It contains a complex of nuclei (six are listed in the *Nomina Anatomica Veterinaria*), of which one group (the **corticomedial nuclei**) receives an essentially olfactory input, and the remainder (the **basolateral nuclei**) belong to the limbic part of the rhinencephalon. The basolateral group becomes progressively more developed in the highest mammals, especially man; not unexpectedly, the corticomedial (olfactory) group is relatively undeveloped in man. The **stria terminalis** carries a large group of efferent fibres from the amygdaloid nuclei to the hypothalamus (and to the septal nuclei (Figs. 21.18, 21.19(d)). This slender but conspicuous bundle runs along the floor at the central part of the lateral ventricle in a groove between the caudate nucleus and thalamus (Fig. 21.20), and is another of the structures that encircles the thalamus (Fig. 21.19(d)); by encircling the thalamus it follows the curvature of the lateral ventricle, and is therefore also related to the temporal horn of the lateral ventricle (Fig. 21.15(c)). The **rostral commissure** (Figs. 21.14(b), 21.19(e)) connects the left and right amygdaloid bodies, as well as other components of the left and right rhinencephalons. The functions of the amygdaloid body are discussed in Section 8.6.

The **septal nuclei** form a small mass of grey matter in the ventrorostral part of the septum pellucidum (Fig. 21.14(b)). Some of them extend rostrally into the adjacent **paraterminal gyrus** (Fig. 21.4). Their numerous connections are shown diagrammatically in Fig. 21.18. Thus they receive an input from olfactory pathways, and project to the habenular nuclei via the **habenular stria**. They have two-way connections with the hippocampus by means of the **precommissural fornix**, and with the hypothalamus through the **medial forebrain bundle**. They receive afferent projections from the amygdaloid body via the stria terminalis. The functions of the septal nuclei are considered in Section 8.6.

Further reading

The authorities given under Principal sources have been extensively consulted during the preparation of most of the chapters in this book. Of these, Brodal's *Neurological anatomy in relation to clinical medicine* has been the main inspiration, lying as it does in the no-man's land between neuroanatomy, neurophysiology, neuropathology, and clinical neurology. The primary literature can be entered through its store of over 4000 references, among which are many reports of experimental observations pertaining to the cat and also to the dog. Guidance to the neurological literature more specifically relating to the domestic mammals is available in the extensive bibliography in de Lahunta's *Veterinary neuroanatomy and clinical neurology.*

The references listed for each individual chapter include the sources of details in the text, and also indicate where additional information can be obtained.

Principal sources

Barr, M. L. and Kiernan, J. A. (1983). *The human nervous system. An anatomical viewpoint* (4th edn). Harper and Row, Philadelphia.

Bowsher, D. (1979). *Introduction to the anatomy and physiology of the nervous system* (4th edn). Blackwell, Oxford.

Brodal, A. (1981). *Neurological anatomy in relation to clinical medicine* (3rd edn). Oxford University Press, New York.

Crosby, E. C., Humphrey, T. and Lauer, E. W. (1962). *Correlative anatomy of the nervous system.* Macmillan, New York.

de Lahunta, A. (1983). *Veterinary neuroanatomy and clinical neurology* (2nd edn). Saunders, Philadelphia.

Jenkins, T. W. (1978). *Functional mammalian neuroanatomy* (2nd edn). Lea and Febiger, Philadelphia.

Meyer, H. (1979). The Brain. In *Miller's anatomy of the dog* (2nd edn, ed. H. E. Evans and G. C. Christensen) pp. 842–902. Saunders, Philadelphia.

Palmer, A. C. (1976). *Introduction to animal neurology* (2nd edn). Blackwell, Oxford.

Williams, P. L. and Warwick, R. (1980). Neurology. In *Gray's anatomy* (36th edn) pp. 802–1215. Churchill Livingstone, Edinburgh.

1. Arterial supply to the central nervous system

Anderson, W. D. and Kubicek, W. (1971). The vertebral–basilar system of dog in relation to man and other mammals. *Am. J. Anat.* **132**, 179–88.

Baldwin, B. A. (1971). Anatomical and physiological factors involved in slaughter by section of the carotid arteries. In *Humane killing and slaughterhouse techniques*, pp. 34–42. UFAW Symposium.

Baldwin, B. A. and Bell, F. R. (1963). The anatomy of the cerebral circulation of the

sheep and ox. The dynamic distribution of the blood supplies by the carotid and vertebral arteries to cranial regions. *J. Anat.* **97**, 203–15.
Baldwin, B. A. and Bell, F. R. (1963). Blood flow in the carotid and vertebral arteries of the sheep and calf. *J. Physiol., Lond.* **167**, 448–62.
Baldwin, B. A. and Bell, F. R. (1963). The effect of temporary reduction in cephalic blood flow on the EEG of sheep and calf. *Electroenceph. clin. Neurophysiol.* **15**, 465–73.
Blackman, N. L., Cheetham, K. and Blackmore, D. K. (1986). Differences in blood supply to the cerebral cortex between sheep and calves during slaughter. *Res. Vet. Sci.* **40**, 252–4.
Bradshaw, P. (1958). Arteries of the spinal cord in the cat. *J. Neurol. Neurosurg. Psychiat.* **21**, 284–9.
Davis, D. D. and Story, H. E. (1943). Carotid circulation in the domestic cat. *Field museum publications* (*Zoology Series*) **28**, 5–47.
Ghoshal, N. G. (1985). Thermoregulatory role of the cranial circulation in cerebral temperature control. *Vlaams Diergeneeskundig Tidjschrift* **54**, 246–61.
Gillilan, L. A. (1958). The arterial blood supply of the human spinal cord. *J. comp. Neurol.* **110**, 75–103.
Gillilan, L. A. (1976). Extra- and intra-cranial blood supply to brains of dog and cat. *Am. J. Anat.* **146**, (3), 237–54.
Griffiths, I. R. (1978). Spinal cord injuries: a pathological study of naturally occurring lesions in the dog and cat. *J. comp. Path* **88**, 303–15.
Innes, J. R. M. and Saunders, L. Z. (1962). Arterial and venous systems. In *Comparative neuropathology*, pp. 762–7. Academic Press, New York.
Jewell, P. A. (1952). The anastomoses between internal and external carotid circulations in the dog. *J. Anat.* **86**, 83–94.
Le Gros Clark, W. E. (1965). The vascular supply of nervous tissue. In *The tissues of the body* (5th edn) pp. 408–10. Clarendon Press, Oxford.
Schmidt-Nielsen, K. (1972). Keeping the head cool. In *How animals work*, pp. 68–70. Cambridge University Press, Cambridge.

2. Meninges and cerebrospinal fluid

Alksne, J. F. and Lovings, E. T. (1972). Functional ultrastructure of the arachnoid villus. *Archs. Neurol.* **27**, 371–7.
de Lahunta, A. and Cummings, J. F. (1965). The clinical and electroencephalographic features of hydrocephalus in three dogs. *J. Am. vet. med. Ass.* **146**, 954–64.
Fletcher, T. F. (1979). Spinal cord and meninges. In *Miller's anatomy of the dog* (2nd edn, ed. H. E. Evans and G. C. Christensen) pp. 935–71. Saunders, Philadelphia.
Jayatilaka, A. D. P. (1965). An electron microscopic study of sheep arachnoid granulations. *J. Anat.* **99**, 635–49.
Livingstone, R. B. (1965). Mechanics of cerebrospinal fluid. In *Physiology and biophysics* (19th edn, ed. T. C. Ruch and H. D. Patton) pp. 935–42. Saunders, Philadelphia.
Malloy, J. J. and Low, F. N. (1974). Scanning electron microscopy of the subarachnoid space in the dog. *J. comp. Neurol.* **157**, 87–107.

Russell, D. S. (1949). Observations on the pathology of hydrocephalus. *Medical Research Council. Special Report* No. 265. HMSO, London.

3. Venous drainage of the spinal cord and brain

Armstrong, L. D. and Horowitz, A. (1971). The brain venous system of the dog. *Am. J. Anat.* **132**, 479–90.

Batson, O. V. (1940). The function of the vertebral veins in metastatic processes. *Ann. Surg.* **112**, 138–49.

Evans, H. E. and Christensen, G. C. (1979). Veins of the central nervous system. In *Miller's anatomy of the dog* (2nd edn, ed. H. E. Evans and G. C. Christensen) pp. 791–801. Saunders, Philadelphia.

Gillilan, L. A. (1968). The arterial and venous blood supplies to the forebrain (including the internal capsule) of primates. *Neurology* **18**, 653–70.

Ghoshal, N. G. (1986). Dural sinuses in the pig and their extracranial venous connections. *Am. J. vet. Res.* **47**, 1165–9.

Magilton, J. H., Swift, C. S., and Ghoshal, N. G. (1981). Experimental evidence of a reciprocal temperature relationship between the parietofrontal region and the orbital emissary vein in the pony. *Am. J. vet. Res.* **42**, 1221–4.

Parker, A. J., Park, R. D., and Stowater, J. L. (1974). Traumatic occlusion of segmental spinal veins. *Am. J. vet. Res.* **35**, 857–9.

Reinhard, K. R., Miller, M. E., and Evans, H. E. (1962). The craniovertebral veins and sinuses of the dog. *Am. J. Anat.* **111**, 67–87.

Towbin, A. (1970). Central nervous system damage in the human fetus and newborn infant. *Diseases of Children* **119**, 529–42.

4. Applied anatomy of the vertebral canal

Griffiths, I. R. (1973). Spinal cord infarction due to emboli arising from the intervertebral discs in the dog. *J. comp. Path.* **83**, 225–32.

Hansen, H. J. (1952). A pathologic–anatomical study on disc degeneration in dog. *Acta orthop. Scandinav.* Supplement XI, pp. 41–117.

Heath, E. H. and Myers, V. S. (1972). Topographic anatomy for caudal anaesthesia in the horse. *Veterinary Medicine, Small Animal Clinician* **67**, 1237–9.

Hoerlein, B. F. (1978). Intervertebral discs. In *Canine neurology* (3rd edn) pp. 470–560. Saunders, Philadelphia.

King, A. S. (1956). The anatomy of disc protrusion in the dog. *Vet. Rec.* **68**, 939–44.

King, A. S. and Smith, R. N. (1964). Degeneration of the intervertebral disc in the cat. *Acta orthop. Scand.* **34**, 139–58.

Littlewood, J. D., Herrtage, M. E., and Palmer, A. C. (1984). Intervertebral disc protrusion in a cat. *J. small anim. Pract.* **25**, 119–27.

Lohse, C. L. and Baba, Y. M. (1982). Comparative anatomy of intervertebral discs and related structures in the cat (*Felis catus*) and Jaguar (*Panthera onca*). *Zbl. vet. Med. C. Anat. Histol. Embryol.* **11**, 334–42.

Mayhew, I. G. (1975). Collection of cerebrospinal fluid from the horse. *Cornell Vet.* **65**, 500–11.

Nixon, J. (1986). Intervertebral disc mechanics: a review. *J. R. Soc. Med.* **79**, 100–4.

Palmer, A. C. (1976). Cerebrospinal fluid. In *Introduction to animal neurology* (2nd edn) pp. 115–20. Blackwell, Oxford.

Smith, R. N. (1966). Anatomy and physiology. In *Intervertebral disc protrusion in the dog* (ed. G. D. Pettit) pp. 1–20. Appleton-Century-Crofts, New York.

Vasseur, P. B., Saunders, G., and Steinback, C. (1981). Anatomy and function of the ligaments of the lower cervical spine in the dog. *Am. J. vet. Res.* **42**, 1002–6.

5. The neuron

Brodal, A. (1981). Regeneration in the central nervous system. Recovery after lesions. In *Neurological anatomy in relation to clinical medicine* (3rd edn) pp. 38–44. Oxford University Press, New York.

Dudel, J. (1978). Excitation of nerve and muscle. In *Fundamentals of neurophysiology* (2nd edn, ed. R. F. Schmidt) pp. 19–71. Springer, Berlin.

Fuortes, M. G. F. (1971). Generation of responses in receptor. In *Handbook of sensory physiology* (ed. W. R. Loewenstein) Vol. 1, pp. 243–63. Springer-Verlag, Berlin.

Ide, C. (1983). Nerve regeneration and Schwann cell basal lamina: observations of the long-term regeneration. *Arch. histol. jap.* **46**, 243–57.

Landowne, D., Potter, L. T., and Terrar, D. A. (1975). Structure–function relationships in excitable membranes. *Ann. Rev. Physiol.* **37**, 485–508.

Loewenstein, W. R. (1971). Mechano-electric transduction in the Pacinian corpuscle. Initiation of sensory impulses in mechanoreceptors. In *Handbook of sensory physiology* (ed. W. R. Loewenstein) Vol. 1, pp. 269–90. Springer-Verlag, Berlin.

Patton, H. D. (1965). Special properties of nerve trunks and tracts. In *Physiology and biophysics* (19th edn, ed. T. C. Ruch and H. D. Patton) pp. 73–94. Saunders, Philadelphia.

Patton, H. D. (1965). Spinal reflexes and synaptic transmission. In *Physiology and biophysics* (19th edn, ed. T. C. Ruch and H. D. Patton) pp. 153–80. Saunders, Philadelphia.

Patton, H. D. (1965). Reflex regulation of movement and posture. In *Physiology and biophysics* (19th edn, ed. T. C. Ruch and H. D. Patton) pp. 181–206. Saunders, Philadelphia.

Sarnat, H. B. and Netsky, M. G. (1974). The coiling reflex. In *Evolution of the nervous system*, pp. 53–71. Oxford University Press, New York.

Schwartz, J. H. (1979). Axonal transport: components, mechanisms, and specificity. *Ann. Rev. Neurosci.* **2**, 467–504.

Woodbury, J. W. (1965). Action potential: properties of excitable membranes. In *Physiology and biophysics* (19th edn, ed. T. C. Ruch and H. D. Patton) pp. 26–72. Saunders, Philadelphia.

6. Nuclei of the cranial nerves

Brodal, A. (1981). The cranial nerves. In *Neurological anatomy in relation to clinical medicine* (3rd edn) pp. 448–571. Oxford University Press, New York.

7. Medial lemniscal system

Bowsher, D. (1979). Specific sensory systems. In *Introduction to the anatomy and physiology of the nervous system* (4th edn) pp. 94–108. Blackwell, Oxford.

Brodal, A. (1981). The somatic afferent pathways. In *Neurological anatomy in relation to clinical medicine* (3rd edn) pp. 46–147. Oxford University Press, New York.

Burgess, P. R. and Clark, F. J. (1969). Characteristics of knee joint receptors in the cat. *J. Physiol. (Lond.)* **203**, 317–35.

Clark, F. J., Landgren, S., and Silfvenius, H. (1973). Projections to the cat's cerebral cortex from low threshold joint afferents. *Acta physiol. Scandinav.* **89**, 504–21.

Ferrell, W. R., Baxendale, R. H., Carnachan, C., and Hart, I. K. (1985). The influence of joint afferent discharge on locomotion, proprioception and activity in conscious cats. *Brain Res.* **347**, 41–8.

Frazer, H. and Palmer, A. C. (1967). Equine inco-ordination and wobbler disease of young horses. *Vet. Rec.* **80**, 338–55.

Gardner, E. (1969). Pathways to the cerebral cortex for nerve impulses from joints. *Acta anat. suppl.* **56**, 203–16.

Iggo, A. (1977). Cutaneous and subcutaneous sense organs. *Br. med. Bull.* **33**, 97–102.

Jones, T. C., Doll, E. R., and Brown, P. G. (1954). The pathology of equine incoordination (ataxia or 'wobbles' of foals). *Proc. Am. vet. med. Ass.*, 91st Annual Meeting, 139–49.

Landgren, S. and Silfvenius, H. (1971). Nucleus Z, the medullary relay in the projection path to the cerebral cortex of group 1 muscle afferents from the cat's hind limb. *J. Physiol.* **218**, 551–71.

Low, F. N. (1976). The perineurium and connective tissue of peripheral nerve. In *The peripheral nerve* (ed. D. N. Landon) pp. 159–87. Chapman and Hall, London.

Mathews, P. B. C. (1977). Muscle afferents and kinaesthesia. *Br. med. Bull.* **33**, 137–42.

Powell, T. P. S. (1977). The somatic sensory cortex. *Br. med. Bull.* **33**, 129–35.

Reynolds, P. J., Talbott, R. E., and Brookhart, J. M. (1972). control of postural reactions in the dog: the role of the dorsal column feedback pathway. *Brain Res.* **40**, 159–64.

Rustioni, A. (1973). Non-primary afferents to the nucleus gracilis from the lumbar cord of the cat. *Brain Res.* **51**, 81–95.

Trotter, E. J., de Lahunta, A., Geary, J. C., and Brasmer, T. H. (1976). Caudal cervical vertebral malformation–malarticulation in Great Danes and Doberman Pinschers. *J. am. vet. med. Ass.* **168**, 917–30.

Webster, K. E. (1977). Somaesthetic pathways. *Br. med. Bull.* **33**, 113–20.

Wright, F., Rest, J. R., and Palmer, A. C. (1973). Ataxia of the Great Dane caused by stenosis of the cervical vertebral canal: comparison with similar conditions in the Basset Hound, Doberman Pinscher, Ridgeback and the thoroughbred horse. *Vet. Rec.* **92**, 1–6.

8. Special senses

Barris, R. W., Ingram, W. R., and Ranson, S. W. (1935). Optic connections of the diencephalon and midbrain of the cat. *J. comp. Neurol.* **62**, 117–53.

Chrisman, C. L. (1980). Vestibular diseases. *Vet. Clin. N. Am.* **10**, (1), 103–29.

Cummings, J. F. and de Lahunta, A. (1969). An experimental study of the retinal projections in the horse and sheep. *Ann. N. Y. Acad. Sci.* **167**, 293–18.

Herron, M. A., Martin, J.. E., and Joyce, J. R. (1978). Quantitative study of the decussating optic axons in the pony, cow, sheep, and pig. *Am. J. vet. Res.* **39**, 1137–9.

Karamanlidis, A. N. and Magras, J. (1974). Retinal projections in domestic ungulates. II. The retinal projections in the horse and the ox. *Brain Res.* **66**, 209–25.

Nyberg-Hansen, R. and Mascitti, T. A. (1964). Sites and mode of termination of fibers of the vestibulospinal tract in the cat. *J. comp. Neurol.* **122**, 369–87.

Skerritt, G. C. and Whitbread, T. J. (1985). Two cases of paradoxical vestibular syndrome in Rough Collies. *J. small Anim. Pract.* **26**, 603–11.

9. Spinocerebellar pathways and ascending reticular formation

Basbaum, A. I. and Fields, H. L. (1984). Endogenous pain control: brainstem spinal pathways and endorphin circuitry. *Ann. Rev. Neurosci.* **7**, 309–38.

Bowsher, D. (1979). The non-specific afferent system. In *Introduction to the anatomy and physiology of the nervous system* (4th edn) pp. 109–17. Blackwell, Oxford.

Bowsher, D. (1983). Pain mechanisms in man. *Resident and Staff Physician* **29**, 26–34.

Brodal, A. (1981). The reticular formation and some related nuclei. In *Neurological anatomy in relation to clinical medicine* (3rd edn) pp. 394–447. Oxford University Press, New York.

Lagerweif, E., Wiegant, V. M., Nelis, P. C., and Ree, J. M. van (1984). The twitch in horses: a variant of acupuncture. *Science* **225**, 1172–3.

Low, J. S. T., Mantle-St. John, L. A., and Tracey, D. J. (1986). Nucleus Z in the rat: spinal afferents from collaterals of dorsal spinocerebellar tract neurons. *J. comp. Neurol.* **243**, 510–26.

Nathan, P. W. (1977). Pain. *Br. med. Bull.* **33**, 149–56.

Nauta, W. J. H. (1963). Central nervous organization and the endocrine system. In *Advances in neuroendocrinology* (ed. A. V. Nalbandov) pp. 5–21. University of Illinois Press, Urbana.

Oscarsson, O. (1965). Functional organization of the spino- and cuneocerebellar tracts. *Physiol. Rev.* **45**, 495–522.

Scheibel, A. B. (1984). The brainstem reticular core and sensory function. In *Handbook of physiology* (2nd edn, ed. I. Darian-Smith) Vol. 3, part 1, pp. 213–56. American Physiological Society, Bethesda.

10. Somatic motor systems: general principles

Brodal, A. (1981). The peripheral motor neuron. In *Neurological anatomy in relation to clinical medicine* (3rd edn) pp. 148–79. Oxford University Press, New York.

Houk, J. and Henneman, E. (1968). Feedback control of movement and posture. In *Medical physiology* (12th edn, ed. V. B. Mountcastle) pp. 1681–96. Mosby, St. Louis.

Patton, H. D. (1966). Reflex regulation of movement and posture. In *Physiology and biophysics* (19th edn, ed. T. C. Ruch and H. D. Patton) pp. 181–206. Saunders, Philadelphia.

Schmidt, R. F. (1978). The monosynaptic reflex arc. In *Fundamentals of neurophysiology* (2nd edn, ed. R. F. Schmidt) pp. 116–24. Springer-Verlag, New York.

11. Pyramidal system

Bowsher, D. (1978). Tone in UMN disorders. In *Mechanisms of nervous disorder: an introduction*, pp. 26–39. Blackwell, Oxford.

Breazile, J. E. and Thompson, W. D. (1967). Motor cortex of the dog. *Am. J. vet. Res.* **28**, 1483–6.

Brodal, A. (1981). Pathways mediating supraspinal influences on the spinal cord. The basal ganglia. In *Neurological anatomy in relation to clinical medicine* (3rd edn) pp. 180–293. Oxford University Press, New York.

Glees, P. (1961). The motor cortex and the pyramidal tract. In *Experimental neurology*, pp. 256–79. Oxford University Press, London.

Palmer, A. C. (1976). Corticospinal system. In *Introduction to animal neurology* (2nd edn) pp. 22–6. Blackwell, Oxford.

Verhaart, W. J. C. (1962). The pyramidal tract, its structure and function in man and animals. *World Neurol.* **3**, 43–53.

Wolsey, C. N. (1958). Organization of somatic sensory and motor areas of cerebral cortex. In *Biological and biochemical bases of behaviour* (ed. H. F. Harlow and C. N. Wolsey) pp. 63–81. University of Wisconsin Press, Madison.

12. Extrapyramidal system

Bowsher, D. (1979). Skeletal motor systems. In *Introduction to the anatomy and physiology of the nervous system* (4th edn) pp. 124–32. Blackwell, Oxford.

Brodal, A. (1981). Pathways mediating supraspinal influences on the spinal cord. The basal ganglia. In *Neurological anatomy in relation to clinical medicine* (3rd edn) pp. 180–293. Oxford University Press, New York.

Buchwald, N. A. and Ervin, F. R. (1957). Evoked potentials and behaviour. A study of responses of subcortical stimulation in the awake, unrestrained animal. *Electroenceph. clin. Neurophysiol.* **9**, 477–96.

Crosby, E. C., Humphrey, T., and Lauer, E. W. (1962). The question of suppressor and facilitatory functions of the cortex. In *Correlative anatomy of the nervous system* pp. 501–4. Macmillan, New York.

DeLong, M. R., Alexander, G. E., Mitchell, S. J., and Richardson, R. T. (1986). The contribution of basal ganglia to limb control. *Prog. Brain Res.* **64**, 161–74.

Forman, D. and Ward, J. (1957). Responses to electrical stimulation of caudate nucleus in cats in chronic experiments. *J. Neurophysiol.* **20**, 230–44.

Laursen, A. M. (1963). Corpus striatum. *Acta physiol. scand.* **59**, suppl. 211, 1–106.

Schmidt, R. F. (1981). Function of the basal ganglia, cerebellum, and motor cortex. In *Fundamentals of neurophysiology* (2nd edn, ed. R. F. Schmidt) pp. 195–204. Springer–Verlag, New York.

Stevens, J. R., Kim, C., and MacLean, P. D. (1961). Stimulation of caudate nucleus. *Archs Neurol.* **4**, 47–54.

13. Extrapyramidal feedback: extrapyramidal disease

Barnes, C. D. and Schadt, J. C. (1979). Release of function in the spinal cord. *Prog. Neurol.* **12**, 1–13.

Bowsher, D. (1978). The basal ganglia and associated nuclei. In *Mechanisms of nervous disorder: an introduction*, pp. 39–43. Blackwell, Oxford.

Brodal, A. (1981). Pathways mediating supraspinal influences on the spinal cord. The basal ganglia. In *Neurological anatomy in relation to clinical medicine* (3rd edn) pp. 180–293. Oxford University Press, New York.

Garcia-Rill, E. (1986). The basal ganglia and the locomotor regions. *Brain Res. Rev.* **11**, 47–63.

Liles, S. L. and Davis, G. D. (1969). Athetoid and choreiform hyperkinesias produced by caudate lesions in the cat. *Science* **164**, 195–7.

Milart, Z. and Chomiak, M. (1971). Pathological changes in the central nervous system of a bull affected with spastic paresis. *Vet. Rec.* **88**, 66–8.

Ruch, T. C. (1966). Transection of the human spinal cord: the nature of higher control. In *Physiology and biophysics* (19th edn, ed. T. C. Ruch and H. D. Patton) pp. 207–14. Saunders, Philadelphia.

Young, S., Brown, W. W. and Klinger, H. (1970). Nigropallidal encephalomalacia in horses caused by ingestion of weeds of the genus Centaurea. *J. Am. vet. med. Ass.* **157**, 1602–5.

15. The cerebellum

Bowsher, D. (1979). The cerebellum. In *Introduction to the anatomy and physiology of the nervous system* (4th edn) pp. 133–47. Blackwell, Oxford.

Bowsher, D. (1978). The cerebellum and posterior fossa. In *Mechanisms of nervous disorder: an introduction*, pp. 58–65. Blackwell, Oxford.

Brodal, A. (1981). The function of the cerebellum as inferred from experimental studies. In *Neurological anatomy in relation to clinical medicine* (3rd edn) pp. 376–88. Oxford University Press, New York.

de Lahunta, A. (1983). Cerebellum. In *Veterinary neuroanatomy and clinical neurology* (2nd edn) pp. 255–78. Saunders, Philadelphia.

Dow, R. S. and Moruzzi, G. (1958). *The physiology and pathology of the cerebellum.* University of Minnesota Press, Minneapolis.

Palmer, A. C. (1976). Cerebellar disease. In *Introduction to animal neurology* (2nd edn) pp. 63–8. Blackwell, Oxford.

16. Autonomic components of the CNS

Baker, B. L. (1962). Hypothalamic functions. In *Correlative anatomy of the nervous system* (ed. E. C. Crosby, T. Humphrey, and E. W. Lauer) pp. 323–42. Macmillan, New York.

Bell, F. R. (1972). Sleep in the larger domestic animals. *Proc. R. Soc. Med.* **65**, 176–7.

Bell, F. R. (1971). Hypothalamic control of food intake. *Proc. Nutr. Soc.* **30**, 103–9.

Brodal, A. (1981). The autonomic nervous system: the hypothalamus. In *Neurological anatomy in relation to clinical medicine* (3rd edn) pp. 688–787. Oxford University Press, New York.

Fox, J. G. and Gutnick, M. J. (1972). Horner's syndrome and brachial paralysis due to lymphosarcoma in a cat. *J. Am. vet. med. Ass.* **160**, 977–80.

Jones, B. R. and Studdert, V. P. (1975). Horner's syndrome in the dog and cat as an aid to diagnosis. *Austral. vet. J.* **51**, 329–32.

Griffiths, I. R. (1970). A syndrome produced by dorso-lateral 'explosions' of the cervical intervertebral discs. *Vet. Rec.* **87**, 737–41.

Olds, J. (1972). Learning and the hippocampus. *Rev. Can. Biol.* **31**, suppl. 215–38.

Palmer, A. C. (1965). The paralysed bladder. In *Introduction to animal neurology* (2nd edn) pp. 49–54. Blackwell, Oxford.

Smith, J. S. and Mayhew, I. G. (1977). Horner's syndrome in large animals. *Cornell Vet.* **67**, 529–42.

Thompson, J. W. (1961). The nerve supply to the nictitating membrane of the cat. *J. Anat.* **95**, 371–84.

17. The cerebral cortex and thalamus

Bowsher, D. (1979). Thalamic projection to telencephalon: association. In *Introduction to the anatomy and physiology of the nervous system* (4th edn) pp. 118–23. Blackwell, Oxford.

Brodal, A. (1981). The thalamus and the thalamocortical projections. In *Neurological anatomy in relation to clinical medicine* (3rd edn) pp. 94–9. Oxford University Press, New York.

de Lahunta, A. (1977). Seizures—convulsions. In *Veterinary neuroanatomy and clinical neurology* (2nd edn) pp. 326–43. Saunders, Philadelphia.

Glees, P. (1961). The thalamus. In *Experimental neurology* pp. 215–36. Oxford University Press, Oxford.
Glees, P. (1961). Evolution of the primate cerebral cortex. In *Experimental neurology* pp. 472–511. Oxford University Press, Oxford.
Palmer, P. (1976). Clinical signs associated with space-occupying lesions of the brain. In *Introduction to animal neurology* (2nd edn) pp. 69–76. Blackwell, Oxford.
Sarnat, H. B. and Netsky, M. G. (1974). Association, instinct, and limbic system. In *Evolution of the nervous system* (2nd edn, 1981) pp. 233–5. Oxford University Press, New York.
Skerritt, G. C. and Stallbaumer, M. F. (1984). Diagnosis and treatment of coenuriasis (gid) in sheep. *Vet. Rec.* **115**, 399–403.
Yeomens, J. S. and Linney, L. (1985). Longitudinal brainstem axons mediating circling: behavioural measurement of conduction velocity distributions. *Behav. Brain Res.* **15**, 121–35.

18. Comparative neuroanatomy

Bowsher, D. (1965). The anatomophysiological basis of somatosensory discrimination. *Int. Rev. Neurobiol.* **8**, 35–75.
Goodrich, E. S. (1958). Peripheral nervous system and sense organs. In *Studies on the structure and development of vertebrates*, pp. 721–85. Dover, New York.
Glees, P. (1961). Comparative neurology. In *Experimental neurology* pp. 474–80. Oxford University Press, Oxford.
King, A. S. and McLelland, J. (1975). Spinal cord, brain. In *Birds: their structure and function*, pp. 237–56. Baillière Tindall, London.
Nauta, W. J. H. and Karten, H. J. (1970). A general profile of the vertebrate brain, with sidelights on the ancestry of cerebral cortex. In *The neurosciences* (ed. F. O. Schmitt) pp. 7–26. Rockefeller University Press, New York.
Romer, A. S. (1962). The nervous system. In *The vertebrate body* (3rd edn) pp. 488–552. Saunders, Philadelphia.

19. Clinical neurology

de Lahunta, A. (1977). Small animal, equine and food animal neurologic examinations. In *Veterinary neuroanatomy and clinical neurology* (2nd edn) pp. 365–406. Saunders, Philadelphia.
McGrath, J. T. (1960). *Neurologic examination of the dog* (2nd edn). Lea and Febiger, Philadelphia.
Palmer, A. C. (1965). The neurological examination based on that of the dog. In *Introduction to animal neurology* (2nd edn) pp. 91–113. Blackwell, Oxford.
Redding, R. W. (1971). Neurological examination. In *Canine neurology* (3rd edn, ed. B. F. Hoerlein) pp. 53–70. Saunders, Philadelphia.
Rooney, J. R. (1971). *Clinical neurology of the horse*. KNA Press Inc., Kennet Square, Pennsylvania.

20. Radiographic anatomy of the head and vertebral column in the diagnosis of disorders of the central nervous system

Adams, W. M. (1982). Myelography. *Vet. Clin. N. Am.* **12**, (2), 295–311.

Hoerlein, B. F. (1978). *Canine neurology* (3rd edn) pp. 103–35. Saunders, Philadelphia.

Kealy, J. K. (1979). *Diagnostic radiology of the dog and cat.* pp. 376–452. Saunders, Philadelphia.

Lee, R. and Griffiths, I. R. (1972). A comparison of cerebral arteriography and cavernous sinus venography in the dog. *J. small anim. Pract.* **13**, 225–38.

McNeel, S. V. (1982). Radiology of the skull and cervical spine. *Vet. Clin. N. Am,* **12**, (2), 259–94.

Oliver, J. E. and Lorenz, M. D. (1983). *Handbook of veterinary neurologic diagnosis,* pp. 110–14. Saunders, Philadelphia.

Spooner, R. L. (1961). Cerebral angiography in the dog. *J. small anim. Pract.* **2**, 243–52.

21. Topographical anatomy of the cerebral nervous system

Barr, M. L. and Kiernan, J. A. (1983). *The human nervous system* (4th edn). Harper and Row, Philadelphia.

Bowsher, D. (1975). *Introduction to the anatomy and physiology of the nervous system* (3rd edn). Blackwell, Oxford.

Brodal, A. (1981). *Neurological anatomy in relation to clinical medicine* (3rd edn). Oxford University Press, New York.

Crosby, E. C., Humphrey, T., and Lauer, E. W. (1962). *Correlative anatomy of the nervous system.* Macmillan, New York.

de Lahunta, A. (1983). *Veterinary anatomy and clinical neurology* (2nd edn). Saunders, Philadelphia.

Dellmann, H. D. and McClure, R. C. (1975). Equine nervous system: central nervous system. In *Sisson and Grossman's anatomy of the domestic animals* (5th edn, ed. R. Getty) Vol. 1, pp. 633–50. Saunders, Philadelphia.

Dellmann, H. D. and McClure, R. C. (1975). Canine neurology: central nervous system. In *Sisson and Grossman's anatomy of the domestic animals* (5th edn, ed. R. Getty) Vol. 2, pp. 1671–86. Saunders, Philadelphia.

Jenkins, T. W. (1978). *Functional mammalian neuroanatomy* (2nd edn). Lea and Febiger, Philadelphia.

Meyer, H. (1979). The brain. In *Miller's anatomy of the dog* (2nd edn, ed. H. E. Evans and G. C. Christensen) pp. 842–902. Saunders, Philadelphia.

Williams, P. L. and Warwick, R. (1980). Neurology. In *Gray's anatomy* (36th edn) pp. 802–1215. Churchill Livingstone, Edinburgh.

Index

Bold type indicates principal page citation.